Synthetic Multidentate Macrocyclic Compounds

Academic Press Rapid Manuscript Reproduction

Synthetic Multidentate Macrocyclic Compounds

edited by

Reed M. Izatt
Department of Chemistry

James J. Christensen
Department of Chemical Engineering

Brigham Young University
Provo, Utah

Academic Press New York San Francisco London 1978
A Subsidiary of Harcourt Brace Jovanovich, Publishers

ACADEMIC PRESS, INC.
111 Fifth Avenue, New York, New York 10003

United Kingdom Edition published by
ACADEMIC PRESS, INC. (LONDON) LTD.
24/28 Oval Road, London NW1 7DX

Library of Congress Cataloging in Publication Data

Main entry under title:

Synthetic multidentate macrocyclic compounds.

Includes bibliographical references and index.
1. Cyclic compounds. 2. Macromolecules.
I. Izatt, Reed McNeil, Date I. Christensen,
James J., Date
QD400.S95 547'.5 77-74052
ISBN 0-12-377650-3

PRINTED IN THE UNITED STATES OF AMERICA

CONTENTS

List of Contributors vii
Preface ix

1. Synthetic Multidentate Macrocyclic Compounds 1
   Charles J. Pedersen

2. Synthesis of Multidentate Compounds 53
   Jerald S. Bradshaw

3. Application of Macrocyclic Polydentate Ligands to Synthetic Transformations 111
   Charles L. Liotta

4. Structural Studies of Synthetic Macrocyclic Molecules and Their Cation Complexes 207
   N. Kent Dalley

5. Kinetic Studies of Synthetic Multidentate Macrocyclic Compounds 245
   Gerard W. Liesegang and Edward M. Eyring

6. Developing the Commercial Potential of Macrocyclic Molecules 289
   Roger A. Schwind, Thomas J. Gilligan, and Edward L. Cussler

Author Index 309
Subject Index 317

LIST OF CONTRIBUTORS

Numbers in parentheses indicate the pages on which authors' contributions begin.

JERALD S. BRADSHAW (53), Department of Chemistry, Brigham Young University, Provo, Utah

EDWARD L. CUSSLER (289), Department of Chemical Engineering, Carnegie-Mellon University, Pittsburgh, Pennsylvania

N. KENT DALLEY (207), Department of Chemistry, Brigham Young University, Provo, Utah

EDWARD M. EYRING (245), Department of Chemistry, University of Utah, Salt Lake City, Utah

THOMAS J. GILLIGAN (289), Department of Chemical Engineering, Carnegie-Mellon University, Pittsburgh, Pennsylvania

GERARD W. LIESEGANG (245), Department of Chemistry, University of Utah, Salt Lake City, Utah

CHARLES L. LIOTTA (111), School of Chemistry, Georgia Institute of Technology, Atlanta, Georgia

CHARLES J. PEDERSEN (1), Research and Development Division, Elastomer Chemicals Department, E. I. du Pont de Nemours and Company, Inc., Wilmington, Delaware

ROGER A. SCHWIND (289), Monsanto Research Corporation, Miamisberg, Ohio

PREFACE

Interest in synthetic multidentate macrocyclic compounds has continually increased since C. J. Pedersen reported the synthesis of the "crown ethers" in 1967. The interest in these compounds is broad, ranging from synthesis of new molecules to investigation of selective metal binding properties to use in catalyzing synthetic organic reactions. This volume is an attempt to bring together selected chapters in which the authors discuss in depth investigations in currently important areas of macrocycle research. The chapters deal mainly with macrocyclic compounds (saturated polyethers and their derivatives), and macrobicyclic compounds (cryptates).

In editing this book we have had in mind those readers interested in becoming acquainted with one or more currently important areas of macrocyclic chemistry. The book is primarily aimed at researchers and students in organic, physical, analytical, and inorganic chemistry, and in chemical engineering. However, we hope it will also be of interest to many in the areas of biology, biochemistry, and physiology. Extensive literature references are found in each chapter.

Chapter 1, by Pedersen, gives a first-hand account of the initial synthesis of the cyclic polyethers. The synthesis of a large number of cyclic polyethers together with substituted polyethers, polyethers containing -S- and -N- groups, and open-chain polyethers is discussed. Macrobicyclic compounds containing -O-, -N-, or -O-, and -N- groups are also described. The binding of metal ions by cyclic polyethers is discussed in terms of spectral changes upon complexation, solubilization in organic solvents, crystalline complexes formed, stoichiometry, and stability. This chapter serves as an introduction from which one can then proceed to the other chapters with a good background in the chemical and physical properties of the macrocyclic polyethers and related compounds.

Bradshaw, in Chapter 2, elaborates on the synthesis of cyclic polyethers, polyether amines, and polyether sulfides. Macropolycyclic compounds containing oxygen, sulfur, and nitrogen donor atoms are also discussed along with chiral macrocyclic polyethers. Included is an extensive table of macrocyclic and macropolycyclic compounds synthesized up to mid-1975. The table includes the formula and ring structure of each macrocyclic together with melting or boiling point, yield, and literature references.

Chapter 3, by Liotta, reviews the large amount of data dealing with the use of macrocyclic polyethers and related ligands in catalyzing synthetic organic reactions and in probing reaction mechanisms. Reactions carried out under both homogeneous and heterogeneous conditions are considered. The advent of macrocyclic polydentate ligands provides a means for solubilizing simple metal salts in nonpolar and dipolar aprotic solvents. Therefore, anions unencumbered by strong solvation forces become potent nucleophiles and bases, and provide the basis for the development of new and valuable reagents for organic synthesis. Liotta surveys the reactions promoted by these "naked" anions, including such information as degree of solubilization of salts by macrocycles, rate constants of reactions, and product yields. The effects of solvent and temperature are also included.

In Chapter 4, Dalley considers the structure of synthetic macrocyclic compounds and their cation complexes. The results discussed in this chapter are based on X-ray crystallographic data and Dalley cautions that these structures may differ significantly from the structures in solution. Information included in addition to type of macrocycle and type and size of anion and cation are bond distances, torsion angle range, distance range between metal–solvent and metal–ligand, coordination type, and structure type. Many drawings or figures showing the confirmation of the complexed and uncomplexed macromolecules are given. Metal–macrocycle complexes are discussed and classified from the point of view of the cavity size of the macrocycle and the diameter of the metal ion.

Chapter 5, by Liesegang and Eyring, turns to the rates of reactions and the mechanism by which synthetic macrocyclic ligands complex substrates in solution. Also discussed is the possibility of simulating the specificity and efficiency of enzymes by using macrocycles as catalytically active sites in synthetic macromolecules. Extensive kinetic data in the form of rate constants for a wide variety of macrocycles and substrates in various solvents at various temperature are given.

In Chapter 6, Schwind, Gilligan, and Cussler indicate several commercial applications of the synthetic macrocyclic ligands. The authors point out that macrocyclic molecules currently show much more potential than accomplishment and that this chapter is more a guide for development than a summary of accomplishment. Four promising areas for development discussed are liquid–liquid extraction, membrane separation, catalysis, and specific ion electrodes. The incorporation of macrocycles in liquid and solid membranes and the resulting cation transport properties of these membranes are detailed.

We would like to acknowledge our indebtedness to all those who have aided in bringing the book to fruition.

J. J. Christensen
R. M. Izatt

Provo, Utah

1 SYNTHETIC MULTIDENTATE MACROCYCLIC COMPOUNDS

Charles J. Pedersen*
Research and Development Division
Elastomer Chemicals Department
E. I. du Pont de Nemours and Company
Wilmington, Delaware

I. Macrocyclic Compounds 1
A. Introduction 1
B. Polyethers 2
C. Salt Complexes of Polyethers 18
D. Other Complexes of Polyethers 30
E. Substituted Derivatives 33
F. Polymeric Polyethers 34
G. Polyethers Containing —S— 34
H. Polyethers Containing —N(R)— 37
II. Macrobicyclic Compounds 40
A. Polyethers (Lanterns) 40
B. Diamines 41
C. Polyether Diamines (Cryptates) 41
III. Open-Chain Polyethers 43
IV. Uses for Polyethers 46
V. Toxicity 47
References 49

I. MACROCYCLIC COMPOUNDS

A. Introduction

This introductory chapter deals with the discovery and properties of a compound systematically named 2,3,11,12-dibenzo-1,4,7,10,13,16-hexaoxacyclo-octa-2,11-diene but nicknamed dibenzo-18-crown-6 for brevity and ease. It was the first multidentate macrocyclic compound synthesized with proven ability to form stable complexes with the alkali and alkaline earth cations. The chapter is also concerned with the early history of the different developments resulting from this discovery.

The useful and unusual property of this polyether led to the preparation

*Present address: 57 Market Street, Salem, New Jersey 08079.

of many others in rapid succession, and eventually to those containing

$$-\mathrm{S}- \quad \text{or} \quad -\overset{\mathrm{R}}{\mathrm{N}}-$$

in place of some of the ether links. Many of these crown compounds were also found to complex the alkali and alkaline earth cations with different degrees of stability. Their salt complexes were prepared, both in solution and as crystals, and the factors involved in their formation and stability were determined.

Some time was devoted to the synthesis of macrobicyclic polyethers, whose complexes were expected to be more stable than those of the crown compounds. A number of open-chain polyethers were prepared and compared to the cyclic compounds as complexing agents. Finally, the potential uses for the crown compounds will be suggested very broadly, and the possible dangers connected with these compounds will be pointed out.

The treatment will be mostly historical, and the different topics will not be dealt with fully nor brought up to date. Specialists will cover many of these topics in detail in the following chapters. The macrocyclic polyethers touched upon here, with a few exceptions, are those containing four or more

$$-\mathrm{O}-, \quad -\mathrm{S}-, \quad \text{and} \quad -\overset{\mathrm{R}}{\mathrm{N}}-$$

links in a ring of 14 or more atoms. They will be called polyethers or crown compounds, interchangeably, in this chapter.

B. Polyethers

1. Discovery of Polyethers

A project was initiated in the fall of 1961 to find new vanadium-containing catalysts for the polymerization of olefins. Most of the contemporary catalysts were obtained by treating inorganic vanadium derivatives with various reagents, such as VCl_3 or $VOCl_2$ with aluminum alkyls.

To be different, it was decided to study the effects of uni- and multidentate phenolic ligands on the catalytic properties of the vanadyl group, VO (Pedersen, 1968a). The quinquedentate ligand selected was bis-[2-(*o*-hydroxyphenoxy) ethyl] ether, and its synthesis was begun in May 1962, according to the route outlined in Fig. 1 (Pedersen, 1971a).

The partially protected catechol was contaminated with about 10% unreacted catechol but this mixture was used for the second step, and the final crude product was an unattractive goo. Initial attempts at purification gave a small quantity (0.4% yield) of white crystals which cried for attention by their silky, fibrous structure and apparent insolubility in hydroxylic solvents.

The reactions of the phenols were followed with an ultraviolet spectrophoto-

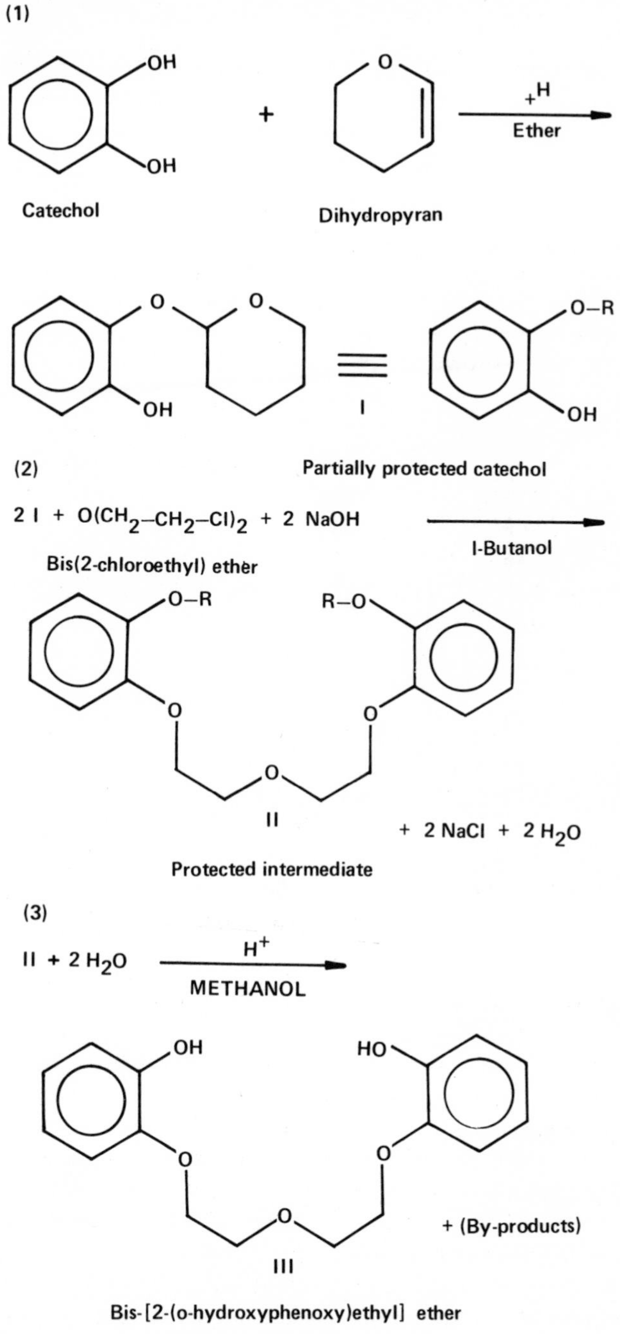

Fig. 1. Synthesis of bis-[2-(*o*-hydroxyphenoxy) ethyl] ether. [Taken by permission from Pedersen (1971a)].

meter because these compounds and their ethers, in neutral methanol solutions, absorb in the region of 275 nm. On treatment with alkali, the absorption curve is not significantly altered if all the hydroxyl groups are covered, but it is shifted to longer wavelengths and higher absorption if one or more hydroxyl groups are still free (Fig. 2).

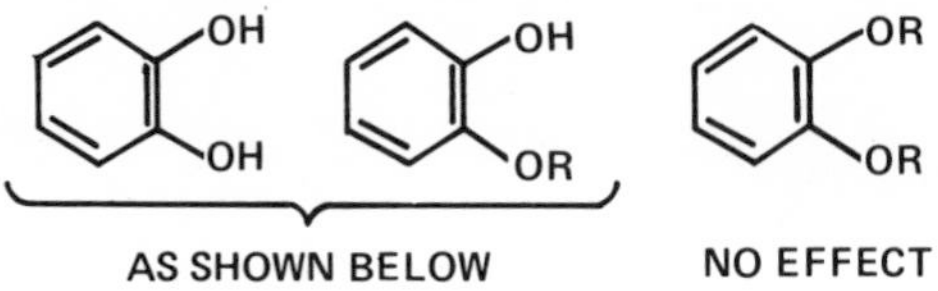

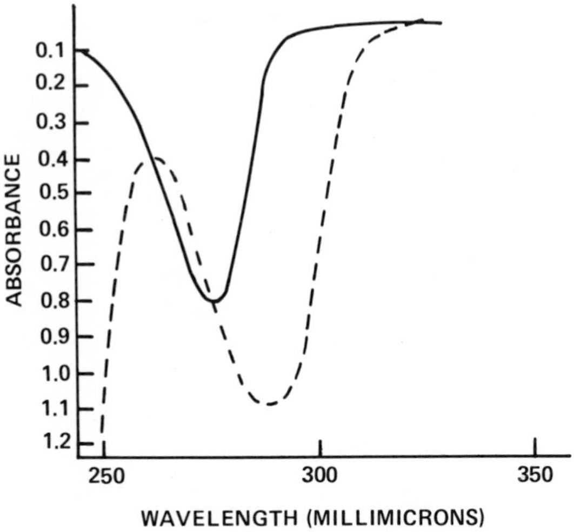

Fig. 2. Effect of NaOH on ultraviolet spectrum (– – –, after addition of NaOH). Neutral sodium salts have no effect on any of these compounds.

The unknown product was very little soluble in methanol and the neutral solution gave an absorption curve characteristic for a phenolic compound. The solution was made alkaline with sodium hydroxide with the expectation that the curve would be either unaffected or shifted to longer wavelengths. The resulting spectrum, however, showed neither effect but the one depicted in Fig. 3 (Pedersen, 1971a). At the same time, it was noticed that the fibrous crystals were freely soluble in methanol in the presence of sodium hydroxide. This seemed strange since the compound did not contain a free phenolic group, a fact confirmed by its infrared and nuclear magnetic resonance (nmr) spectra. It was then found that the compound was soluble in methanol containing any soluble sodium salt. Thus the increased solubility was due not to alkalinity but to the sodium ions; there was, however, no obvious explanation for this behavior, since the compound's elementary analysis corresponded with that

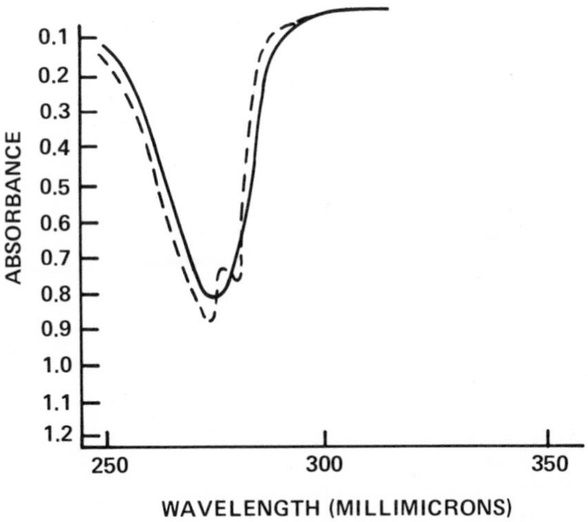

Fig. 3. Effect of NaOH on ultraviolet spectrum of dibenzo-18-crown-6 (Fig. 4, XXVIII) (– – –, after addition of NaOH). Soluble sodium salts have the same effect.

for 2,3-benzo-1,4,7-trioxacyclononane (Fig. 4, I), a plausible product from the reaction of catechol and bis(2-chloroethyl) ether in the presence of sodium hydroxide. Its molecular weight, however, was found to be exactly twice that of the foregoing compound and, thus, the true structure of the first aromatic crown compound, dibenzo-18-crown-6 (Fig. 4, XXVIII), was recognized on July 5, 1962. It was found later that dibenzo-18-crown-6 can be synthesized in 45-80% yield without resorting to high dilution (see Table II).

Face to face with a molecular model of the compound with its empty gaping hole in the middle (Fig. 5) (Pedersen, 1967b), it was easy to see that the sodium ion had fallen into the hole and was held there by the electrostatic attraction between its positive charge and the negative dipolar charge on the six oxygen atoms symmetrically arranged around it in the polyether ring. Tests showed that the ammonium ion and the ions of the alkali and alkaline earth metals, and some others, behaved like the sodium ion.

Excitement, which had been gradually rising during the investigation, turned to elation when it was realized that, at long last, a neutral compound had been found capable of forming stable complexes with the salts of the alkali metals. This reaction was the keener because work on the crown compounds was carried out in total ignorance of the important researches on the naturally occurring macrocyclic antibiotics. The first encounter with these products was in April 1967, in an article by Mueller and Rudin (1967). It contained pictures of the molecular models of macrocyclic fungal metabolites, such as valinomycin (Brockmann and Schmidt-Kastner, 1955; Shemyakin *et al.*, 1963), nonactin (Corbaz *et al.*, 1955; Dobler *et al.*, 1969), and enniatin A (Plattner and Nager, 1947; Plattner *et al.*, 1963), which were isolated between 1947 and 1955, and their structures proven between 1963 and 1969. These and other neutral macrocyclic antibiotics, some of which antedate the crown compounds, form stable complexes with the alkali cations (Ovchinnikov *et al.*, 1974).

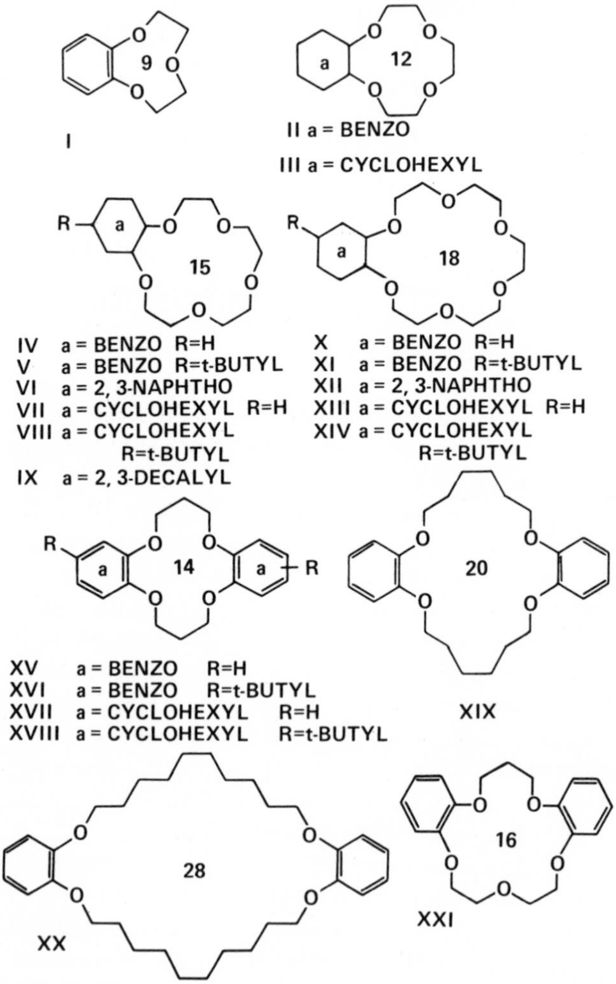

Fig. 4. Structural formulas of polyethers synthesized. The number associated with each structure refers to the total number of atoms in the particular ring where the number is located. [Taken (in part) by permission from Pedersen (1967b, 1970b).]

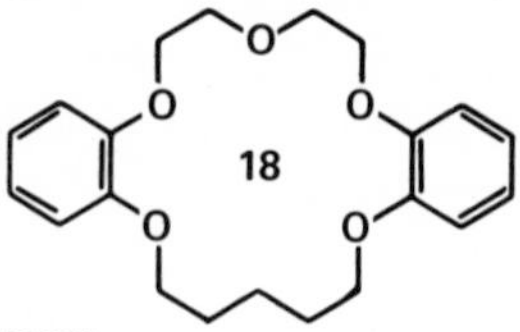

XXII

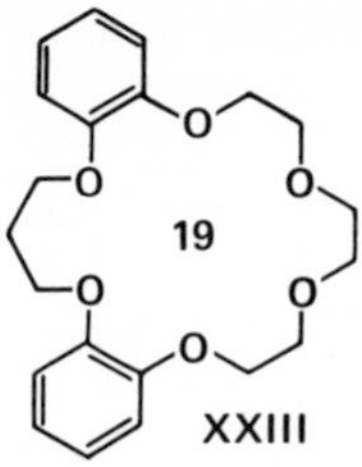

XXIII

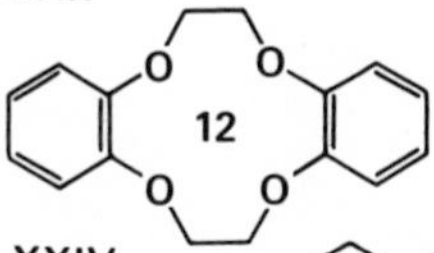

XXIV

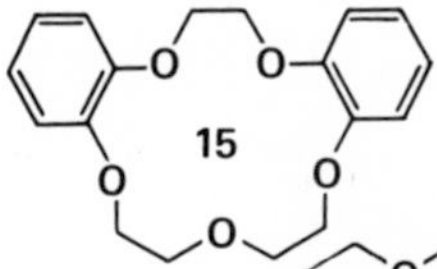

XXV

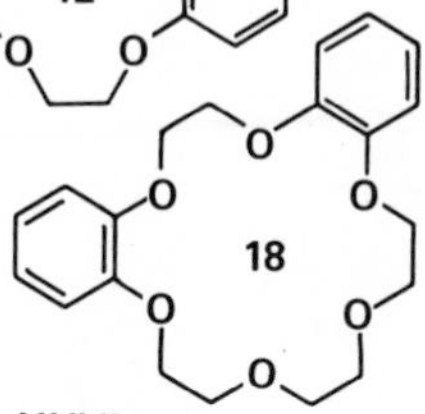

XXVI

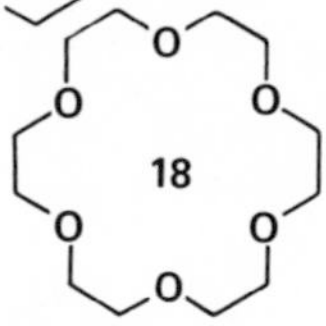

XXVII

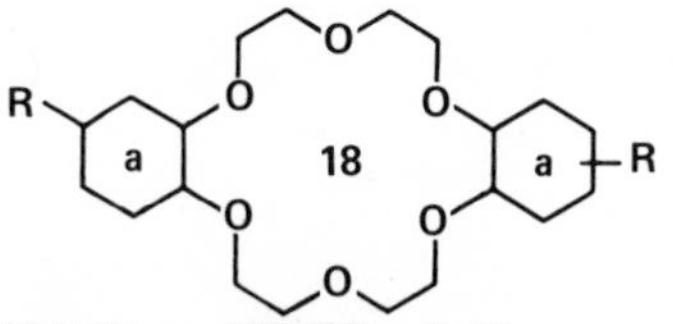

XXVIII a = BENZO R=H
XXIX a = BENZO R=t-BUTYL
XXX a = 2, 3-NAPHTHO
XXXI a = CYCLOHEXYL R=H
XXXII a = CYCLOHEXYL R=t-BUTYL

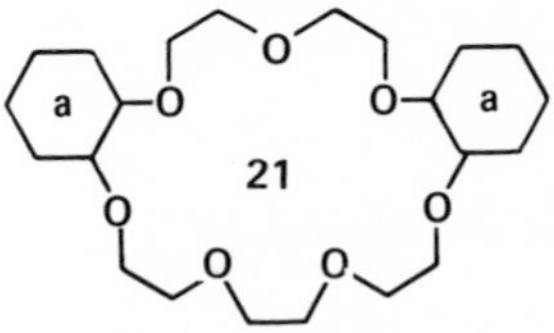

XXXIII a = BENZO
XXXIV a = CYCLOHEXYL

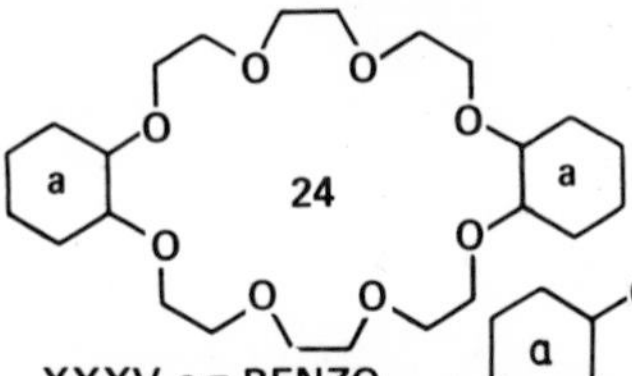

XXXV a = BENZO
XXXVI a = CYCLOHEXYL
XXXVII a = 2, 3-NAPHTHO

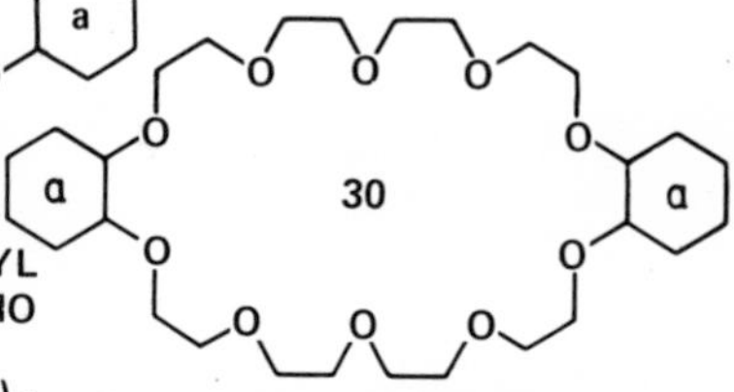

XXXVIII a = BENZO
XXXIX a = CYCLOHEXYL

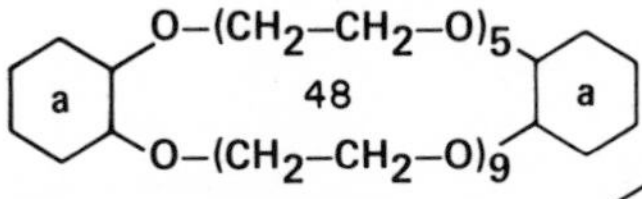

XL a = BENZO
XLI a = CYCLOHEXYL

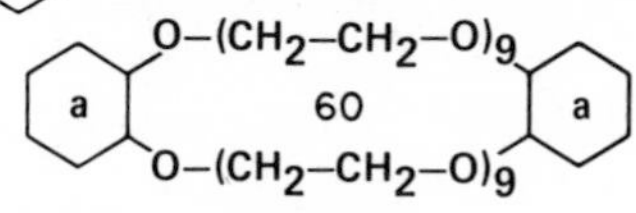

XLII a = BENZO
XLIII a = CYCLOHEXYL

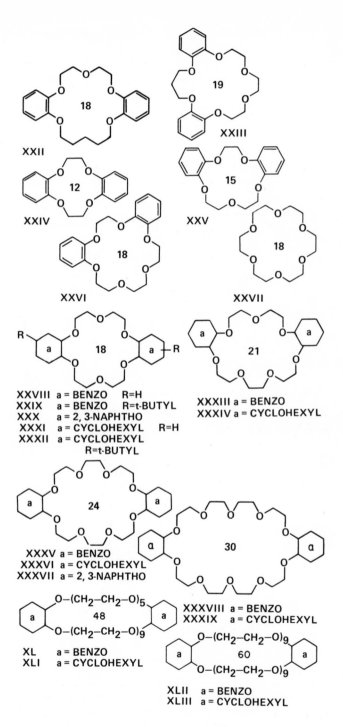

Fig. 4. Continued

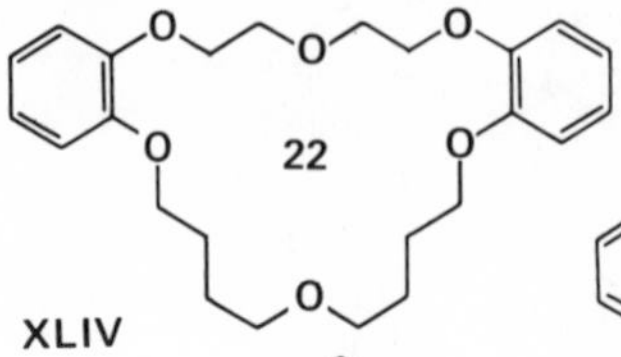

XLIV

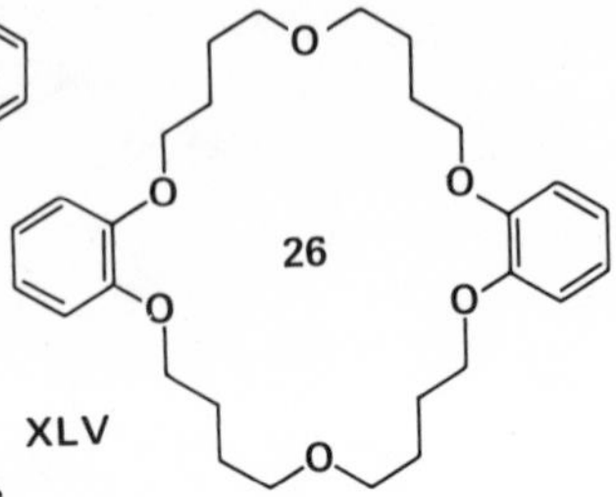

XLV

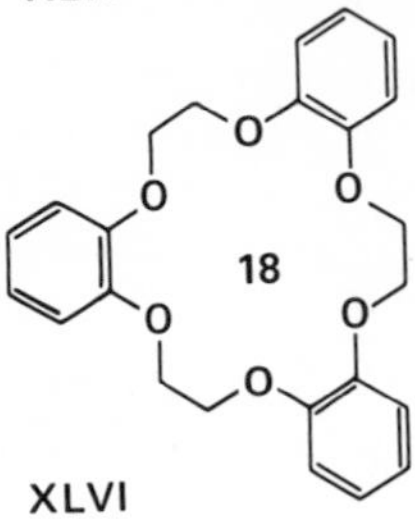

XLVI

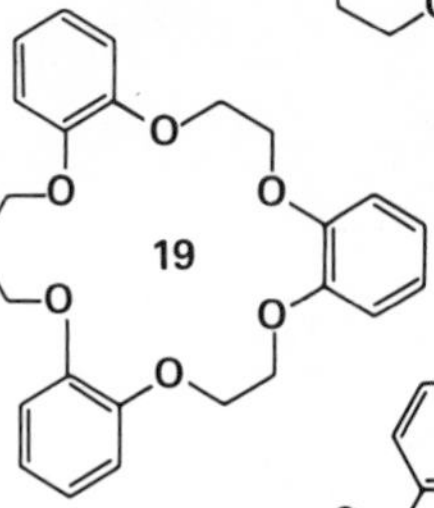

XLVII

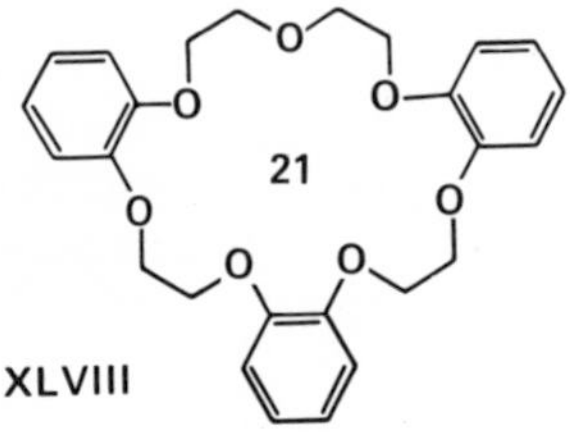

XLVIII

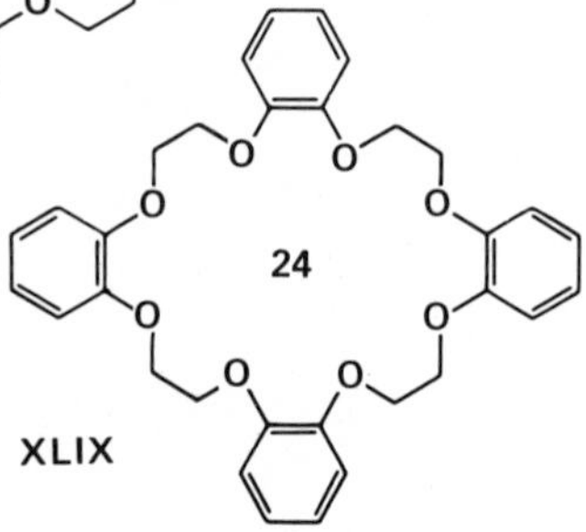

XLIX

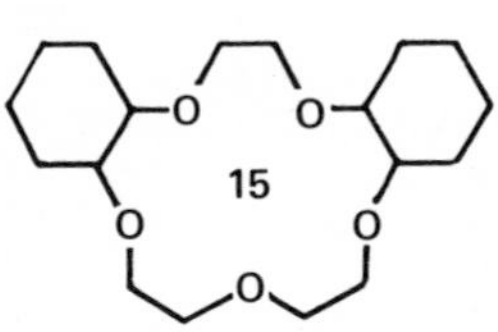

L

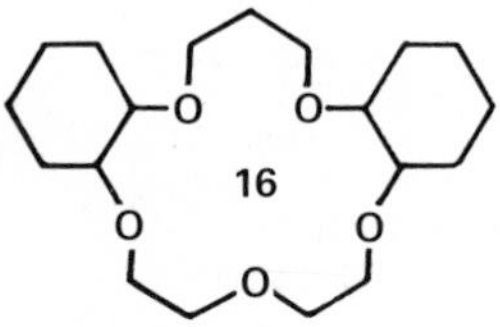

LI

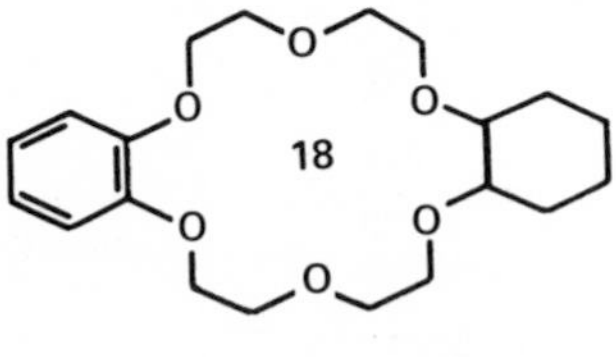

LII

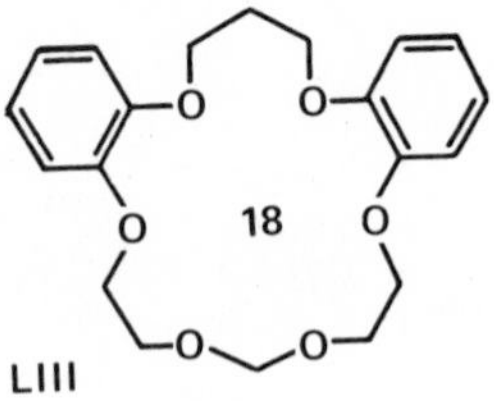

LIII

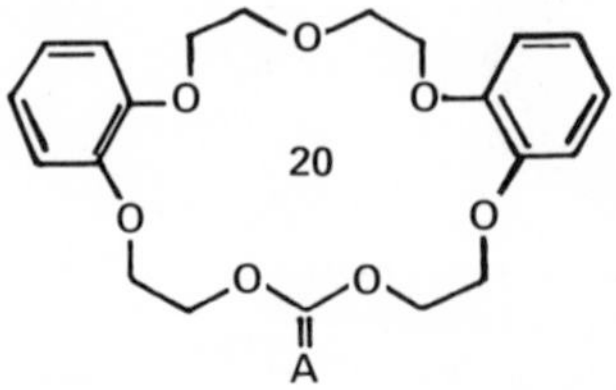

LIV A = 2 Hydrogen
LV A = Pentamethylene
LVI A = Oxygen

Fig. 4. Continued

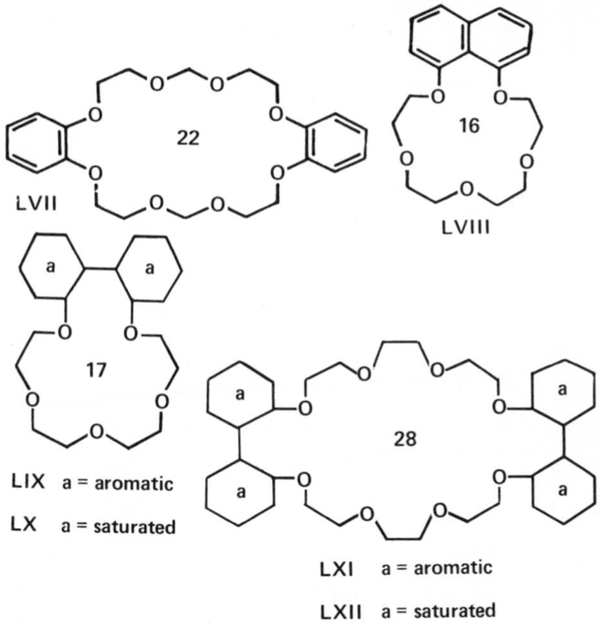

Fig. 4. Continued

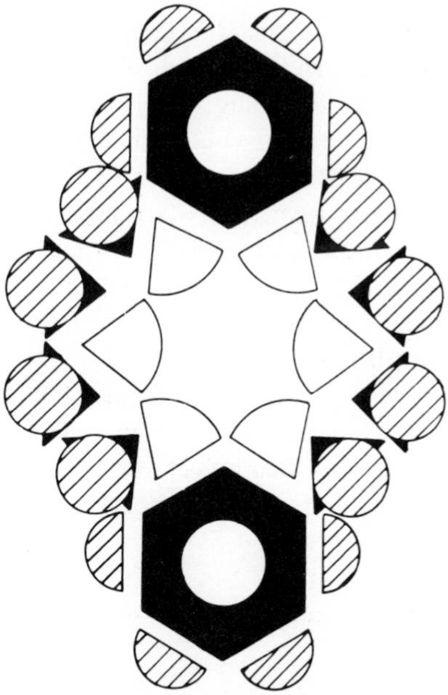

Fig. 5. Courtauld model of dibenzo-18-crown-6 (Fig. 4, XXVIII). [Taken by permission from Pedersen (1967b).]

2. Previously Reported Polyethers

Only a few publications prior to 1967 referred to polyethers, and none of them considered the possibility of their forming complexes with the salts of the alkali and alkaline earth metals. Of the previously reported compounds, the most closely related to the crown compounds were 2,2,7,7,12,12,17,17-octamethyl-21,22,23,24-tetraoxaquatrene, $C_{28}H_{32}O_4$, made by condensing acetone with furan (Ackman *et al.*, 1955); 2,3,12,13,22,23-tribenzo-1,4,11,14,21,24-hexaoxacyclotriaconta-2,12,22-triene (tribenzo-30-crown-6), $C_{36}H_{48}O_6$, and 2,3,12,13-dibenzo-1,4,11,14-tetraoxacycloeicosa-2,12-diene (dibenzo-20-crown-4), $C_{24}H_{32}O_4$ (Fig. 4, XIX) (Luttringhaus and Sichert-Modrow, 1956). Cyclic tetramers of ethylene oxide (Stewart *et al.*, 1957) and propylene oxide (Down *et al.*, 1957, 1959) had also been synthesized previously.

3. Polyethers Synthesized

Following the demonstration of the unusual ability of dibenzo-18-crown-6 to complex the alkali and alkaline earth cations, the preparation of all sorts of polyethers was undertaken. The work was spurred by intense curiosity regarding factors involved in the stability of the salt complexes, such as the relative sizes of the hole and the cation, and the number and symmetrical arrangement of the oxygen atoms in the polyether ring.

Nine new compounds were prepared and identified during 1962 and, by the end of 1968, 60 authentic crown compounds were on hand, containing 12-60 atoms in the polyether ring, including 4-20 oxygen atoms. Many of these compounds were found to be useless as complexing agents but they helped to define the effective ones (Pedersen, 1967a, b, 1970b, 1971c, 1972b).

Two of the five theoretical isomers of dicyclohexyl-18-crown-6 have been isolated so far from the products obtained by hydrogenating dibenzo-18-crown-6 (Mercer and Truter, 1973). Using column chromatography, D. J. Sam* was the first to separate the isomers: isomer A, cis-syn-cis (meso), melting at 61-62°C, and isomer B, cis-anti-cis (dl), melting at 69-70°C.

Biphenyl-17-crown-5 (Fig. 4, LIX), and dibiphenyl-28-crown-8 (Fig. 4, LXI) were synthesized from 2,2′-dihydroxybiphenyl in 1968,† and they were hydrogenated to the corresponding saturated polyethers.**

Parsons (1975) reacted catechol with the ditosylates of 1,1′-oxydipropan-2-ol and with 2,2′-oxydipropan-1-ol to prepare tetramethyl-dibenzo-18-crown-6, and he separated the resulting ten isomers.

Kyba *et al.*, (1973) synthesized chiral crown compounds based on 2,2′-

*Private communication, D. J. Sam, E. I. du Pont de Nemours and Company.
†Unpublished results of R. E. Evans.
**Unpublished results.

dihydroxy-1,1′-binaphthyl. They have been used with interesting results in a new field recently developed and named "host-guest" chemistry (Cram and Cram, 1974).

The code numbers and the structural formulas of the crown compounds are shown in Fig. 4. The digits within the diagram indicate the total number of atoms in the polyether ring. Their empirical formulas, molecular weights, and melting points are listed in Table I (Pedersen, 1967b, 1970b).

TABLE 1

Data on Macrocyclic Polyethers[a]

Compound[b]	Formula	Calcd. mol. wt.	M.P. (°C)[c]
I	$C_{10}H_{12}O_3$	180	67-69
II	$C_{12}H_{16}C_4$	224	44-45.5
III	$C_{12}H_{22}O_4$	230	Below 26
IV	$C_{14}H_{20}O_5$	268	79-79.5
V	$C_{18}H_{28}O_5$	324	43.5-44.5
VI	$C_{18}H_{22}O_5$	318	117-119
VII	$C_{14}H_{26}O_5$	274	Below 26
VIII	$C_{18}H_{34}O_5$	330	Below 26
IX	$C_{18}H_{32}O_5$	328	Below 26
X	$C_{16}H_{24}O_6$	312	43-44
XI	$C_{20}H_{32}O_6$	368	35-37
XII	$C_{20}H_{26}O_6$	362	110-111.5
XIII	$C_{16}H_{30}O_6$	318	Below 26
XIV	$C_{20}H_{38}O_6$	374	Below 26
XV	$C_{18}H_{20}O_4$	300	150-152
XVI	$C_{26}H_{36}O_4$	412	149-152
XVII	$C_{18}H_{32}O_4$	312	153.5-155.5
XVIII	$C_{26}H_{48}O_4$	424	A glass at 26°
XIX	$C_{24}H_{32}O_4$	384	139-141
XX	$C_{32}H_{45}O_4$	496	137-138.5
XXI	$C_{19}H_{22}O_5$	330	117-118
XXII	$C_{21}H_{26}O_5$	358	157-158
XXIII	$C_{21}H_{26}O_6$	374	84.5-86
XXIV	$C_{16}H_{16}O_4$	272	208-209
XXV	$C_{18}H_{20}O_5$	316	113.5-115
XXVI	$C_{20}H_{24}O_6$	360	117-118
XXVII	$C_{12}H_{24}O_6$	264	39-40
XXVIII	$C_{20}H_{24}O_6$	360	164
XXIX	$C_{28}H_{40}O_6$	472	135-137
XXX	$C_{28}H_{28}O_6$	460	224-246
XXXI	$C_{20}H_{36}O_6$	372	A 61-62.5
			B 69-70
XXXII	$C_{28}H_{52}O_6$	484	Below 26
XXXIII	$C_{22}H_{28}O_7$	404	106.5-107.5
XXXIV	$C_{22}H_{40}O_7$	416	Below 26
XXXV	$C_{24}H_{32}O_8$	448	103-104[d]

TABLE 1, Cont.

Compound[b]	Formula	Calc. mol. wt.	M.P. (°C)[c]
XXXVI	$C_{24}H_{44}O_8$	460	Below 26
XXXVII	$C_{32}H_{36}O_8$	548	190-191.5
XXXVIII	$C_{28}H_{40}O_{10}$	536	106-107.5
XXXIX	$C_{28}H_{52}O_{10}$	548	Below 26
XL	$C_{40}H_{64}O_{16}$	800	Below 26
XLI	$C_{40}H_{76}O_{16}$	812	Below 26
XLII	$C_{48}H_{80}O_{20}$	976	Below 26
XLIII	$C_{48}H_{92}O_{20}$	988	Below 26
XLIV	$C_{24}H_{32}O_6$	416	82-83
XLV	$C_{28}H_{40}O_6$	472	125-127
XLVI	$C_{24}H_{24}O_6$	408	190-192
XLVII	$C_{25}H_{26}O_6$	422	147-149
XLVIII	$C_{26}H_{28}O_7$	452	98.5-100
XLIX	$C_{32}H_{32}O_8$	544	150-152
L	$C_{18}H_{32}O_5$	328	62-63
LI	$C_{19}H_{34}O_5$	342	Below 26
LII	$C_{20}H_{30}O_6$	366	Below 26
LIII	$C_{20}H_{24}O_6$	348	118
LIV	$C_{21}H_{26}O_7$	390	151-152
LV	$C_{26}H_{34}O_7$	458	162
LVI	$C_{21}H_{24}O_8$	404	122-125
LVII	$C_{22}H_{28}O_8$	420	166-167
LVIII	$C_{18}H_{22}O_5$	318	112-115
LIX	$C_{20}H_{24}O_5$	344	106-109
LX	$C_{20}H_{36}O_5$	356	Below 26
LXI	$C_{36}H_{40}O_8$	600	151-152
LXII	$C_{36}H_{64}O_8$	624	Below 26

[a] Data taken from Pedersen (1967b, 1970b).
[b] Numbers refer to compounds in Fig. 4.
[c] Melting points were taken on a Fisher-Johns apparatus and are uncorrected. The melting point of XXVIII, 164°, however, was obtained by another method and is corrected. This compares with 163-165° on the aluminum block. All the hydrogenation products except XVII, XXXI, and XVIII (a glass) are viscous liquids at room temperature.
[d] The melting point of this compound was given as 113-114° in Pedersen (1967b). P. E. Stott of Parish Chemical Company informed the author that the melting point is 102-103°. On reexamination, the author's original sample of dibenzo-24-crown-8 was found to melt at 103-104°.

4. Polyether (Crown) Nomenclature

The opening paragraph of this chapter makes it clear that the systematic names of the polyethers are not practical for their identification. The epithet "crown" was applied to the first-discovered member of this class of polyethers because its molecular model looked like one and, with it, cations could be crowned and uncrowned, just like the heads of some royalty. From this beginning the crown names were developed (Pedersen, 1967b).

The names consist of, in order:

(a) the number and kind of hydrocarbon rings;
(b) the total number of atoms in the polyether ring;
(c) the class name, crown; and
(d) the number of oxygen atoms in the polyether ring.

In most cases, if the first number in the name divided by the second equals three, the compound is a derivative of a cyclic polyethylene oxide. Figure 4, LIII, is one of the exceptions.

TABLE II

Some Crown Compounds: Synthesis and Yields

Compound[a]	Synthesis	
	Method	Yield (%)
Benzo-12-crown-4 (II)	V	4
Benzo-15-crown-5 (IV)	V	62
Benzo-18-crown-6 (X)	V	60
Dibenzo-12-crown-4 (XXIV)	W	11
Dibenzo-14-crown-4 (XV)	W	27
Dibenzo-20-crown-4 (XIX)	W	4
Dibenzo-15-crown-5 (XXV)	W	43
Dibenzo-18-crown-6 (XXVIII)	W	80
asym-Dibenzo-18-crown-6 (XXVI)	W	25
Tribenzo-18-crown-6 (XLVI)	W	28
Dibenzo-21-crown-7 (XXXIII)	W	36
Dibenzo-18-crown-6 (XXVIII)	X	45
Dibenzo-24-crown-8 (XXXV)	X	38
Tetrabenzo-24-crown-8 (XLIX)	X	18
Dibenzo-30-crown-10 (XXXVIII)	X	Over 6[b]
Dibenzo-28-crown-4 (XX)	Y	3
Dicyclohexyl-18-crown-6 (XXXI)	Z	45-67
Benzo-cyclohexyl-18-crown-6 (LII)	Z	8

[a]Numbers refer to compounds in Table I and Fig. 4.
[b]Benzo-15-crown-5, IV, was also formed in this preparation.

The placements of the hydrocarbon rings and the oxygen atoms are as symmetrical as possible and the others are indicated by the prefix *asym*. Because the names are trivial ones contrived for brevity, some liberties have been taken with them, such as the use of cyclohexyl for 1,2-hexahydrobenzo. It is possible to make the names more precise by using positional numbers, but such elaboration carried too far would be contrary to their main purpose. The crown nomenclature, though not foolproof, serves to identify the better-complexing crown compounds simply and with reasonable accuracy, but structural formulas are recommended for positive identification. Some examples of this new nomenclature are given in Table II (Pedersen, 1967b, 1970b).

5. Methods of Synthesizing Polyethers

The four different ways of preparing the aromatic crown compounds are represented by the stoichiometric equations shown in Fig. 6, where R, S, T, U,

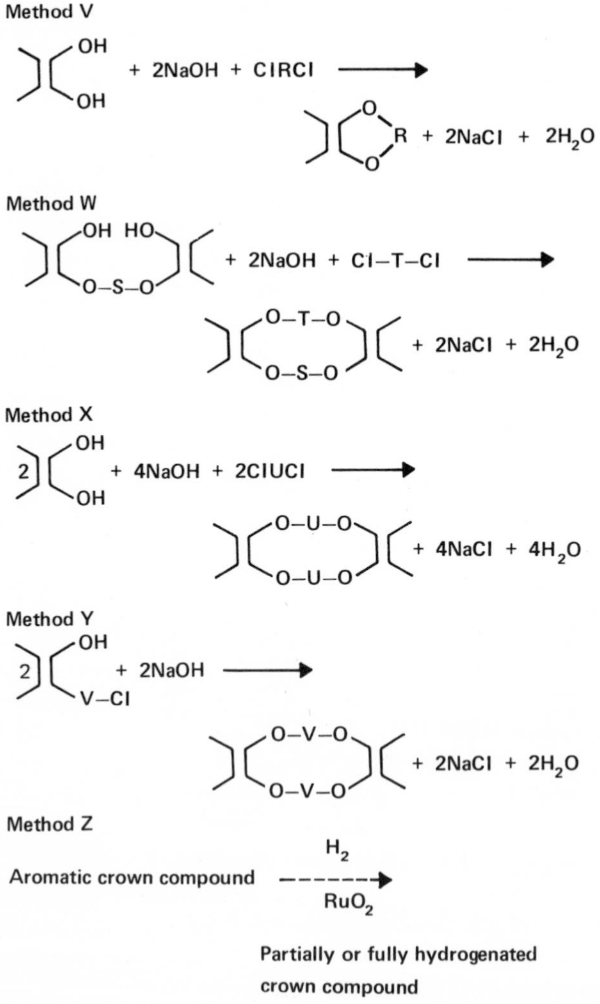

Fig. 6. Methods of syntheses [taken by permission from Pedersen (1967b)].

and V are divalent organic groups, generally of the type $-(CH_2-CH_2-O)_n-CH_2-CH_2-$. In Method W, S and T may or may not be identical (Pedersen, 1967b, 1970b). This method is the most versatile for the synthesis of crown compounds containing two or more aromatic groups, particularly if they have an odd number of oxygen atoms in the polyether ring. The starting phenolic intermediates for this method are prepared by adapting the reactions shown in Fig. 1. Benzyl is also a convenient group for protecting phenolic hydroxyl, and it can be removed later by catalytic hydrogenation.

The condensations are usually run in 1-butanol under nitrogen at reflux temperature for 12-24 hr. After evaporating the solvent and water, the reaction products are dissolved in chloroform and extracted with 5% aqueous sodium hydroxide to eliminate phenolic compounds; then the chloroform is removed and the aromatic crown compound is recovered by extraction with a saturated hydrocarbon, such as *n*-heptane.

Partially or fully saturated crown compounds are prepared from the corresponding aromatic polyethers by Method Z, typically in 1-butanol at 100°C and 7-10 atm over a ruthenium catalyst in a glass or stainless steel vessel. These products are best recovered by column chromatography over alumina (Pedersen, 1970b).

Later, the yield of 18-crown-6, which originally had been 2%, was raised to as high as 93% by others (Dale and Kristiansen, 1971, 1972; Greene, 1972; Gokel *et al.,* 1974). Satisfactory yields of 21-crown-7, 24-crown-8, and 18-crown-O_5N_1 were also obtained (Greene, 1972).

An assortment of crown compounds, methods of their synthesis, and yields are given in Table II (Pedersen, 1967b, 1970b). It is evident that the yields for the aromatic compounds are highest for rings 15-crown-5 and 18-crown-6, followed by rings 14-crown-4, 21-crown-7, and 24-crown-8, and the lowest yields are for compounds which do not form stable complexes: II, XIX, XX, and XXIV.

The high yield, 45% (see Table II), of dibenzo-18-crown-6 obtained by reacting equimolar proportions of catechol and bis(2-chloroethyl) ether without resorting to high dilution is remarkable. It was concluded that the "coordination template effect" is operating and the probability of the ring-closing step is greatly increased by the sodium ion (Pedersen, 1971a). This hypothesis had been previously proposed to explain the formation of macrocyclic compounds, containing four nitrogen atoms in the ring, in good yields in the presence of transition metal cations but not in their absence (Taylor *et al.,* 1966).

The results of further experiments appear to support this assumption. The yield of dibenzo-18-crown-6 is higher when it is prepared with sodium or potassium hydroxide than when lithium or tetramethylammonium hydroxide is used. Lithium and quaternary ammonium ions are not strongly complexed by the crown compound. The best complexers are obtained in highest yields and the noncomplexers in lowest yields (Table II; Pedersen, 1971a). Furthermore, some open-chain polyethers were found to complex alkali cations (see Section III).

Greene (1972) ascribes the high yields of the single-ring polyethers which are strong complexers to a "template effect" involving a cation which is not too bulky guiding together the termini of an intermediate, thus increasing the rate of cyclization with respect to polymerization.

6. Identification of Polyethers

The proof of structure of the free crown compounds is based on elementary composition, molecular weight, ultraviolet, infrared, and nmr spectra. These properties of each of the authentic compounds are consistent with their proposed structures. Occasionally they were determined by X-ray analysis and high-resolution mass spectroscopy.

The ultraviolet spectra of crown compounds containing benzo groups have absorption bands, in methanol, at 273-275 nm. The extinction coefficient, ϵ, of these compounds ranges from 2100 to 8400 cm^{-1} $mole^{-1}$, depending on the number of benzo groups (from one to four). The saturated crown compounds do not absorb in this region (Pedersen, 1967b).

The infrared spectra of dibenzo-18-crown-6 and dicyclohexyl-18-crown-6 are shown in Fig. 7. They establish the absence of any hydroxyl group, and indicate the presence of ether linkages by strong, broad bands centering near 8.1 μ for aromatic-O-aliphatic, and 8.85 μ for aliphatic-O-aliphatic (Pedersen, 1967b).

The nmr spectra of the crown compounds, obtained in deuteriochloroform, confirm the absence of terminal groups, such as hydroxyl or alkoxy (Pedersen, 1967b).

The structure of the partially hydrogenated compound, benzo-cyclohexyl-18-crown-6 (Fig. 4, LII), was definitely established by high-resolution mass spectroscopy (Pedersen, 1970b).

The crystallographic structures of uncomplexed dibenzo-18-crown-6 (Bright and Truter, 1970a, b) and dibenzo-30-crown-10 (Fig. 4, XXXVIII) (Bush and Truter, 1970) have been determined by X-ray analysis.

7. Properties of Polyethers

The aromatic polyethers are colorless crystalline compounds and, for a given polyether ring, the melting point rises with the number of benzo groups (Table I). These compounds, particularly those containing more than one benzo group, are nearly insoluble in water and sparsely soluble in alcohol and many other common solvents at room temperature. They are readily soluble in methylene chloride and chloroform.

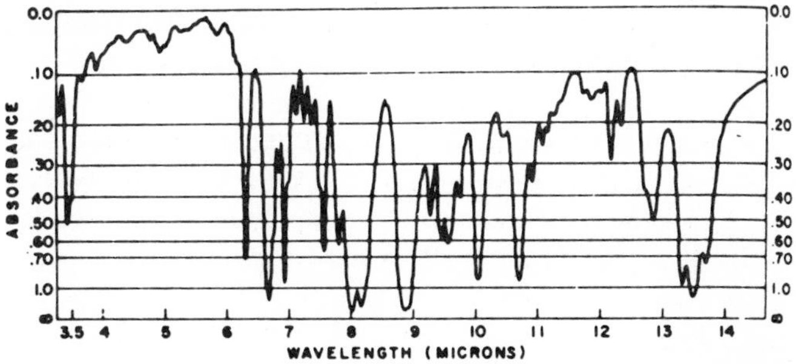

Infrared spectrum of dibenzo-18-crown-6 (XXVIII), KBr pellet.

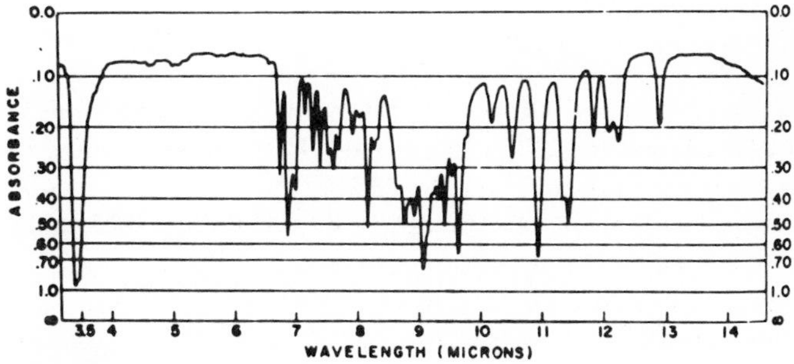

Infrared spectrum of dicyclohexyl-18-crown-6 (XXXI) mp 68.5–69.5°, KBr pellet.

Fig. 7. Infrared spectra [taken by permission from Pedersen (1967b)].

The saturated crown compounds are colorless, viscous liquids or solids of low melting point. They are very much more soluble in all solvents than their aromatic precursors; many of them dissolve even in petroleum ether and yet are appreciably soluble in water.

The polyethers are thermally stable; for instance, dibenzo-18-crown-6 can be distilled at 380°C at atmospheric pressure. They must be protected from oxygen, however, at high temperatures. The polyether ring is destroyed by reagents which cause the scission of aromatic and aliphatic ethers. The aromatic crown compounds react like veratrole and undergo many substitution reactions (see Section I.E), and form polymeric products (see Section I.F) (Pedersen, 1967b).

Another property of the crown compounds quickly aroused the interest of the biological community. Their superficial resemblance to macrocyclic antibiotics, and their shared ability to complex alkali cations, prompted several investigators of biological and artificial membranes to use crown compounds in experiments designed to clarify the mechanism for the transport of salts through the membranes. (Although less efficient than the natural products, some crown compounds decrease the electrical resistance of cell membranes.) Their results were ready for presentation at a symposium in 1968 (Eisenman and Ciani, 1968; Lardy, 1968; Pressman, 1968; Tosteson, 1968).

C. Salt Complexes of Polyethers

The first observation of complex formation between a crown compound and the sodium ion was described in Section I.B.1. This unusual phenomenon will now be considered in greater detail (Pedersen, 1967a, b, 1970a). Selective complexing of cations by different types of macrocyclic compounds was reviewed by Christensen *et al.* (1971).

From the beginning, complexing was detected or measured by:

(a) the changes in the ultraviolet spectra of the aromatic crown compounds;
(b) the changes in the solubilities of salts and crown compounds in different types of solvents;
(c) the isolation of crystalline complexes; and, somewhat later,
(d) two-phase liquid extraction.

1. Ultraviolet Spectrum

The change in the ultraviolet spectrum of dibenzo-18-crown-6 in methanol on addition of the sodium ion is illustrated in Fig. 3. The effects of other cations are shown in Fig. 8 (Pedersen, 1967b). Note that the salts are of different acids. Anions have pronounced effects in the solubilization of inorganic salts in aprotic solvents and in the formation of crystalline complexes, but they have little effect under the conditions used for obtaining these spectra: very low concentration of crown compound in methanol and a 50-fold excess of salt.

If a salt causes no significant change in the spectrum, it is considered not to complex (there are exceptions: cadmium chloride does not alter the spectrum of dibenzo-18-crown-6 but the two form a crystalline complex, hence, a positive test is more reliable than a negative one; Pedersen, 1967b); the greatest change is assumed to be caused by the salt forming the most stable complex, and the others are assumed to lie between these limits. The present discussion is restricted to the alkali cations and crown compounds having rings of 14-30 atoms, including 4-10 oxygen atoms. The conclusions drawn regarding complex formation under these conditions and assumptions are summarized in Table III.

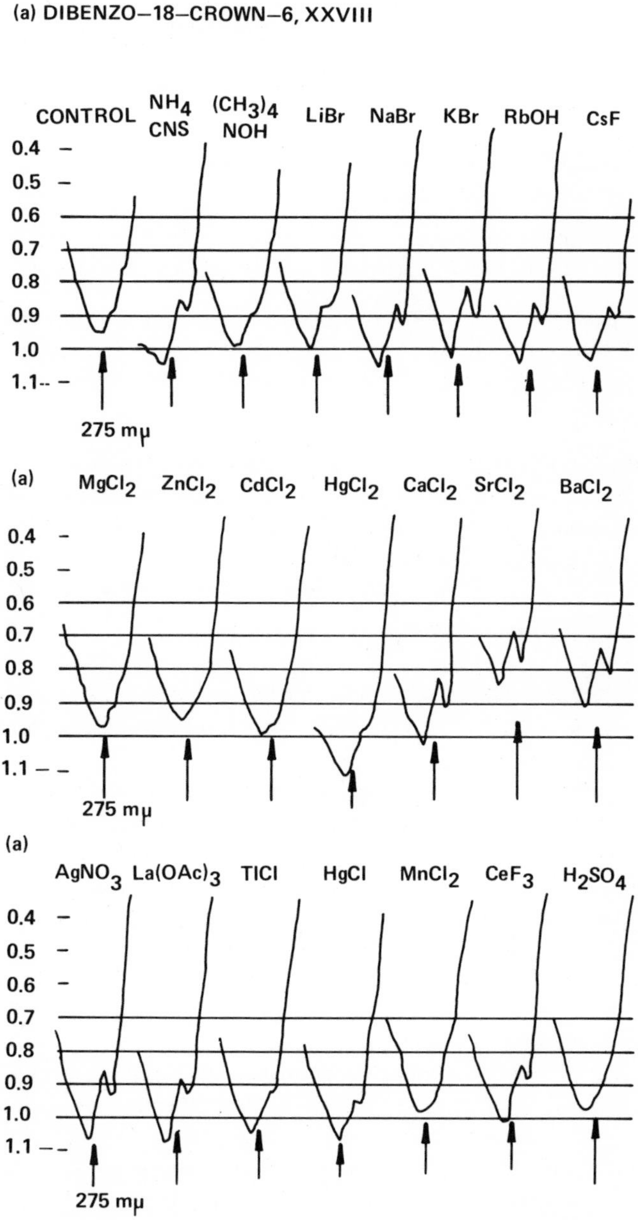

Fig. 8. Effects of salts on ultraviolet spectrum of dibenzo-18-crown-6. Solvent: methanol; concentration of dibenzo-18-crown-6: 1.86×10^{-4} mole/liter; concentration of salt: 50-fold of the foregoing. [Taken (in part) by permission from Pedersen (1967b).]

TABLE III

Complex Formation According to Ultraviolet Spectra

Compound		Li^+	Na^+	K^+	Rb^+	Cs^+
Dibenzo-14-crown-4	(XV)	++[a]	+	+?	–	0
Benzo-15-crown-5	(IV)	+	++[b]	+	–	–
Dibenzo-18-crown-6[c]	(XXVIII)	+	++	+++	++	+
Dibenzo-21-crown-7	(XXXIII)	0	0	+	–	++
Dibenzo-30-crown-10	(XXXVIII)	–	0	++[b]	–	+

[a]– not tested; 0 no interaction; + slight interaction; ++ marked interaction; +++ very strong interaction.

[b]A significant decrease in absorbance without the development of a second peak.

[c]The cation selectivity of the 18-crown-6 polyethers is shifted to $Na^+ \gtrsim K^+ > Cs^+ > Li^+$ in tetrahydrofuran, indicating that selectivity is dependent on the solvent (Wong *et al.*, 1970).

This method was used extensively for the qualitative detection of complexing. Although it can be used indirectly for saturated polyethers and for the quantitative estimation of complex formation (Pedersen, 1967b), solvent extraction is simpler and more accurate for these purposes (see Section I.C.3).

2. Solubilization of Inorganic Salts in Organic Solvents

The simplest manifestation of complexing by polyethers is the solubilization of inorganic salts in organic solvents. This property is also potentially the most useful.

Polyethers and complexable salts mutually increase their solubilities in solvents wherein the complexes are soluble. Sometimes these effects are remarkable; the solubility of dibenzo-18-crown-6 in methanol is 0.001 mole/liter, but when excesses of the polyether and potassium thiocyanate are mixed together in this solvent, a solution is obtained containing 0.107 mole/liter of the complex, a 100-fold increase (Pedersen, 1967b).

The solubilization of alkali metal salts in aprotic solvents is still more impressive, the most spectacular being that of solid potassium permanganate in aromatic hydrocarbons.

When a pinch of powdered potassium permanganate crystals is lightly sprinkled on the surface of benzene in a tall vessel, the particles settle rapidly to the bottom and leave no sign whatever of their passage through the colorless liquid. If this procedure is repeated with a benzene solution of dicyclohexyl-18-crown-6, the particles descend like a flock of tiny peacocks, each trailing a tail of flashing purple. Eventually a solution is obtained exceeding 0.02 M in permanganate (Pedersen, 1967a). Such solutions have been used for oxidizing organic compounds in benzene (Sam and Simmons, 1972).

Still more unexpected is the solubilization of potassium hydroxide in aromatic hydrocarbons to give solutions which are up to 0.2 *N* in base. In this case, potassium hydroxide pellets cannot be dissolved directly in the aromatic solvent containing dicyclohexyl-18-crown-6. It has to be done by solvent exchange with a methanol solution (Pedersen, 1967b) or by direct addition of 1% methanol (Pedersen and Frensdorff, 1972). Such solutions, which contain methoxide as well as hydroxide anions, saponify the hindered esters of 2,4,6-trimethylbenzoic acid by normal acyl-oxygen fission. This power was attributed to bare, unsolvated hydroxide anions, which can attack the carbonyl group of the hindered esters much more easily than the ordinary solvated hydroxide anions (Pedersen, 1967b). The chemistry of "naked" anions was established recently and has expanded rapidly (Liotta *et al.,* 1974; Liotta and Harris, 1974). It is concerned with reactions of activated anions obtained by complexing salts with crown compounds.

Only a few transition metal salts can be solubilized in organic solvents by complexing directly with a crown compound (see Section I.D), but many of them can be solubilized indirectly by a reaction represented by the equation

$$MX_m + nKX + nP \longrightarrow nPK^+ + [MX_{m+n}]^{n-}$$

where M is the transition metal, X the halide, m the valence of M, $n = 1, 2,$ or 3, P the crown compound, PK^+ the crown-complexed potassium ion, and $[MX_{m+n}]^{n-}$ the complex anion.

Vanadium trichloride, $FeCl_3$, $CoCl_2$, $NiCl_2$, $PdCl_2$, $PtCl_2$, and $PtCl_4$ were solubilized in *o*-dichlorobenzene by reaction with KCl and dicyclohexyl-18-crown-6. Rhodium trichloride did not respond to this treatment.* Cobaltous chloride and $PdCl_2$ were also made soluble in benzene by this method (Pedersen, 1967b). Some interesting insight into the nature of the complex anions might be gained by a systematic study of the reactions of dicyclohexyl-18-crown-6, potassium halides, and the stable halides of the transition metals.

Nearly any salt with a complexable cation can be solubilized in an organic solvent with a medium to high dielectric constant, but salts of "hard" anions (Pearson, 1966), such as fluoride and sulfate, cannot be solubilized in a solvent with a low dielectric constant, since the complex reverts to its components (Pedersen, 1967b). Solubilization by polyethers, however, is not a simple function of solvent polarity. Specific interactions, such as solvation of anion or competition between solvent and polyether, evidently overcome more general polarity effects in some cases (Pedersen and Frensdorff, 1972).

*Unpublished results.

3. Solvent Extraction

This method is more generally useful because it can be applied to all crown compounds, and complexing efficiencies can be ranked numerically (Pedersen, 1968b). When an aqueous solution of an alkali metal salt (hydroxides are here considered to be salts) containing a very low concentration of the picrate of the same cation is mixed with an equal volume of an immiscible organic solvent (methylene chloride is usually a good choice), nearly all the picrate is present in the yellow aqueous phase and the organic phase remains colorless. If a polyether is added to the system by dissolving in the organic solvent, the complexed picrate transfers to the organic phase, the extent depending on the effectiveness of the polyether as a complexing agent for the cation. If it is ineffective, the organic phase will be colorless; if it is very effective, most of the color will be in the organic phase. The efficiencies of the crown compounds will lie between these two limits, and can be expressed as percentage extracted. The concentration of picrate, which has an absorption maximum near 360 nm, is determined with a spectrophotometer, and the test concentrations are usually made low enough so that they can be read in the instrument without dilution. Some extraction results are given in Table IV (Pedersen, 1968b).

TABLE IV

Extraction Results[a,b]

	Picrate extracted (%)			
Polyether	Li^+	Na^+	K^+	Cs^+
Dicyclohexyl-14-crown-4 (XVII)	1.1	0	0	0
Cyclohexyl-15-crown-5 (VII)	1.6	19.7	8.7	4.0
Dibenzo-18-crown-6[c] (XXVIII)	0	1.6	25.2	5.8
Dicyclohexyl-18-crown-6[c] (XXXI)	3.3	25.6	77.8	44.2
Dicyclohexyl-21-crown-7 (XXXIV)	3.1	22.6	51.3	49.7
Dicyclohexyl-24-crown-8 (XXXVI)	2.9	8.9	20.1	18.1

[a]Two-phase liquid extraction: methylene chloride and water. Concentration of picric acid: 7×10^{-5} *M*; concentration of polyethers: 7×10^{-5} *M*; concentration of metal hydroxides: 0.1 *M*.

[b]Data taken from Pedersen (1968b).

[c]See Table III, footnote *c*.

Frensdorff systematized extraction by establishing the equilibria governing the partition of alkali-picrate complexes of two macrocyclic polyethers between water and immiscible solvents, and analyzing them in terms of the underlying molecular processes. He also deduced equilibrium constants for the dissociation of complex cation-picrate ion pairs in methylene chloride, and found that complex ion pairs are dissociated considerably more than the ion pairs of the corresponding salts (Frensdorff, 1971b).

4. Crystalline Complexes

The first crystalline salt complex of a crown compound was obtained in 1962. Many more of them have been prepared since then by warming together different proportions of polyethers and salts, or doing this in a solvent, such as methanol, and removing as much of the solvent as necessary to recover the crystals. Examples of salt complexes are given in Table V (Pedersen, 1967b, 1970a; Poonia and Truter, 1973).

TABLE V

Crystalline Complexes of Polyethers

Crystalline complex	Mole ratio[a]	M.P. (°C)	Reference[b]
Benzo-15-crown-5 (IV)			
NaI	1 : 1	152.5-156	*1*
KCNS	2 : 1	176	*2*
Dibenzo-15-crown-5 (XXV)			
KCNS	2 : 1	143-144	*2*
Dibenzo-18-crown-6 (XXVIII)			
KCNS	1 : 1	248-249	*1*
NH_4CNS	1 : 1	187-189	*1*
RbCNS	1 : 1	184-185	*2*
RbCNS	2 : 1	175-176	*2*
CsCNS	2 : 1	146-147	*2*
CsCNS	3 : 2	145-146	*2*
Dicyclohexyl-18-crown-6 (XXXI)			
KI_3	1 : 1	194-195	*2*
CsI_3	3 : 2	112-114	*2*
Dibenzo-24-crown-8 (XXXV)			
KCNS	1 : 1	113-114	*1*
KCNS	1 : 2	132[c]	*3*
Dibenzo-30-crown-10 (XXXVIII)			
KCNS	1 : 1	176-177.5	*1*
NaCNS	1 : 2	176[d]	*3*

[a](Polyether) : (salt).
[b]References *1* Pedersen (1967b), *2* Pedersen (1970a), *3* Poonia and Truter (1973).
[c]With decomposition.
[d]Darkens at 145°C.

The solid salt complexes of polyethers are colorless unless the anion is colored. Many of them form well-defined crystals, usually melting above the melting point of the polyether and sometimes of the salt. Although salts with high lattice energy, such as fluorides, nitrates, and carbonates, form complexes with crown compounds in alcoholic solutions, they cannot be isolated in the solid state because one or the other of the uncomplexed components precipitates when the solutions are concentrated. Most complexes are stable in water containing excess of salt, but they decompose in pure water. There are some exceptions: the KI_3 complex of dicyclohexyl-18-crown-6 is stable indefinitely in contact with water (Pedersen, 1967b, 1970a). It is likely that the salts of other very "soft" acids (Pearson, 1966) will behave in this way.

5. Stoichiometry, Structure, and Stability of Salt Complexes

"The stoichiometry of the complexes is one molecule of polyether per single ion regardless of the valence." This statement occurs in the first two publications on the crown compounds (Pedersen, 1967a, b), but later work showed that the stoichiometry is not as simple as originally assumed.

A systematic overestimation of the diameters of the holes of the crown compounds, and an unintentional disregard of a few contradictory data, delayed the preparation of crystalline complexes with molar excesses of the polyethers. When such experiments were carried out, complexes with molar ratios [(polyether) : (salt)] of 3 : 2 and 2 : 1 were obtained in addition to 1 : 1 (Pedersen, 1970a). Later, complexes with a ratio of 1 : 2 were prepared (Poonia and Truter, 1973).

Dibenzo-18-crown-6, one of the compounds containing the most versatile polyether ring, complexes the cations of the elements listed, with their ionic

TABLE VI

Complexable Cations and Their Diameters in Angstrom Units[a]

Group I		Group II		Group III		Group IV	
Li	1.36						
Na	1.94						
K	2.66	Ca	1.98				
Cu(I)	1.92	Zn	1.48				
Rb	2.94	Sr	2.26				
Ag	2.52	Cd	1.94				
Cs	3.34	Ba	2.68	La	2.30		
Au(I)	2.88	Hg(II)	2.20	Tl(I)	2.80	Pb(II)	2.40
Fr[b]	3.52	Ra	2.80				
NH_4	2.86						

[a]Data taken from Pedersen (1967b). More recent values are given in Dalley, this volume, Chapter 4, Table IV.

[b]Not tested.

TABLE VII

Diameters of Holes in Angstrom Units[a]

Macrocyclic polyethers	Revised diameters
All 14-crown-4 (e.g., XV)	1.2[b]-1.5[c]
All 15-crown-5 (e.g., IV)	1.7-2.2
All 18-crown-6 (e.g., XXVIII)	2.6-3.2[d]
All 21-crown-7 (e.g., XXXIII)	3.4-4.3

[a]Data taken from Pedersen (1970a). For values based on X-ray crystallographic data, see Dalley, this volume, Chapter 4, Table IV.
[b]According to Corey-Pauling-Koltun atomic models.
[c]According to Fisher-Hirschfelder-Taylor atomic models.
[d]The original estimate was 4Å.

diameters, in Table VI (Pedersen, 1967b; Pedersen and Frensdorff, 1972). The transition metals are not included here (see Section I.D).

The estimated diameters of the holes of crown compounds whose polyether rings can be substantially planar are given in Table VII (Pedersen, 1970a).

The existence of a well-defined crystalline salt complex was judged by the height and sharpness of its melting point, and its composition was determined by chemical analysis. If these guides are dependable, the stoichiometry of the complexes in the crystalline state is readily resolved (see Table V), but it is not so simple for their solutions.

There is considerable evidence for the 1 : 1 complex in solution: one atom of K^+ solubilizes 1 mole of dibenzo-18-crown-6 in methanol; one atom of K^+ and 1 mole of dibenzo-18-crown-6 develop fully the characteristic ultraviolet spectrum of the complex in methanol (Pedersen, 1967b); and conductometric titration of potassium chloride with dicyclohexyl-18-crown-6 in methanol and in chloroform-methanol showed convincingly that the stoichiometry is 1 : 1 in both cases (Frensdorff, 1971b). The original concept of a 1 : 1 complex is shown in Fig. 9 (Pedersen, 1967a).

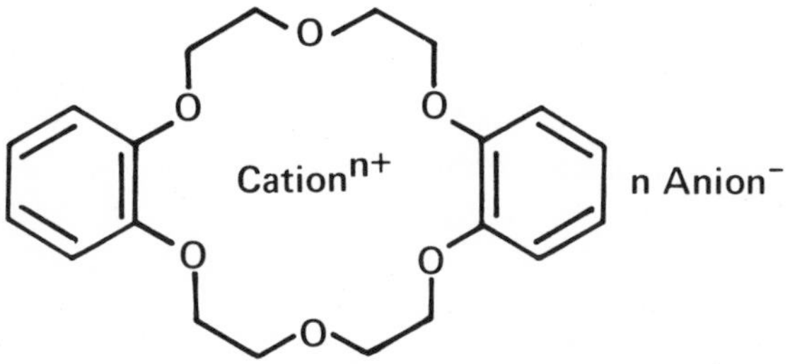

Fig. 9. 1 : 1 Salt complex of dibenzo-18-crown-6 [taken by permission from Pedersen (1967a)].

On the other hand, it was found that 1 mole of cesium thiocyanate solubilized 1.2 moles of dibenzo-18-crown-6 in methanol (Pedersen, 1967b). Pressman observed in his biological study that a rubidium cation could interact with more than one molecule of dicyclohexyl-18-crown-6.* During his work on stability constants, Frensdorff (1971a) obtained evidence of 2 : 1 complexing in methanol solutions for K^+ with cyclohexyl-15-crown-5, and Cs^+ with cyclohexyl-15-crown-5, cyclohexyl-18-crown-6, and dibenzo-18-crown-6.

Considering all these facts relating to stoichiometry, especially the data in Table V, the structures depicted in Fig. 10 were thought to represent roughly the 1 : 1, 2 : 1, and 3 : 2 complexes. The two-layered one was named a "sandwich" complex, and the three-layered one a "club sandwich" complex† (Pedersen, 1970a).

The structures of these complexes are no longer subjects for speculation. Truter and her associates pioneered the crystal structure analysis of the complexes by X-ray diffraction. They published the first results in 1970 (Bright and Truter, 1970a, b), and they have continued to elucidate the structures of many of them.

Their results are in substantial accord with the concept of the 1 : 1 complex: a circular, evenly spaced, planar arrangement of oxygen atoms with a cation located equidistant from them in the same plane. In $(\text{dibenzo-18-crown-6})_1$: $(RbCNS)_1$, however, the cation is out of the plane of the oxygen atoms by 0.08 Å (Bright and Truter 1970a, b). There are small deviations, too, in $(\text{benzo-15-crown-5})_1$: $(NaI)_1H_2O$, $(\text{dibenzo-18-crown-6})_1$: $(NaBr)_1 2H_2O$ (Bush and Truter, 1970, 1971), and $(\text{dicyclohexyl-18-crown-6, isomer B})_1$: $(NaBr)_1 2H_2O$ (Mercer and Truter, 1973).

The existence of a sandwich structure was confirmed by the crystal structure of $(\text{benzo-15-crown-5})_2$: $(KI)_1$, which has the cation located between two parallel rings consisting of five oxygen atoms (Mallinson and Truter, 1972). Such complexes constitute a new group in the class of sandwich compounds previously reported (Taylor *et al.*, 1966; Su and Weiher, 1968).

$(\text{Dibenzo-18-crown-6})_3$: $(RbCNS)_2$ is the only 3 : 2 complex analyzed by X ray so far (Bright and Truter, 1970a, b). It is not a club sandwich complex. In this crystal, one extra molecule of uncomplexed ligand is present for every two molecules of 1 : 1 complex in which the cation is equidistant from six coplanar oxygen atoms. The bond lengths and angles are the same for the uncomplexed and complexed dibenzo-18-crown-6, but the conformation is different. This, however, does not rule out the existence of a club sandwich

*Private communication. B. C. Pressman, Johnson Research Foundation University of Pennsylvania, Philadelphia, Pennsylvania.

†The term "club sandwich" was first proposed by R. Pariser, E. I. du Pont de Nemours and Company, Wilmington, Delaware.

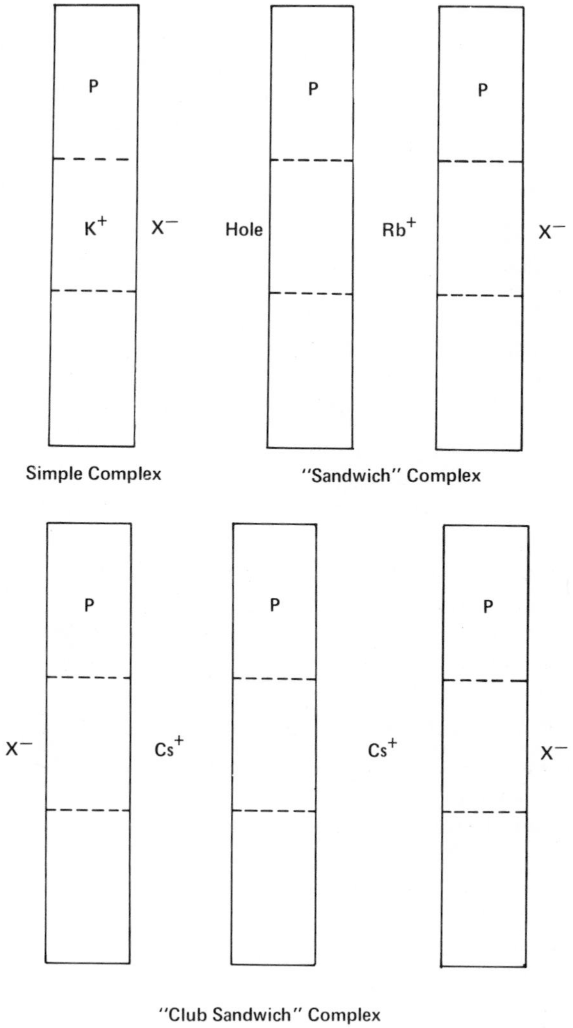

Fig. 10. Rough approximations of complexes; P represents dibenzo-18-crown-6.

structure, and the final word on it will have to wait for an X-ray analysis of a 3 : 2 complex of cesium.

The structure of dibenzo-30-crown-10, as determined by the X-ray crystallographic method, is shown in Fig. 11 (Bush and Truter, 1970). It forms a 1 : 1 complex with potassium iodide in which the ligand is wrapped around the K^+ "like the seam of a tennis ball" (Dobler *et al.,* 1969). The potassium ion is enclosed in a coordination sphere formed by the ten ether-oxygen atoms and

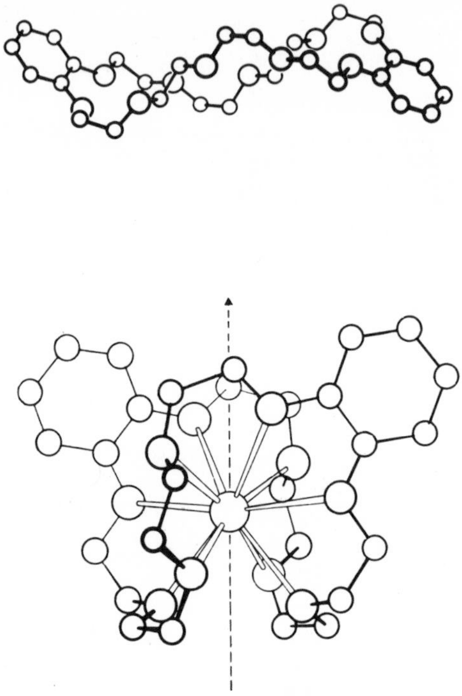

Fig. 11. Crystalline dibenzo-30-crown-10 and its K^+ complex. The halves of the molecule (top) are related by a center of symmetry. The arrow (bottom) shows a crystallographic twofold axis. [Taken by permission from Truter and Pedersen (1971).]

there are no interactions other than those from the ligand. The uncomplexed and complexed ligands have the same bond lengths and angles but a different conformation (Truter and Pedersen, 1971).

The thermal stability of crystalline complexes is attested by these examples: (dibenzo-18-crown-6)$_1$: (KCNS)$_1$ melts sharply at 248-249°C, which is higher than the melting points of the ligand, 164°C, and the salt, 173°C; and (dicyclohexyl-18-crown-6)$_1$: $(KI_3)_1$ melts at 194-195°C to a clear, red-brown liquid without giving off any noticeable iodine vapor, whereas iodine is observable above the uncomplexed salt even at room temperature.

The earliest measurements of the stability constants or the equilibrium constants for complexing were reported by Izatt and associates (1969, 1971), who used a calorimetric titration technique. They were also determined by spectroscopic methods (Wong *et al.,* 1970), and were determined indirectly from conductance and potential measurements on phospholipid bilayers (McLaughlin *et al.,* 1970). It was, however, Frensdorff (1971a) who made the comprehensive study of the stability constants for the complexes of many varieties of synthetic macrocyclic multidentate compounds with univalent cations in water and in methanol by potentiometry with cation-selective electrodes (Eisenman, 1967). His stability constants in methanol are given in Table VIII. These and other constants are fully analyzed (Frensdorff, 1971a). Note the exceptionally high stability of the potassium complex of dibenzo-30-crown-10, whose unusual structure is depicted in Fig. 11.

TABLE VIII

Stability Constants in Methanol at 25°[a]

Polyether	Log K_1'[b]		
	Na^+	K^+	Cs^+
Tetramethyl-12-crown-4[c]	1.41		
Dicyclohexyl-14-crown-4 (XVII)	2.18[d]	1.30[d]	
Cyclohexyl-15-crown-5 (VII)	3.71	3.58[e]	2.78[d,e]
18-Crown-6 (XXVII)	4.32	6.10	4.62
Cyclohexyl-18-crown-6 (XIII)	4.09	5.89	4.30[e]
Dicyclohexyl-18-crown-6 (isomer A, XXXI)	4.08	6.01	4.61
Dicyclohexyl-18-crown-6 (isomer B, XXXI)	3.68	5.38	3.49
Dibenzo-18-crown-6 (XXVIII)	4.36	5.00	3.55[e]
21-Crown-7		4.41	5.02
Dibenzo-21-crown-7 (XXXIII)	2.40[d]	4.30[d]	4.20
24-Crown-8		3.48	4.15
Dibenzo-24-crown-8 (XXXV)		3.49	3.78[d]
Dibenzo-30-crown-10 (XXXVIII)	2.0[f]	4.60[d]	
Dibenzo-60-crown-20		3.90[d]	
Pentaglyme[g]	1.52[d]	2.20[d]	

[a]Data taken from Frensdorff (1971a).
[b]K_1' in liters/mole; uncertainty ±0.04 unless noted otherwise.
[c]Cyclic tetramer of propylene oxide: Down *et al.* (1959).
[d]Uncertainty ±0.08.
[e]Evidence of 2 : 1 complexing.
[f]Uncertainty ±0.2.
[g]$CH_3(OCH_2CH_2)_5OCH_3$.

The conclusion that the relative sizes of the holes and the cations control the stoichiometry, structure, and stability of the complexes is inescapable. In every case, a cation is smaller than its best hole, but it can also be complexed by polyethers with smaller or much larger holes, although with diminished stability. If the hole is large enough, a 1 : 1 complex is formed. [Crystalline (benzo-15-crown-5)$_2$: [NaB(phenyl)$_4$]$_1$ has just been announced (Parsons and associates, 1975). The hole of this ligand is larger than Na^+.] The complex which is formed when the hole is too small for the cation depends on whether it is recovered as crystals or remains in solution. Crystals will be 2 : 1 or 3 : 2 (Pedersen, 1970a), but in solution this is not sufficient, although it may be a necessary condition for 2 : 1 complex formation (Frensdorff, 1971a). The rubidium ion in relation to the 18-crown-6 ring is a borderline case, and both 1 : 1 and 2 : 1 complexes are obtained, at least in the crystalline state (Pedersen, 1970a).

The ultraviolet spectral data (Table III) and the extraction results (Table IV) indicate that Li^+ is best for the 14-crown-4 ring, Na^+ for 15-crown-5, K^+ for 18-crown-6, and K^+ and Cs^+ about equally for 21-crown-7. The stability constants permit even finer distinctions. Of all the complexes of univalent cations and crown compounds tested, the completely symmetrical $(18\text{-crown-}6)_1 : (K^+)$ (Fig. 12) is the most stable. The factors involved in this top stability are (1) the largest circular ring possible with the greatest number of evenly distributed oxygen atoms of high basicity with their maximum charge density directed toward the center of the ring; and (2) a cation that fits the hole in such a way as to make the total attraction between it and the oxygen atoms a maximum and that has relatively slight tendency to solvation. Any deviation from this model will decrease the stability of the complex. Steric hindrance in or near the hole will also interfere with complexing (Pedersen, 1967b).

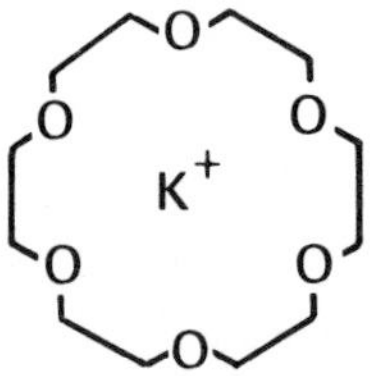

Fig. 12. Potassium complex of 18-crown-6.

D. Other Complexes of Polyethers

The simple ions which are normally complexed by crown compounds are listed in Table VI. This section deals with a miscellaneous group of complexes or pseudo-complexes, the exact natures of some of which are yet unknown.

The dicyclohexyl-18-crown-6 complexes of $[(\text{ion})^+(\text{solvated electron})^-]$, where ion^+ is K^+ or Cs^+ and solvent is tetrahydrofuran, have been prepared. Solutions of these complexes in the solvent are stable for several hours at room temperature even in the absence of excess metal. The solution of the complex was stable for only 5-10 min at room temperature when potassium was reacted in diethyl ether (Dye *et al.*, 1970).

Crystalline oxonium perchlorate complexes of dicyclohexyl-18-crown-6 were characterized by infrared spectroscopy. $(\text{Dicyclohexyl-18-crown-6})_1 : (OH_3{}^+$

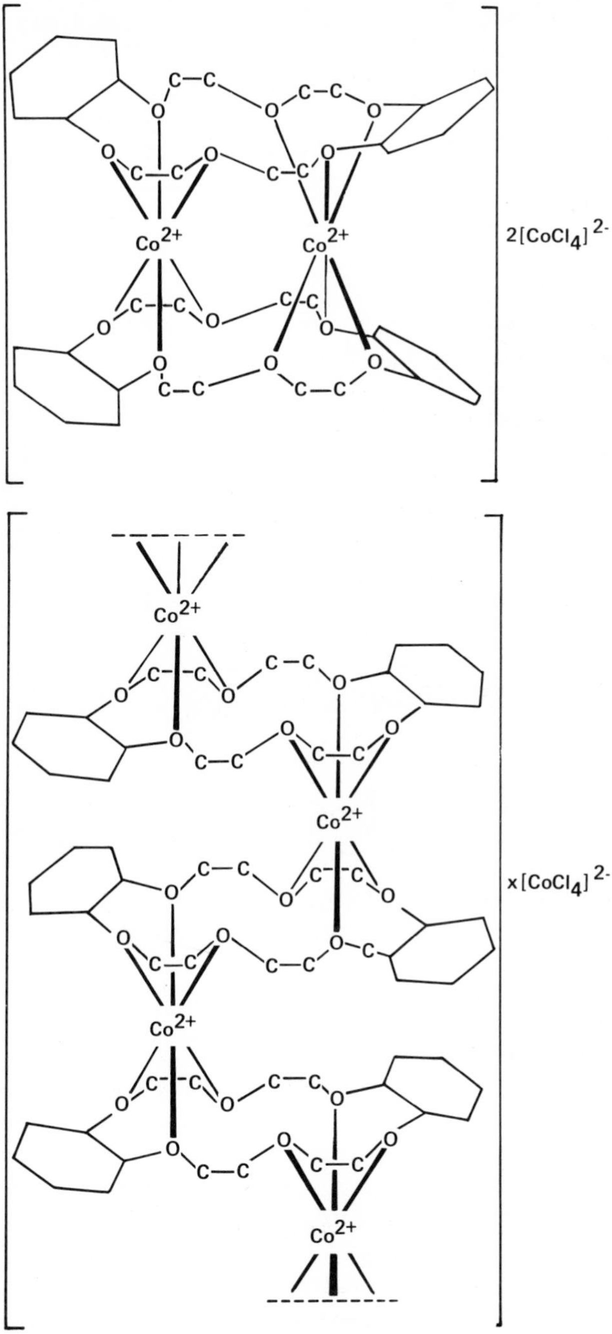

Fig. 13. Crystalline complex of dicyclohexyl-18-crown-6 and $CoCl_2$ [taken by permission from Su and Weiher (1968)].

$ClO_4^-)_1$ melts at 143-144°C, is stable to moisture, and remained unchanged at room temperature for over two years (Izatt *et al.,* 1972).

Complex formation between cyclic polyethers and metal cations, such as Ti^{3+}, V^{3+}, Fe^{3+}, Co^{2+}, and Zn^{2+}, can be observed in organic solvents of low dielectric constants, ϵ = 2-9, by change of colors, change of solubility, or spectral changes of the reactants. The blue crystalline complex [(dicyclohexyl-18-crown-6)$_2$: $Co_2]^{4+}2[CoCl_4]^{2-}$ was isolated from acetic acid, and melted sharply at 238-239°C. In more polar solvents, such as nitromethane, the complex dissociated into the original components. Two probable structures of the complex are shown in Fig. 13. The cation moiety involves a sandwich structure with two Co^{2+} ions squeezed between two polyether molecules (Su and Weiher, 1968). This was the first sandwich complex of a polyether reported as such.

Some substituted derivatives of the ammonium ion are complexed by polyethers. According to their effects on the ultraviolet spectrum of dibenzo-18-crown-6 in methanol, compounds containing $-NH_3^+$, and probably

$$-\overset{|}{C}=NH_2^+,$$

form complexes, but not those containing

$$-\overset{|}{N}H_2^+ \quad \text{or} \quad -\overset{|}{\underset{|}{N}}H^+.$$

Apparently, the first two groups can intrude sufficiently into the polyether ring of dibenzo-18-crown-6 to affect its spectrum but, due to steric hindrance, the last two cannot do so (Pedersen, 1967b).

The idea of forming a thiourea inclusion compound (Schiessler and Flitter, 1952; Powell, 1954) of a salt complex of a polyether was an intriguing one, since it would create a doubly wrapped salt. Many crown compounds and their salt complexes were treated with varying proportions of thiourea in methanol.

It was found that among the compounds tested, ten with polyether rings containing 5-8 oxygen atoms form crystalline adducts with thiourea, the thiourea/polyether mole ratio ranging from 6/1 to 1/1. The salt complexes of two of them, for example, (benzo-15-crown-5)$_2$(KCNS)$_1$ and (dibenzo-18-crown-6)$_1$(NaCNS)$_1$, also form crystalline adducts, the thiorea/complex mole ratio being 1/1 for those with sharp melting points. Thiourea, on the other hand, tends to displace ammonium, rubidium, and cesium salts from their complexes. Some compounds related to thiourea also give crystalline adducts with polyethers, particularly dicyclohexyl-18-crown-6.

The exact nature of these adducts is not known. A polyether ring of sufficient size and the absence of steric hindrance are two of the requisites for their formation. Whatever forces are involved, they are not strong enough to displace potassium from its complexes with the 18-crown-6 polyethers (Pedersen, 1971d, 1975).

A solid adduct of 18-crown-6 with acetonitrile is used to purify the polyether in its preparation from triethyleneglycol and triglycol dichloride, using potassium hydroxide in 10% aqueous tetrahydrofuran. Its stoichiometry depends on conditions, and 18-crown-6 of high purity is recovered on evaporating the acetonitrile (Gokel *et al.*, 1974).

Variously colored complexes of dibenzo-18-crown-6 and bromine in the form of crystalline powder were obtained from chloroform. The molar ratios, polyether/Br_2, ranged from 0.9 to 1.85. The equilibrium constants of complexation in carbon tetrachloride-ethyl bromide solutions were determined by paramagnetic resonance (pmr) spectroscopy (Shchori and Jagur-Grodzinski, 1972a).

E. Substituted Derivatives of Polyethers

Only a few substituted derivatives of crown compounds were prepared in the early days because there was little call for them. Many will be made in the future to introduce the polyether rings into substances designed for specific purposes, such as dyes, products for agriculture, drugs, polypeptides, elastomers, and other polymers useful for conversion into fibers and films.

There are three general methods of preparing substituted crown compounds:

(1) using intermediates carrying the substituents;
(2) running substitution reactions on the polyethers; and
(3) converting the substituents on the polyethers into others.

In the first case, the substituents must withstand the reaction conditions for the synthesis of the crown compounds and not interfere with the reaction. Hydroxyl and hydrocarbon groups are permissible on aliphatic intermediates, and unreactive nonfunctional groups on the aromatic rings. Functional groups, such as carboxyl and extra phenolic hydroxyl groups, must not be present unless they can be protected.

The second method is more applicable to aromatic polyethers than the others because the former will undergo substitution reactions characteristic for aromatic ethers, such as veratrole, provided the conditions do not lead to the scission of the ether linkage (Pedersen, 1967b). They have been halogenated (Shchori and Jagur-Grodzinski, 1972b), nitrated, (Feigenbaum and Michel, 1971), and sulfonated.*

*Unpublished results.

By the third method, carboxyl group has been obtained from methyl,[†] and amino from nitro (Feigenbaum and Michel, 1971). If the substituents do not poison the catalyst and are not removed, substituted aromatic crown compounds can be transformed into the corresponding saturated compounds by catalytic hydrogenation. Some substituents will be reduced simultaneously, such as nitro to amino group.

F. Polymeric Polyethers

Metal-complexing polymers containing chemically bonded polyethers have been prepared. Free radical and anionic polymerization of vinylbenzo-15-crown-5 and vinylbenzo-18-crown-6 gave products with degree of polymerization *n* greater than 300 (Kopolow *et al.*, 1971). Bis(aminobenzo)-18-crown-6 was condensed with isophthaloyl and terephthaloyl chlorides. A specimen of such a polymer complexed (in mole percent): Li^+ 24, Na^+ 54, K^+ 64, Cs^+ 0, Ca^{2+} 0, and Ba^{2+} 0. It is noteworthy that the barium ion was not complexed under these conditions (Feigenbaum and Michel, 1971). Dibenzo-18-crown-6 was converted into polymers by reaction with chloral, bromal, or glyoxalic acid in presence of a strong acid catalyst (Takekoshi and Webb, 1974).

A resin containing 88% by weight of crosslinked dibenzo-18-crown-6 was obtained by mixing a chloroform solution of the crown compound with glacial acetic acid containing paraformaldehyde and a little concentrated sulfuric acid. The resin picked up 151% of KI_3 on immersion in an aqueous solution of this salt, and still held 129% after an overnight contact with lots of water. It is likely that other salts with very "soft" anions will be taken up, but salts with "hard" anions, such as potassium chloride, acetate, and sulfate, were readily displaced from the resin by water. Benzo-15-crown-5 was also converted into a formaldehyde resin.*

G. Polyethers Containing —S—

The preparation of polyether sulfides was begun in 1964 and continued sporadically as time and suitable intermediates became available. In general, the compounds were prepared by refluxing in 1-butanol under nitrogen vicinal mercaptophenols and cyclic dithiols with equivalent proportions of terminally substituted ether dichlorides and sodium hydroxide. The use of $Cl-(CH_2-CH_2-S)_n-CH_2-CH_2-Cl$ was considered but discarded because these dichlorides are dangerously vesicant, especially if *n* is more than 1. The polyether sulfides which were synthesized are depicted in Fig. 14, and some of their properties are listed in Table IX (Pedersen, 1971b, 1974)

[†]Private communication. D. J. Sam, E. I. du Pont de Nemours and Company, Wilmington, Delaware.

*Unpublished results.

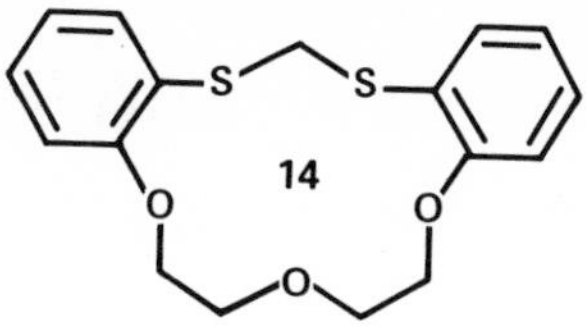

LXIII

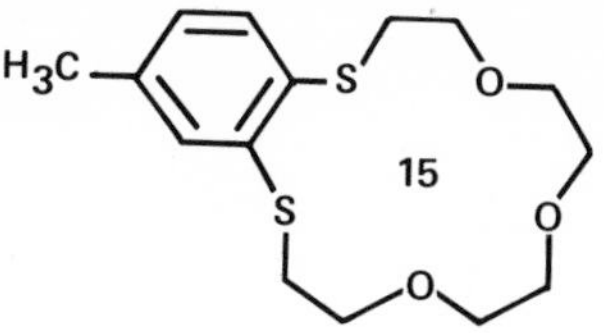

LXIV

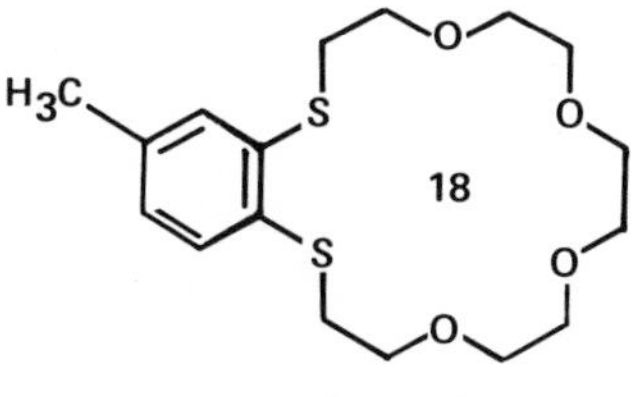

LXV

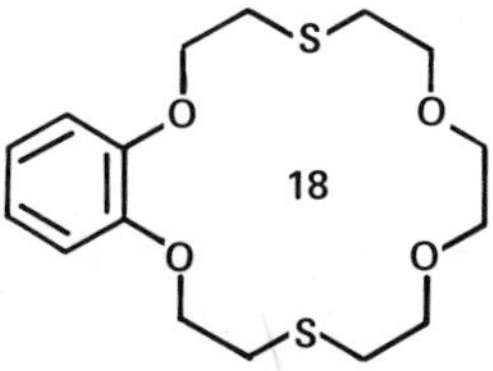

LXVI

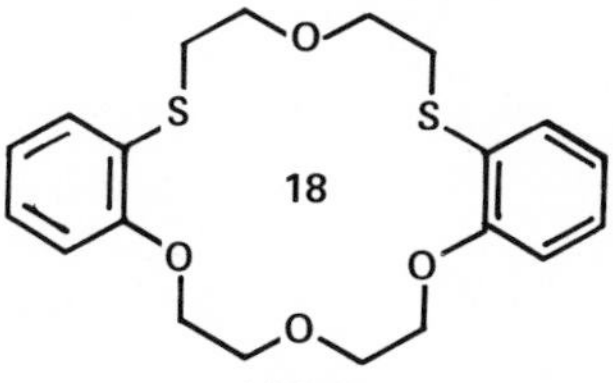

LXVII

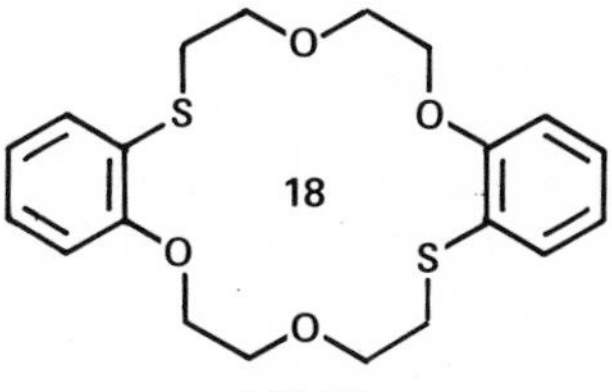

LXVIII

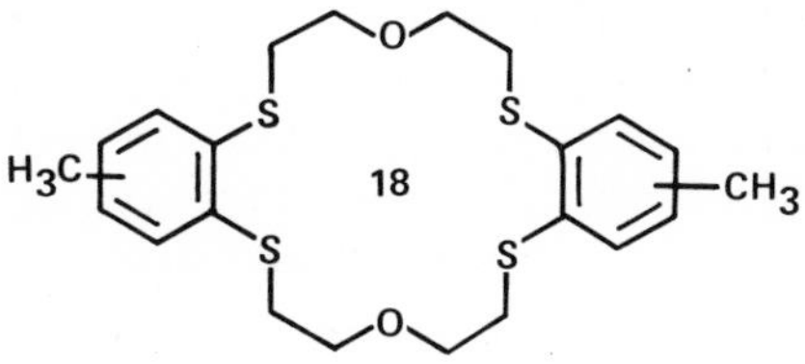

LXIX

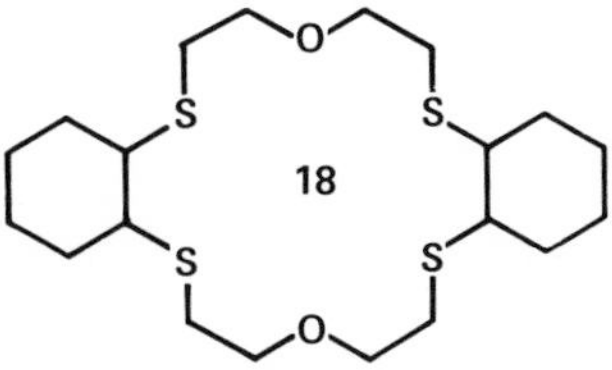

LXX

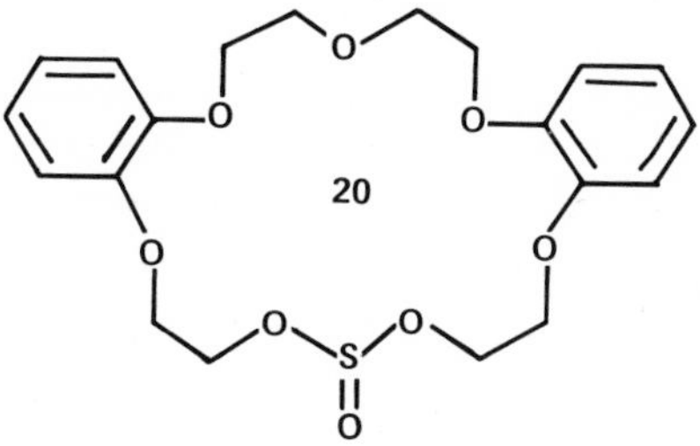

LXXI

Fig. 14. Structural formulas of polyether suflides synthesized [taken by permission from pedersen (1971b)].

TABLE IX

Data on Polyether Sulfides[a]

Compound[b]	Formula	Calc. mol. wt.	M.P. (°C)
LXIII	$C_{17}H_{18}O_3S_2$	334	150-153
LXIV	$C_{15}H_{22}O_3S_2$	314	Viscous oil
LXV	$C_{17}H_{26}O_4S_2$	358	Viscous oil
LXVI	$C_{16}H_{24}O_4S_2$	344	91
LXVII	$C_{20}H_{24}O_4S_2$	392	143-144
LXVIII	$C_{20}H_{24}O_4S_2$	392	114-115
LXIX	$C_{22}H_{28}O_2S_4$	452	147
LXX	$C_{20}H_{36}O_2S_4$	436	Viscous oil
LXXI	$C_{20}H_{24}O_8S$	424	133

[a]Data taken from Pedersen (1971b).
[b]Numbers refer to compounds in Fig. 14.

All tests indicate that the polyether sulfides are excellent for complexing silver but not sodium or potassium (Pedersen, 1971b; Frensdorff, 1971a).

The strong complexing power of the crown compounds for the alkali and alkaline earth cations is destroyed by substituting —S— for —O—, probably because the symmetrical distribution of the negative charge around the hole of the polyethers is perturbed by the larger sulfur atom, its lower electronegativity, and the different bond angles associated with the sulfur atom. All these factors tend to weaken the electrostatic attraction for these cations.

Frensdorff concluded that the powerful complexing of silver by the polyether sulfides provides the first evidence of complexing by other than purely electrostatic forces.

Crystalline complexes of silver nitrate with LXVII and LXX (Fig. 14) were obtained, but no crystalline complex of potassium thiocyanate with LXVII was formed when an attempt was made to prepare it by the method which gave crystalline (dibenzo-18-crown-6)$_1$: (KCNS)$_1$ (Pedersen, 1971b).

Some reactions can be carried out on the polyether sulfides which cannot on crown compounds. For example, LXX was oxidized to the tetrasulfoxide with hydrogen peroxide, and to the tetrasulfone with an excess of the same reagent.*

Polyether sulfides have been reviewed recently (Christensen *et al.,* 1974; Bradshaw and Hui, 1974).

*Unpublished results.

H. Polyethers Containing $-\overset{R}{N}-$

M. H. Bromels collaborated in the work on polyethers with one or more —O— links replaced by

$$-\overset{R}{N}-$$

groups, and she prepared most of the compounds shown in Fig. 15 (Pedersen and Bromels, 1974). Some of their properties are listed in Table X.

TABLE X

Data on Polyethers Containing $-\overset{R}{N}-$[a]

Compound[b]	Formula	Calc. mol. wt.	M.P. (°C)
LXII	$C_{14}H_{21}NO_4$	267	100-101
LXXIII	$C_{16}H_{25}NO_5$	311	Below 26
LXXIV	$C_{20}H_{25}NO_5$	359	152-153
LXXV	$C_{22}H_{29}NO_5$	387	130-132
LXXVI	$C_{28}H_{41}NO_5$	471	109-110
LXXVII	$C_{50}H_{68}N_2O_{10}$	756	122-124
LXXVIII	$C_{22}H_{27}NO_6$	401	198
LXXIX	$C_{20}H_{26}N_2O_4$	358	175-176
LXXX	$C_{34}H_{38}N_2O_4$	538	176

[a]Data taken from Pedersen and Bromels (1974).
[b]Numbers refer to compounds in Fig. 15.

LXXII and LXXIII (Fig. 15) were prepared by first reacting 2-aminophenol with tetraglycol dichloride or pentaglycol dichloride to form (2-hydroxylphenyl)-NH—$(CH_2-CH_2-O)_n-CH_2-CH_2-Cl$:HCl ($n$ = 3 or 4), then adding sodium hydroxide to close the ring by reacting the remaining chloride with the phenolic hydroxyl group.*

Frensdorff (1971a) found that in methanol, the logarithm of the stability constant, K_1' for the potassium complexes of dibenzo-18-crown-6, LXXVI, LXXIV, and LXXIX (Fig. 15) are 5.00, 4.10, 3.20, and 1.63, respectively. In water, log K_1' for the silver complex of 18-crown-6, 1.6, is increased to 7.8 for the compound with two opposing oxygen atoms replaced by

$$-\overset{H}{N}-$$

*Unpublished results.

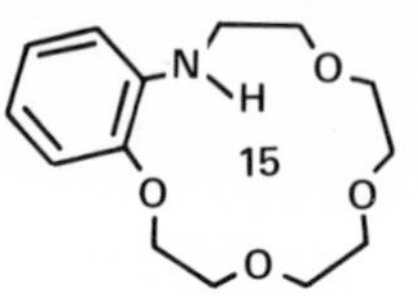

LXXII

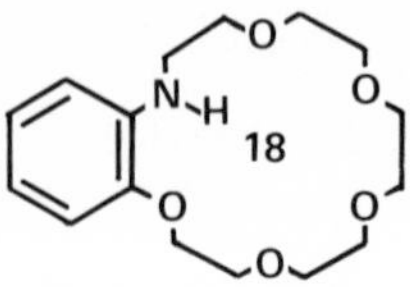

LXXIII

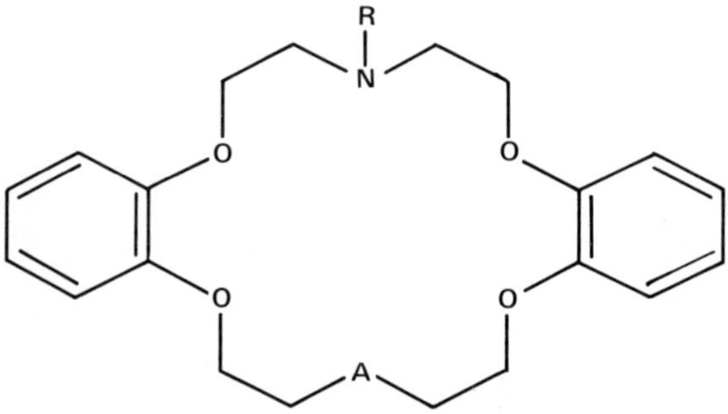

Compound	R	A
LXXIV	H	O
LXXV	C_2H_5	O
LXXVI	n–C_8H_{17}	O
LXXVII[a]	$(CH_2)_{10}$	O
LXXVIII	CH_3–C(O)	O
LXXIX	H	N–H
LXXX	CH_2–C_6H_5	N–CH_2–C_6H_5

Fig. 15. Polyethers containing $-\overset{R}{N}-$. [a]Two polyether rings connected.

He concluded that covalent bonding between silver and amino group was involved here rather than simple electrostatic attraction.

According to ultraviolet spectral data, complexing power for cations is destroyed by acids or by acetylating the amino group.*

There is no advantage in replacing oxygen atoms in the polyether ring with

$$-\overset{R}{N}-$$

*Unpublished results.

unless the trivalency of nitrogen or sensitivity to acid is the desired property.

Compound LXXII was converted to the *N*-nitroso derivative, which was recovered as white fibrous crystals melting at 95-96°C. It was isomerized to the ring-nitroso compound, which was a yellow solid.†

Polyethers containing

R
–N–

have been reviewed recently (Christensen *et al.*, 1974).

Polyethers containing

R
–N–

differ from those which do not in several ways. The complexing power of the former is dependent on pH, whereas that of the latter is not. Compounds containing two amino links, such as LXXIX, can serve as intermediates for the synthesis of bicyclic compounds.

The speculative reaction shown in Fig. 16 brings up an interesting possibility. A molecular model of LXXXI indicates that the amino hydrogen juts into the polyether ring and either prevents the formation of complexes or greatly reduces their stability. On the other hand, the quinoid structure, LXXXII, resulting from the removal of two hydrogen atoms, having an unobstructed hole, might form complexes as well as the oxygen analogue, benzo-18-crown-6. If these assumptions are correct, the redox potential of LXXXI should respond to the presence of ions complexable by the quinoid compound, and their concentrations could be determined directly by measuring the redox potential. This effect might be made visible by replacing the phenolic hydroxyl group by a conjugate dye moiety. These ideas have not yet been tested.

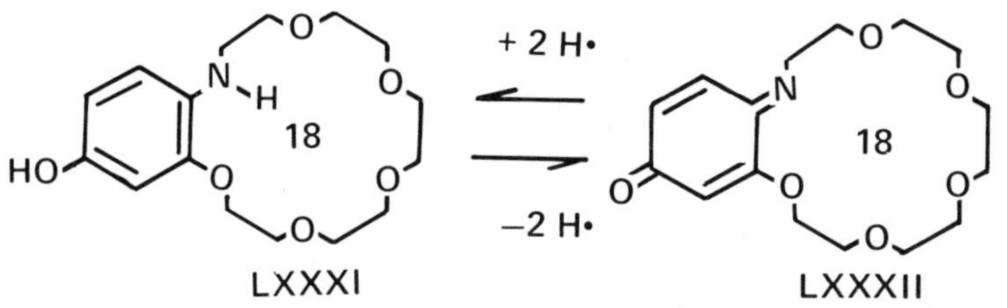

Fig. 16. Speculative compounds.

†Unpublished results.

II. MACROBICYCLIC COMPOUNDS

By the end of a year's work preparing and testing many crown compounds, it became evident that nearly all their salt complexes are decomposed by water. There seemed to be at least two ways of thwarting the attack of water molecules from the open sides of the complex by placing obstacles in the way. One was to attach a lid to the crown compound in such a position that it would shield the complexed ion with a protective cover. The other was to convert the polyether into a bicyclic compound to improve its complexing power, like a hat made windproof by a strap under the chin. The latter course of action was chosen.

A. Polyethers (Lanterns)

The first compound selected for synthesis, shown in Fig. 17,* might be considered the three-dimensional form of XV and expected to hold the smallest cations in its cavity. Its molecular model was readily constructed, and its resemblance to a hanging lantern suggested the epithet "lantern" for this type of compound. Besides, it could be lighted like a lantern by putting a cation in it as if it were a flaming candle.

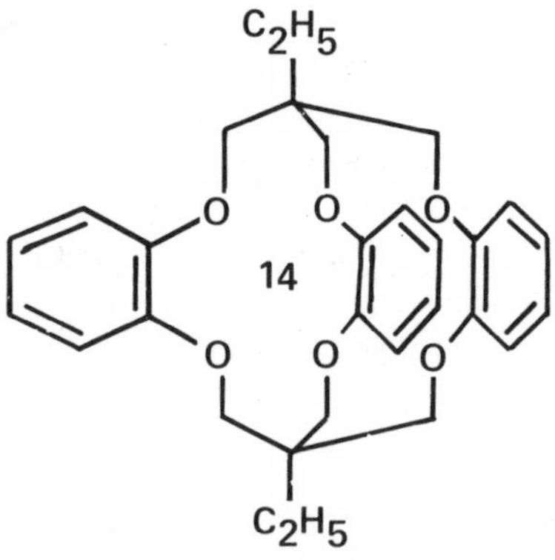

LXXXIII

Fig. 17. Lantern.

At least on paper, LXXXIII could be prepared by the overall reaction

$$2C_2H_5{-}C(CH_2{-}Cl)_3 + 3\ \text{Catechol} + 6NaOH \longrightarrow C_{30}H_{34}O_6 + 6NaCl + 6H_2O$$

It did not take long to find out, however, that organic halides with a neopentyl

*Unpublished matter.

structure are very unreactive, and compounds with bridgehead carbon atoms are not easy to synthesize. Some other tack had to be taken, and an interesting development elsewhere pointed the new direction.

B. Diamines

Although these compounds are not polyethers, they form a link with the important class of bicyclic polyether diamines whose salt complexes are called cryptates (see Section II.C).

Simmons and Park (1967) announced the preparation of bicyclic diamines with bridgehead nitrogen atoms. They later (Simmons and Park, 1968) published their synthesis by a two-step application of a procedure by Stetter and Marx (1957).

These diamines undergo the newly discovered geometrical "out-in" isomerism (Park and Simmons, 1968a), and encapsulate halide ions by a process called "katapinosis" (Fig. 18, LXXXVI) (Park and Simmons, 1968b). The ability to encapsulate Cl^- (diameter 3.62 Å), Br^- (3.90 Å), or I^- (4.32 Å) depends upon the size of the cavity in the molecule of the diamine. When each of the three chains connecting the two nitrogen atoms consists of eight atoms, none of these halides is accommodated; when nine atoms each, Cl^- and Br^- are taken in but not I^-; when ten atoms each, all of them are readily incorporated.

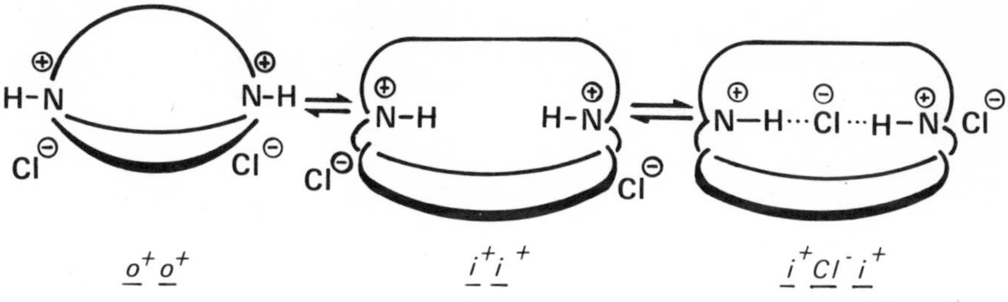

LXXXIV LXXXV LXXXVI

Fig. 18. Encapsulation of chloride ion [taken by permission from Park and Simmons (1968)].

C. Polyether Diamines (Cryptates)

By combining the structural elements of the polyethers and the bicyclic diamines, the following compounds were conceived: the structure in Fig. 17 with each bridgehead C_2H_5-C replaced by N; $N(-CH_2-CH_2-O-CH_2-CH_2-O-CH_2-CH_2-)_3N$; and $N(-CH_2-CH_2-O-CH_2-CH_2-O-CH_2-CH_2-O-CH_2-CH_2-)_3N$.

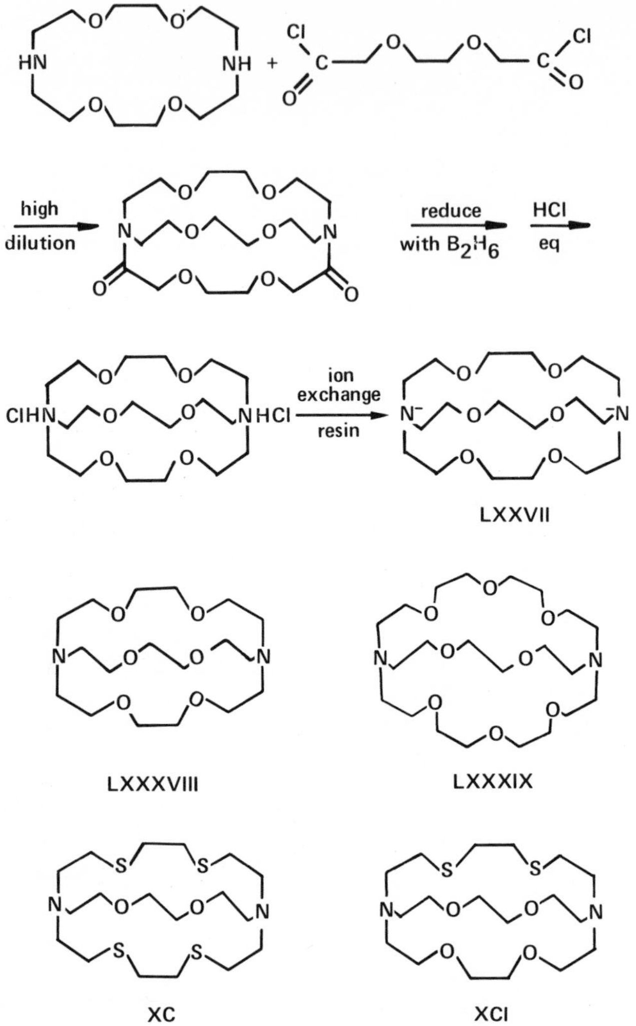

Fig. 19. Synthesis of bicyclic polyether diamines [taken by permission from Truter and Pedersen (1971)].

In the meantime, the preparation of such compounds had been undertaken independently in France, and Lehn and his collaborators announced their admirable synthesis of diazapolyoxamacrobicyclic compounds in 1969 (Dietrich *et al.*, 1969a). At the same time, they described the salt complexes of these compounds, which they named cryptates, in another article (Dietrich *et al.*, 1969b). The method of synthesis and a variety of bicyclic polyether diamines prepared by them are shown in Fig. 19 (Truter and Pedersen, 1971). They have been fully described only recently (Lehn, 1975).

As expected, these compounds are potent complexing agents for many salts. For example, an aqueous solution of LXXXVII slowly dissolves barium sulfate; a benzene solution takes up potassium permanganate; and a chloroform solution dissolves potassium thiocyanate in seconds. The rate of uptake of salts is related to the nature of the anion, the presence of traces of water, and the temperature (Dietrich *et al.*, 1969b).

Bromels prepared a bicyclic polyether diamine by reacting LXXIX with *o*-phenylene($-O-CH_2-CH_2-O-$tosyl$)_2$ (Pedersen and Bromels, 1974).

III. OPEN-CHAIN POLYETHERS

Although it was taken for granted from the beginning that the cyclic polyethers were better complexing agents than open-chain polyethers of equal chain length, several examples of the latter were eventually synthesized for comparison. They were prepared by reacting derivatives of catechol with appropriate organic chlorides in the presence of sodium hydroxide.* Some of them are shown in Fig. 20.

Structurally, these open-chain polyethers are derived from crown compounds by breaking a bond in their polyether rings: XCII from XXV; XCIII from X; XCIV and XCV from XXVI; XCVI from XLVIV; and XCVII is a hydrolytic product from LIV. Data on these compounds are given in Table XI.

TABLE XI

Data on Open-Chain Polyethers

Compound[a]	Formula	Calc. mol. wt.	M.P. (°C)
XCII	$C_{18}H_{22}O_5$	318	79-80
XCIII	$C_{16}H_{26}O_6$	314	Below 26[b]
XCIV	$C_{20}H_{26}O_6$	362	83-85
XCV	$C_{20}H_{26}O_6$	362	92-95
XCVI	$C_{24}H_{26}O_6$	410	124-126
XCVII[c]	$C_{20}H_{30}O_9$	414	73-91

[a]Numbers refer to compounds in Fig. 20.
[b]Boiling point: 346°C/752 mm.
[c]Recovered as the dihydrate, which gradually loses water on melting.

*Unpublished results.

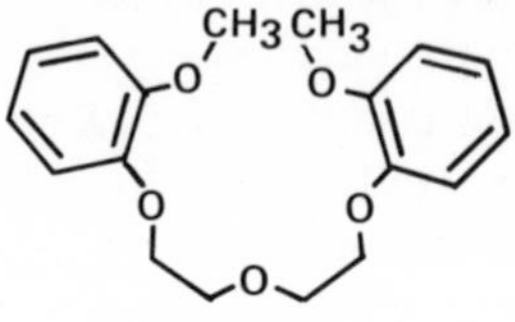

XCII

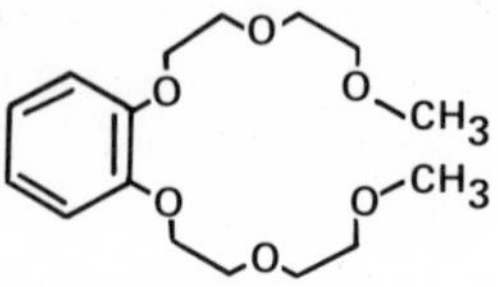

XCIII

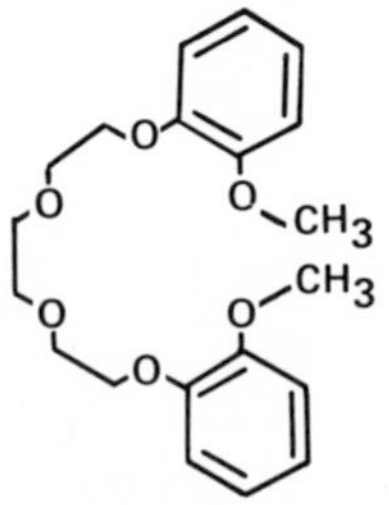

XCIV

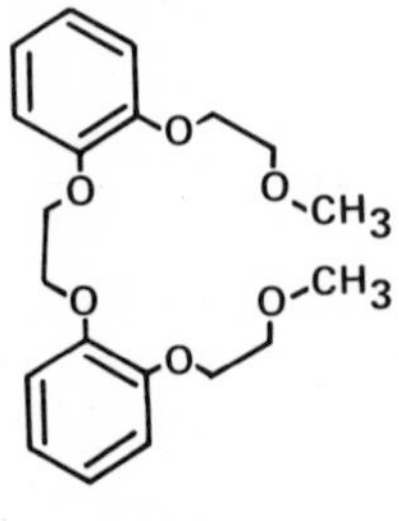

XCV

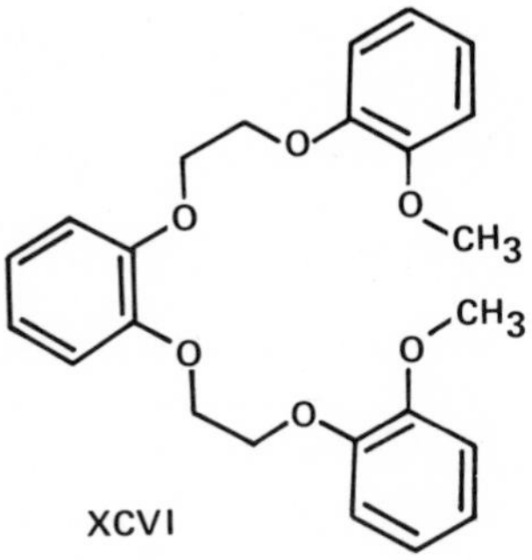

XCVI

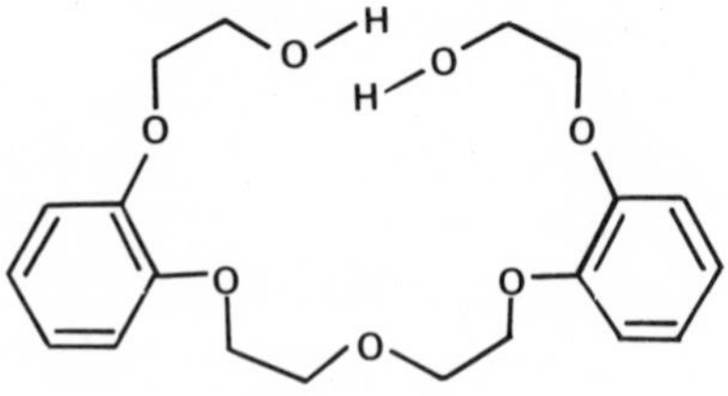

XCVII

Fig. 20. Open-chain polyethers.

The relative complexing powers of some open-chain polyethers and dicyclohexyl-18-crown-6, XXXI (Fig. 4), as measured by two-phase liquid extraction, are shown in Table XII.*

TABLE XII

Comparison with Open-Chain Polyethers[a]

Polyether[b]	Mole ratio (polyether picric acid)	K^+ extracted (%)
XXXI	1	95
	10	Over 99
$CH_3-(O-CH_2-CH_2)_4-OCH_3$	10	5
$CH_3-(O-CH_2-CH_2)_5-OCH_3$	1	1
	10	28
XCII	10	12
XCVI	10	6
$CH_3-(O-CH_2-CH_2)_6-OCH_3$	10	47
$CH_3-(O-CH_2-CH_2)_7-OCH_3$	1	15
	10	75

[a]Two-phase liquid extraction: methylene chloride and water. Concentration of picric acid: 2×10^{-4} M; concentration of KOH: 1 M.
[b]Numbers refer to compounds in Figs. 4 and 20.

The crown compound is more effective than the others, but the efficiency of the open-chain compounds rises steadily with increasing chain length. The replacement of aliphatic ether links with aromatic ones decreases the basicity of the oxygen atoms, thereby diminishing the complexing power.

Even if the open-chain polyethers are less effective complexing agents than the cyclic polyethers, crystalline salt complexes of two of them have been obtained:† $(XCIII)_1 : (KCNS)_1$ melting at 76-78°C; $(XCIV)_1 : (KCNS)_1$, 123-125°C; and $(XCIV)_1 : (NH_4CNS)_1$, 101.5-109.5°C.

Hurd and Sims (1949) reported that phenacyl kojate forms crystalline complexes with sodium salts $(\text{phenacyl kojate})_2 : (NaX)_1$.** The chloride melts at 158°C and the bromide at 197°C. The structures of the complexes of phenacyl kojate with alkali metal salts have been determined by X-ray analysis (Phillips and Truter, 1974, 1975a, b).

Cotton and Wilkinson (1962) mentioned two crystalline etherates of alkali

*Unpublished results.
†Unpublished results.
**C. D. Hurd called the author's attention to these complexes on May 25, 1970.

metal salts:

$$[Na(CH_3{-}O{-}{-}CH_2{-}CH_2{-}O{-}CH_2{-}CH_2{-}O{-}CH_3)_2][Ta(CO)_6]$$

and

$$[K(CH_3{-}O{-}CH_2{-}CH_2{-}O{-}CH_2{-}CH_2{-}O{-}CH_3)_3][MO(CO)_5I].$$

Facts disclosed in this section support the "template effect" (see Section I.B.5). Compound XCVII, obtained by acid hydrolysis of LIV, which might tend to close the polyether ring temporarily by hydrogen bonding, can be converted into cyclic compounds with a variety of reagents: cyclohexanone, 16% yield; phosgene, 48% (Pedersen, 1970b); and thionyl chloride, 33% (Pedersen, 1971b).

Compounds having the structures

$RO{-}(CH_2{-}CH_2{-}O)_n{-}CH_2{-}COOH$ and 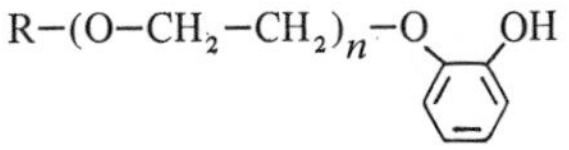

where R = H or alkyl, and n = 3-5, form salts which are very soluble in organic solvents. The solubility of potassium palmitate, $CH_3{-}(CH_2)_{14}{-}COOK$, in benzene at room temperature is no higher than 0.0002 M, but the solubility of $RO{-}(CH_2{-}CH_2{-}O)_3{-}CH_2{-}COOK$ in benzene is over 2 M.*

The high solubility of these salts in aprotic solvents must be due to their structures, which probably consist of a cation near the carboxylate or phenoxide group, and the ethereal tail wrapped around the ion to provide a lipophilic exterior. This configuration is maintained by the attraction between the cation and the oxygen atoms of the ether links.

IV. USES FOR POLYETHERS

It is not possible to list in this space all the different purposes for which the crown compounds have been tested or suggested for testing. Their uses can only be considered in the broadest aspects.

The crown compounds are useful because they complex the salts of the alkali and alkaline earth elements and some others, thereby profoundly altering their properties.

Besides this, they have unique qualities which permit their use under all sorts of conditions and assure their survival in many chemical reactions: insensitivity to the full range of pH; and inertness to a great variety of reagents, depending on whether they are aromatic or saturated. In addition, they are able to complex ions of large molecules if they have a small positively charged group, such as $-NH_3^+$, in an accessible position.

*Unpublished results.

The melting point, hygroscopicity, and electrical and other properties of crystalline salt complexes differ from those of the corresponding salts, and such changes might be useful. Crown compounds can act as salt scavengers in solid systems.

The solubilization of many salts by crown compounds in organic solvents, especially aprotic solvents, opens up a wider field of application. Ionic reactions, whether chemical or electrochemical, can now be run in media once considered impossible. This new technique can be used in any branch of experimental chemistry. "Naked" anions made by this method catalyze and promote many organic reactions, and alter the course of others.

"The use of crown ethers as additives to organic reactions is limited solely by the chemist's imagination and will almost certainly continue to yield new and exciting results" (Baillargeon, 1974).

Complexation of silver brings up photography, and of gold and radium the recovery of precious metals. Ion selectivity will permit the separation of salts of different metals and the construction of ion-selective electrodes.

Having such profound effects on sodium and potassium ions, which are so vital to life processes, it seems inevitable that, either as tools in biological studies or as integral parts of drugs and agricultural chemicals, the crown compounds will eventually contribute directly to the betterment of life.

V. TOXICITY

When dealing with new compounds with unusual properties, their careless handling must be avoided in case they turn out to be injurious. Moreover, this must be continued for years to guard against possible dangers arising from long-delayed harmful action.

Crown and related compounds, by interacting strongly with the alkali and alkaline earth cations, would be expected to influence the vital processes in which the cations participate. Although the physiological effects of crown compounds have not yet been studied extensively, the available results call for care in handling them (Pedersen, 1967b).

The approximate lethal dose for dicyclohexyl-18-crown-6 for ingestion by rats was 300 mg/kg. In a 10-day subacute oral test, the compound did not exhibit any cumulative oral toxicity when administered to male rats at a dose level of 60 mg kg^{-1} day^{-1}. It should be noted that dosage at the ALD level caused death in 11 min, but that a dose of 200 mg/kg was not lethal in 14 days.

Dicyclohexyl-18-crown-6 produced some generalized corneal injury, some iritic injury, and conjunctivitis when introduced as a 10% solution in propylene glycol. Although tests are not complete, there may be permanent injury to the eye even if the eye is washed after exposure.

Dicyclohexyl-18-crown-6 is readily absorbed through the skin of test animals. It caused fatality when absorbed at the level of 130 mg/kg. Primary skin irritation tests run on this compound indicate that it should be considered a very irritating substance.

It has been reported (Leong *et al.*, 1974; Leong, 1975) that the inhalation of the vapors of 12-crown-4 (ethylene oxide cyclic tetramer) by rats caused testicular atrophy besides other injuries.

Except for the lithium ion, this compound is a relatively poor complexing agent for the alkali cations. If its injurious effects are due to its interaction with physiologically active ions, the more effective crown compounds should be treated with caution.

ACKNOWLEDGMENTS

The author's work on the crown compounds described in this chapter was carried out while he was a research associate in the Research and Development Division of the Elastomer Chemicals Department of E. I. du Pont de Nemours and Company, during the consecutive directorship of Dr. A. S. Carter and Dr. H. E. Schroeder. Without their sympathetic interest and support, the program would not have prospered. The author also acknowledges the following:

Dr. Rudolph Pariser, laboratory director, for his constant encouragement and the creation of a lovely pin;

Dr. H. K. Frensdorff who made the first comprehensive studies of the distribution equilibria and stability constants of the complexes, for his many courtesies during the years and his advice on the preparation of this chapter;

Mrs. Marilyn H. Bromels for collaborating on the nitrogen crowns and for providing the pearl for the crown with the feminine touch;

Dr. R. N. Greene for improving the synthesis of the crown compounds and promoting their use in organic chemistry;

Dr. G. S. Wich for providing information for the preparation of this chapter;

The experts of the analytical divisions whose many skills sped the research by the prompt identification of reaction products;

All the laboratory colleagues who listened with friendly ears and an open mind;

Drs. B. C. McKusick and H. E. Simmons whose early interest in the crown compounds helped to promote the testing of these compounds;

Dr. D. J. Sam who undertook extensive work with the crown compounds in the early days;

The late Sir Ronald Nyholm who brought the crown compounds and the author to London, for his influential interest and great kindness;

Professor Mary R. Truter and her group who first elucidated the structures of crown complexes by X-ray analysis, for an enduring association which began in London;

Professor D. H. Busch for suggesting that the first public announcement of the crown compounds be made at the Tenth International Conference on Coordination Chemistry in Nikko, Japan (1967), and helped to make this possible;

Professor B. C. Pressman for pioneering the use of crown compounds in biological studies;

Professor George Eisenman for generously sharing his expert knowledge of ion-selective electrodes and the results of his many biological experiments with the crown compounds;

Professor D. J. Cram who has involved the crown compounds in his ingenious and pioneering work in "host-guest" chemistry, for a private viewing of some of his results;

Professors J. J. Christensen and R. M. Izatt who began the physical chemical investigation of crown compounds, for inviting the author to write this chapter;

Professors G. A. Berchtold, J. H. Richards, J. D. Roberts, and G. W. Watt for valuable consultations;

The many investigators who requested samples for use in their studies, for adding support to the belief that all this was not just a flash in the flask;

The authors and publishers who kindly gave permission to reproduce facts and figures from their publications.

Finally, last but not least, thanks to Mr. T. T. Malinowski, whose blithe spirit, mechanical ingenuity, and ever-willing help hastened the project to beat the deadline of retirement of April 3, 1969.

REFERENCES

Ackman, R. G., Brown, W. H., and Wright, G. F. (1955). *J. Org. Chem.* **20**, 1147-1158.

Baillargeon, M. J. (1974). Org. Chem. Seminar, Univ. Illinois, Urbana, Illinois, December 12, pp. 1-9.

Bradshaw, J. S., and Hui, J. Y. K. (1974). *J. Heterocycl. Chem.* **11**, 649-673.

Bright, D., and Truter, M. R. (1970a). *Nature (London)* **255**, 176-177.

Bright, D., and Truter, M. R. (1970b). *J. Chem. Soc. B* 1544-1550.

Brockmann, H., and Schmidt-Kastner, G. (1955). *Chem. Ber.* **88**, 57-61.

Bush, M. A., and Truter, M. R. (1970). *J. Chem. Soc. Chem. Commun.* 1439-1440.

Bush, M. A., and Truter, M. R. (1971). *J. Chem. Soc. B* 1440-1446.

Christensen, J. J., Hill, J. O., and Izatt, R. M. (1971). *Science,* **174**, 459-467.

Christensen, J. J., Eatough, D. J., and Izatt, R. M. (1974). *Chem. Rev.* **74**, 351-384.

Corbaz, R., Ettlinger, L., Gaumann, E., Keller-Schierlein, W., Kradolfer, F., Neipp, L., Prelog, V., and Zahner, H. (1955). *Helv. Chim. Acta* **38**, 1455-1448.

Cotton, F. A., and Wilkinson, G. (1962). *In* "Advanced Inorganic Chemistry," p. 318. Wiley (Interscience), New York.

Cram, D. J., and Cram., J. M. (1974). *Science* **183**, 803-809.

Dale, J., and Kristiansen, P. O. (1971). *J. Chem. Soc. Chem. Commun.* 670-671.

Dale, J., and Kristiansen, P. O. (1972). *Acta Chem. Scand.* **26**, 1471-1478.

Dietrich, B., Lehn, J. M., and Sauvage, J. P. (1969a). *Tetrahedron Lett.* 2885-2888.

Dietrich, B., Lehn, J. M., and Sauvage, J. P. (1969b). *Tetrahedron Lett.* 2889-2892.

Dobler, M., Dunitz, J. D., and Kilbourn, B. T. (1969). *Helv. Chim. Acta* **52**, 2573-2583.

Down, J. L., Lewis, J., Moore, B., and Wilkinson, G. (1957). *Proc. Chem. Soc.* 209-210.

Down, J. L., Lewis, J., Moore, B., and Wilkinson, G. (1959). *J. Chem. Soc.* 3767-3773.

Dye, J. L., DeBacker, M. G., and Nicely, V. A. (1970). *J. Am. Chem. Soc.* **92**, 5226-5228.

Eisenman, G. (ed.) (1967). *In* "Glass Electrodes for Hydrogen and Other Cations." Dekker, New York.

Eisenman, G., and Ciani, S. (1968). *Fed. Proc.* **27**, 1289-1304.

Feigenbaum, W. M., and Michel, R. H. (1971). *J. Polym. Sci.* (A-2) 9, 817-820.

Frensdorff, H. K. (1971a). *J. Am. Chem. Soc.* **93**, 600-606.

Frensdorff, H. K. (1971b). *J. Am. Chem. Soc.* **93**, 4684-4688.

Gokel, G. W., Cram, D. J., Liotta, C. L., Harris, H. P., and Cook, F. L. (1974). *J. Org. Chem.* **39**, 2445-2446.

Greene, R. N. (1972). *Tetrahedron Lett.* 1793-1796.
Hurd, C. D., and Sims, R. J. (1949). *J. Am. Chem. Soc.* **71**, 2440-2443.
Izatt, R. M., Rytting, J. H., Nelson, D. P., Haymore, B. L., and Christensen, J. J. (1969). *Science* **164**, 443-444.
Izatt, R. M., Nelson, D. P., Rytting, J. H., Haymore, B. L., and Christensen, J. J. (1971). *J. Am. Chem. Soc.* **93**, 1619-1623.
Izatt, R. M., Haymore, B. L., and Christensen, J. J. (1972). *J. Chem. Soc. Chem. Commun.* 1308-1309.
Kopolow, S., Hogen-Esch, T. E., and Smid, J. (1971). *Macromolecules* **4**, 359-360.
Kyba, E. P., Siegel, M. G., Sousa, L. R., Sogah, G. D. Y., and Cram, D. J. (1973). *J. Am. Chem. Soc.* **95**, 2691-2692.
Lardy, H. A. (1968). *Fed. Proc.* **27**, 1278-1282.
Lehn, J. M. (1975). U. S. Patent 3,888,877, June 10.
Leong, B. K. J. (1975). *Chem. Eng. News* **53** (4), 5.
Leong, B. K. J., Ts'o, T. O. T., and Chenoweth, M. B. (1974). *Toxicol. Appl. Pharmacol.* **27**, 342.
Liotta, C. L., and Harris, H. P. (1974). *J. Am. Chem. Soc.* **96**, 2250-2052.
Liotta, C. L., Harris, H. P., McDermott, M. Gonzalez, T., and Smith, K. (1974). *Tetrahedron Lett.* 2417-2420.
Luttringhaus, A., and Sichert-Modrow, I. (1956). *Makromol. Chem.* **18-19**, 511-521.
Mallinson, P. R., and Truter, M. R. (1972). *J. Chem. Soc. Perkin Trans. 2* 1818-1823.
McLaughlin, S. G. A., Szabo, G., Eisenman, G., and Ciani, S. (1970). *Biophys. Soc. Abstr.* **10**, 96a.
Mercer, M., and Truter, M. R. (1973). *J. Chem. Soc. Dalton Trans.* 2215-2220.
Mueller, P., and Rudin, D. O. (1967). *Biochem. Biophys. Res. Commun.* **26**, 398-404.
Ovchinnikov, Yu. A., Ivanov, V. T., and Shkrob, A. M. (1974). *In* "Membrane-Active Complexones." Elsevier, Amsterdam.
Park, C. H., and Simmons, H. E. (1968a). *J. Am. Chem. Soc.* **90**, 2429-2431.
Park, C. H., and Simmons, H. E. (1968b). *J. Am. Chem. Soc.* **90**, 2431-2432.
Parsons, D. G. (1975). *J. Chem. Soc. Perkin Trans.* **1**, 245-250.
Parsons, D. G., Truter, M. R., and Wingfield, J. N. (1975). *Inorg. Chim. Acta* **14**, 45-48.
Pearson, R. G. (1966). *Science* **151**, 172-177.
Pedersen, C. J. (1967a). *J. Am. Chem. Soc.* **89**, 2495-2496.
Pedersen, C. J. (1967b). *J. Am. Chem. Soc.* **89**, 7017-7036.
Pedersen, C. J. (1968a). U. S. Patent 3,361,778, January 2.
Pedersen, C. J. (1968b). *Fed. Proc.* **27**, 1305-1309.
Pedersen, C. J. (1970a). *J. Am. Chem. Soc.* **92**, 386-391.
Pedersen, C. J. (1970b). *J. Am. Chem. Soc.* **92**, 391-394.
Pedersen, C. J. (1971a). *Aldrichim. Acta* **4**, 1-4.
Pedersen, C. J. (1971b). *J. Org. Chem.* **36**, 254-257.
Pedersen, C. J. (1971c). U. S. Patent 3,562,295, February 9.
Pedersen, C. J. (1971d). *J. Org. Chem.* **36**, 1690-1693.
Pedersen, C. J. (1972a). U. S. Patent 3,686,225, August 22.
Pedersen, C. J. (1972b). U. S. Patent 3,687,978, August 29.
Pedersen, C. J. (1974). U. S. Patent 3,856,813, December 24.
Pedersen, C. J. (1975). U. S. Patent 3,873,569, March 25.
Pedersen, C. J., and Frensdorff, H. K. (1972). *Angew. Chem.* **84**, 16-26; *Int. Ed. English* **11**, 16-25.
Pedersen, C. J., and Bromels, M. H. (1974). U. S. Patent 3,847,949, November 12.
Phillips, S. E. V., and Truter, M. R. (1974). *J. Chem. Soc. Dalton Trans.* 2517-2520.

Phillips, S. E. V., and Truter, M. R. (1975a). *J. Chem. Soc. Dalton Trans.* 1066-1070.
Phillips, S. E. V., and Truter, M. R. (1975b). *J. Chem. Soc. Dalton Trans.* 1071-1077.
Plattner, P. A., and Nager, U. (1947). *Experientia* **3**, 325-326.
Plattner, P. A., Vogler, K., Studer, R. O., Quitt, P., and Keller-Schierlein, W. (1963). *Experientia* **19**, 71-72.
Poonia, N. S., and Truter, M. R. (1973). *J. Chem. Soc. Dalton Trans.* 2062-2065.
Powell, H. M. (1954). *J. Chem. Soc.* 2658-2663.
Pressman, B. C. (1968). *Fed. Proc.* **27**, 1283-1288.
Sam, D. J., and Simmons, H. E. (1972). *J. Am. Chem. Soc.* **94**, 4024-4025.
Schiessler, R. W., and Flitter, D. (1952). *J. Am. Chem. Soc.* **74**, 1720-1723.
Shchori, E., and Jagur-Grodzinski, J. (1972a). *Israel J. Chem.* **10**, 935-940.
Shchori, E., and Jagur-Grodzinski, J. (1972b). *J. Am. Chem. Soc.* **94**, 7957-7962.
Shemyakin, M. M., Aldanova, N. A., Vinogradova, E. I., and Feigina, M. Yu. (1963). *Tetrahedron Lett.* 1921-1925.
Simmons, H. E., and Park, C. H. (1967). *Organ. Symp., 20th, Burlington, Vermont,* June 20.
Simmons, H. E., and Park, C. H. (1968). *J. Am. Chem. Soc.* **90**, 2428-2429.
Stetter, H., and Marx, J. (1957). *Justus Liebigs Ann. Chem.* **607**, 59-66.
Stewart, D. G., Waddam, D. Y., and Borrows, E. T. (1957). British Patent 785,229, October 23.
Su, A. C. L., and Weiher, J. F. (1968). *Inorg. Chem.* **7**, 176-177.
Takekoshi, T., and Webb, J. (1974). U. S. Patent 3,824,215, July 16.
Taylor, L. T., Vergez, S. C., and Busch, D. H. (1966). *J. Am. Chem. Soc.* **88**, 3170-3171.
Tosteson, D. C. (1968). *Fed. Proc.* **27**, 1269-1277.
Truter, M. R., and Pedersen, C. J. (1971). *Endeavour* **30**, 142-146.
Wong, K. H., Koniser, G., and Smid, J. (1970). *J. Am. Chem. Soc.* **92**, 666-670.

2 SYNTHESIS OF MULTIDENTATE COMPOUNDS

Jerald S. Bradshaw
Chemistry Department
Brigham Young University
Provo, Utah

I.	Introduction	53
II.	Macrocyclic Polyethers (Crowns)	54
III.	Macrocyclic Polyether-Amines	56
IV.	Macrocyclic Polyether-Sulfides	57
V.	Macropolycyclic Compounds	57
VI.	Chiral Macrocyclic Polyethers	58
VII.	Comprehensive List of Polyethers, Polyether-Amines, Polyether-Sulfides, and Cryptates	60
	References	107

I. INTRODUCTION

This chapter deals with the synthesis of multidentate compounds. Most of the compounds included belong to the so-called crown class of compounds first prepared by Pedersen (1967) (see Chapter 1, Section I.B.4, for crown compound nomenclature). A few closely related compounds are also included in Table I at the end of the chapter. No porphyrin-type compounds are included. A review of various types of nitrogen-containing ligands has been published by Black and Hartshorn (1973).

Extensive reviews have been written concerning the crown compounds. A limited discussion of the synthesis of crown compounds is given in some of these (Christensen *et al.,* 1974; Lehn, 1973; Pedersen and Frensdorff, 1972; Vögtle and Neumann, 1973). Extensive discussions of complexation by crown and related cyclic compounds are given in the following reviews: Christensen *et al.* (1971, 1974); Izatt *et al.* (1973); Kapoor and Mehrotra (1974); Kappenstein (1974); Lehn (1973); Lindoy (1975); Pedersen and Frensdorff (1972). Cram and Cram (1974) discuss the correlation of structures for host and guest

*Contribution No. 84 from the Thermochemical Institute. Brigham Young University. Appreciation is expressed to the U.S. Public Health Service for NIH grant GM18811 and the National Science Foundation for NSF grant GP-33536X.

molecules in complexation. Truter (1973) reviews the structure determination work that has been done on crown compounds and their complexes. An extensive review of all types of macrocyclic sulfur compounds has also been published (Bradshaw and Hui, 1974).

Table I contains a reasonably complete list of all crown compounds reported before July 1, 1975. The synthetic techniques are not given; however, references are provided. Short discussions of synthetic methods are given in Sections II-VI. These discussions are followed by Section VII, Table I.

II. MACROCYCLIC POLYETHERS (CROWNS)

Pedersen (1967) first reported the synthesis of macrocyclic polyethers to be used for metal ion complexation. Chapter 1 deals with his work and gives the background on his synthetic methods as well as crown nomenclature. Pedersen used three main methods to prepare his crown compounds (see Scheme 1 and Chapter 1). Most of the compounds were prepared using processes (1) and (2)

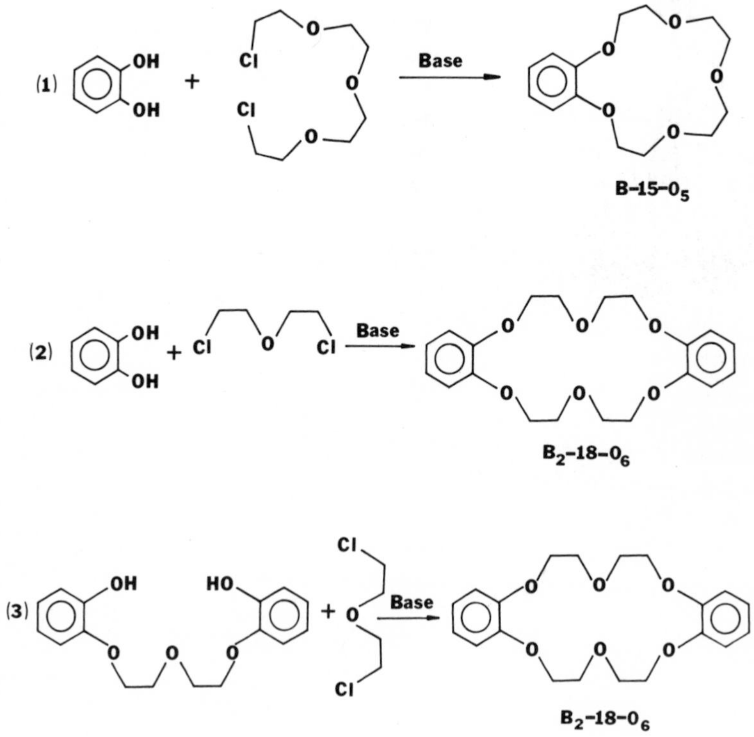

Scheme 1

from catechol or an oligoethylene glycol and the appropriate dihalide (Scheme 1). The product yields for crown compounds prepared by Pedersen (1967, 1971) were generally low except where five or six oxygens were included in the macrocyclic rings. These were found to be the most effective metal-complexing agents; and, presumably, a "template effect" of the cation aided in their synthesis (Greene, 1972). An excellent procedure for preparing dibenzo- and dicyclohexo-18-crown-6 (B_2-18-O_6 and C_2-18-O_6) is given by Pedersen (1973).

The benzo crown compounds were reduced to make the cyclohexo derivatives. The most popular crown compound, dicyclohexo-18-crown-6 (C_2-18-O_6) prepared in this fashion proved to be a mixture of two of the five possible isomers—the cis-syn-cis and the cis-anti-cis forms (Frensdorff, 1971). These can be separated easily by differential complexation (Izatt *et al.*, 1975). Stoddart and Wheatley (1974) have reported the synthesis of two other isomers, namely the trans-anti-trans and trans-syn-trans forms.

Since Pederson's initial paper was published (1967), hundreds of additional crown compounds have been prepared. The majority of these compounds were synthesized using Pedersen's procedures from the glycol (or catechol) and a dihalide (Scheme 1). Others have used the ditosylate derivatives of the oligoethylene glycols rather than the dihalides (Ashby *et al.*, 1974; Greene, 1972). Yields have often been greatly improved by using the proper-sized cation in a template effect (Greene, 1972). An excellent method to isolate 18-crown-6 (18-O_6) by means of an acetonitrile-18-O_6 complex has recently been reported by Cram and Liotta and their co-workers (Gokel *et al.*, 1974). Dale *et al.*, (1974) have been able to isolate all possible cyclic oligomers of ethylene oxide by using the reaction of ethylene oxide with boron trifluoride.

A great variety of multidentate crown ethers are available (see Table I). Some of the crown ethers contain the furan system; an example is Fur-15-O_5 (Kyba *et al.*, 1973a; Parsons, 1975; Reinhoudt and Gray, 1975; Timko and Cram, 1974). The pyridine ring has also been incorporated into the polyether ring, as in compound Py-18-O_5N (Gokel *et al.*, 1975b; Newcomb *et al.*, 1974; Newkome and Robinson, 1973; Newkome *et al.*, 1975; Vögtle and Weber, 1974; Vögtle *et al.*, 1974).

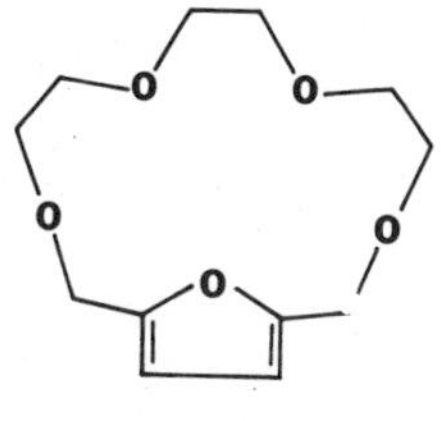

Fur-15-O_5

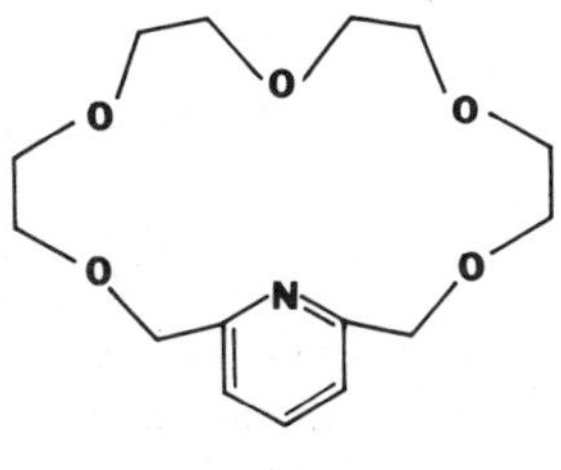

Py-18-O_5N

A new class of multidentate compounds, the ether esters, has recently been reported by Bradshaw *et al.*, (1975). As an example, they reacted tetraethylene glycol with malonyl dichloride to make k_2-16-O_5 in an excellent yield. These compounds have structures which are somewhat similar to valinomycin, a naturally occurring cyclic polyamide ester, and nonactin, a cyclic polyester ether.

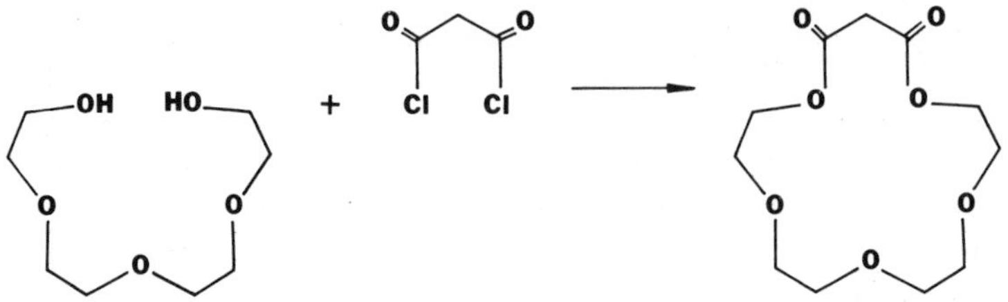

k_2-16-O_5

III. MACROCYCLIC POLYETHER AMINES

The aza crown compounds have been prepared by a different process than those mentioned in Section II. Macrocyclic diamides were first prepared from diacid chlorides and diamines. The diamide was then reduced to form the aza crown compounds (see Scheme 2). Some of these compounds contain sulfur and/or oxygen atoms in addition to the nitrogen atoms (Dietrich *et al.*, 1969, 1970, 1973a; Hogberg and Cram, 1975; King and Krespan, 1974; Lockhart *et al.*, 1973; Pelissard and Louis, 1972; Vögtle *et al.*, 1974). Dye *et al.*, (1973) reported the use of a flow synthesis technique to prepare the cyclic diamides.

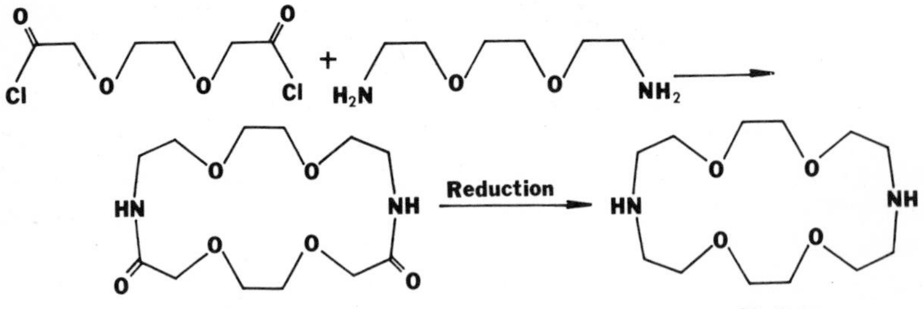

k_2-18-O_4N_2 18-O_4N_2

Scheme 2

Richman and Atkins (1974) prepared a number of aza crown compounds by reacting the sodium salts of bis-*p*-toluenesulfonamides with dihalides or ditosylates. All secondary amine groups were protected by forming the *p*-toluenesulfonamide. Yields of 45-80% were reported for this reaction.

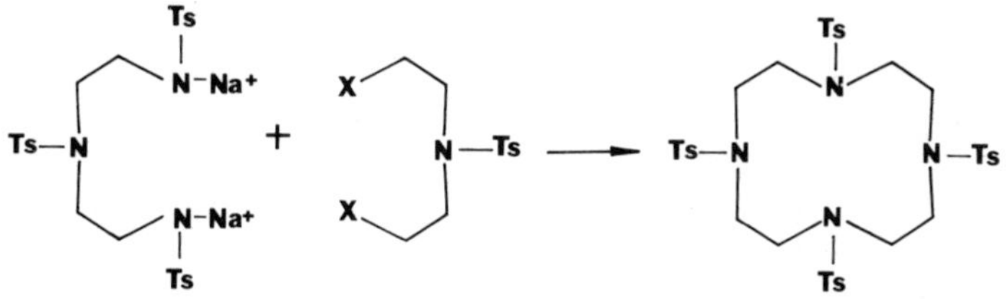

Ts_4-12-N_4

IV. MACROCYCLIC POLYETHER SULFIDES

Sulfur is the most common heteroatom (other than oxygen) in crown compounds. The thia crown compounds are easily prepared from dimercaptans or sodium sulfide and the dihalide (Bradshaw *et al.*, 1973, 1974, 1976; Dann *et al.*, 1961; Pedersen, 1971). The yields for these reactions are often very good (25-30%). A review of the preparation of all types of macrocyclic sulfur compounds has been published (Bradshaw and Hui, 1974).

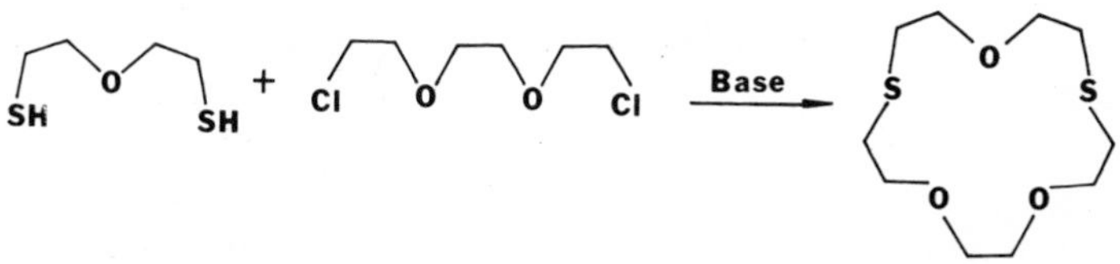

$15\text{-}O_3S_2{}^{7,13}$

V. MACROPOLYCYCLIC COMPOUNDS

The aza crown compounds discussed in Section III have been used by Lehn and his co-workers and by others to prepare many bi- and tricyclic ring systems which are called cryptates (Cheney and Lehn, 1972; Cheney *et al.*, 1972; Coxon and Stoddart, 1974; Dietrich *et al.*, 1969, 1970, 1973b, 1974; Lehn and Montavan, 1972; Lehn *et al.*, 1973; Lutringhaus, 1937). The aza crown compounds (such as 18-$O_2N_2S_2$) are reacted with a diacid chloride to make an amide (k_2-$(18)_3$-$O_4N_2S_2$) which is reduced to make the cryptate (Scheme 3). More than 50 cryptate compounds have been prepared (see Table I). Some of these compounds, such as $(18)_2(30)_2$-$O_{10}N_4$ (Lehn *et al.*, 1973) and $(18)_4$-O_6N_4 (Graf and Lehn, 1975), are macropolycyclic.

Coxon and Stoddart (1974) have prepared macrobicyclic compounds without using the aza crown compounds. Their compounds contain a carbon bridgehead [$(HyMe)_2$-$(20)_3$-O_9].

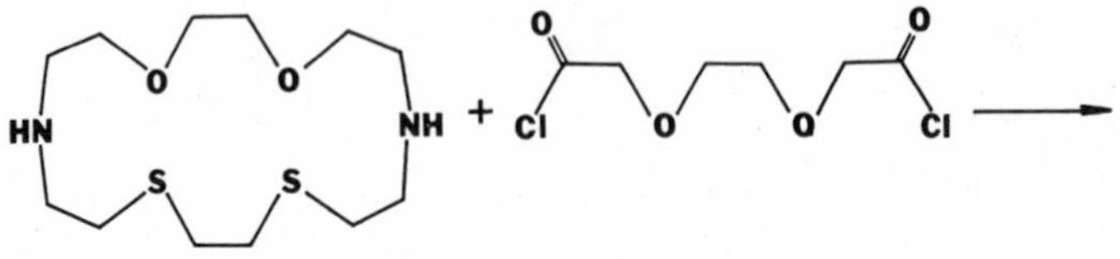

$18\text{-}O_2N_2S_2$

$k_2\text{-}(18)_3\text{-}O_4N_2S_2$ $(18)_3\text{-}O_4N_2S_2$

Scheme 3

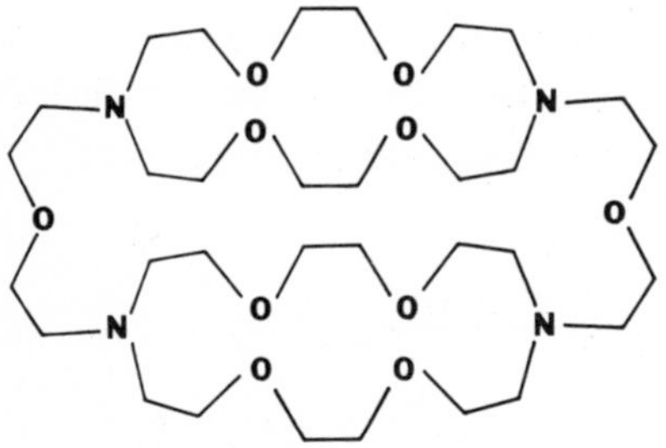

$(18)_2(30)_2\text{-}O_{10}N_4$

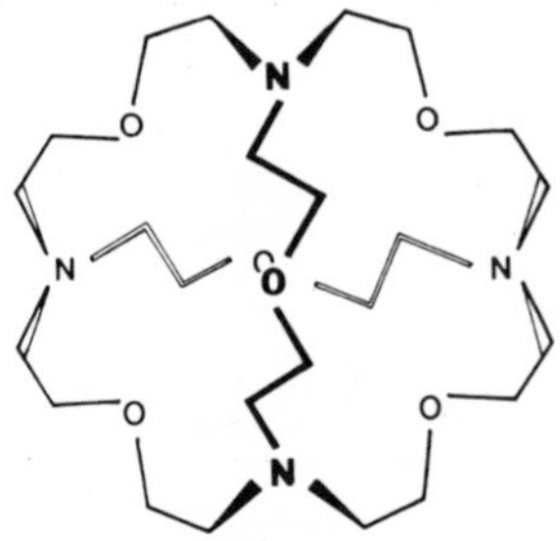

$(18)_4\text{-}O_6N_4$

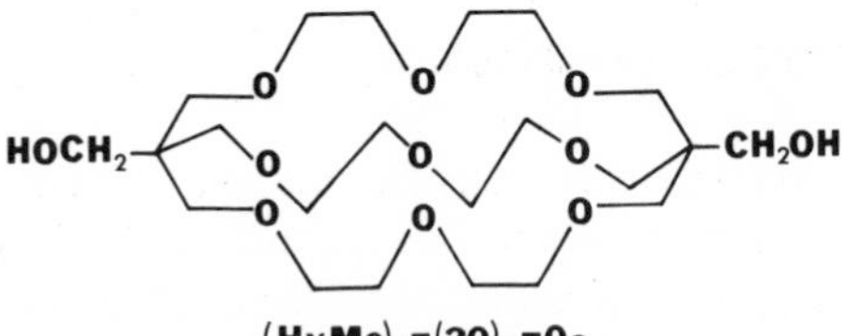

$(HyMe)_2\text{-}(20)_3\text{-}O_9$

VI. CHIRAL MACROCYCLIC POLYETHERS

Cram and his co-workers have synthesized a number of crown compounds containing biphenyl, binaphthyl, and paracyclopanyl moieties (Cram and Cram, 1974; deJong *et al.*, 1975; Dotsevi *et al.*, 1975; Gokel *et al.*, 1975a, b; Helgeson *et al.*, 1973a, b, 1974a, b; Kyba *et al.*, 1973a, b; Madan and Cram, 1975; Newcomb *et al.*, 1974; Newcomb and Cram. 1975; Timko and Cram,

1974). Many of these compounds have chiral centers and have been used to partially resolve optically active compounds (see Cram and Cram, 1974, for a review). The binaphthol system was particularly suited for the chiral crown compounds because it could be resolved into the R and S forms. When one of the optically active forms of binaphthol was treated with diethylene glycol ditosylate, optically active Np_2-11-O_3 and Np_4-22-O_6 were formed (see Scheme 4) (Kyba *et al.*, 1973b).

Two chiral crown compounds were also prepared by Wudl and Gueta (1972).

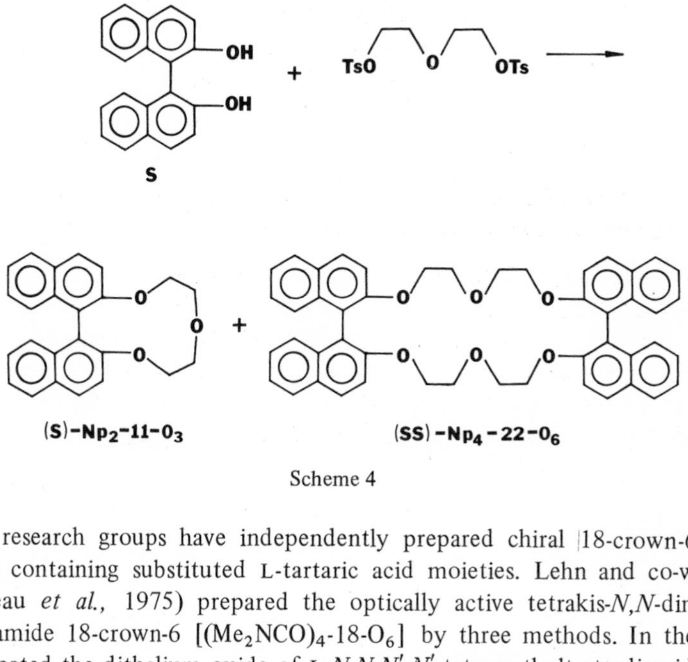

Scheme 4

Two research groups have independently prepared chiral 18-crown-6 compounds containing substituted L-tartaric acid moieties. Lehn and co-workers (Girodeau *et al.*, 1975) prepared the optically active tetrakis-*N,N*-dimethylcarboxamide 18-crown-6 [$(Me_2NCO)_4$-18-O_6] by three methods. In the first, they treated the dithalium oxide of L-*N,N,N′,N′*-tetramethyltartardiamide with diethylene glycol diiodide to give a 20% yield of $(Me_2NCO)_4$-18-O_6. The other methods utilized a partially assembled compound much like process (3) in Scheme 1. The Lehn group also prepared a crown compound containing two L-tartaric acid and two pyridine moieties.

Stoddart and co-workers (Curtis *et al.*, 1975) converted the acid groups of L-tartaric acid to benzyloxymethyl groups before cyclization to a crown compound. They prepared L,L-$(PhCH_2OCH_2)_4$-18-O_6 in 12% yield along with a 9-crown-3 compound. The benzyl group was removed by hydrogenation to give tetrakis-hydroxymethyl compounds. D-Mannitol was also used by the Stoddart group to prepare an 18-crown-6 compound containing four 1′,2′-dihydroxyethyl groups [$(Hy_2Et)_4$-18-O_6].

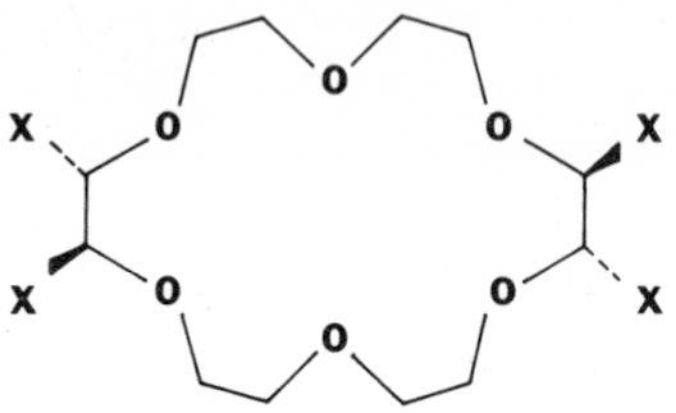

$(Me_2NCO)_4$-18-O_6

$(PhCH_2OCH_2)_4$-18-O_6

$(Hy_2Et)_4$-18-O_6

VII. COMPREHENSIVE LIST OF POLYETHERS, POLYETHER AMINES, POLYETHER SULFIDES, AND CRYPTATES

Table I contains a comprehensive list of crown compounds and cryptates prepared up to about July 1975. The table is organized by the number of ring atoms, type of heteroatom(s) in the ring, and the substituents on the ring members. For example, 18-O_6 is 18-crown-6 and MeB-18-O_6 is methylbenzo-18-crown-6. Each compound is correlated with the structure that appears above the data. The substituents are abbreviated as follows:

Ac	aceto	Acet	acetic acid	am	acetomido
B	benzo	Bz	benzene	C	cyclohexo
carb	carboxy	Cyn	cyano		
DMD	4′-(2′,2′-dimethyl-1′,3′-dioxolyl)				
Fur	furan	Hy	hydroxy	HyMe	hydroxymethyl
k	keto	Me	methyl	MeOMe	methoxymethyl
Mo	morpholinyl	Ms	methane sulfonyl	nB	nitrobenzo
Nor	norborneno	Np	naphtho	P	propyl
Ph	phenyl	pip	piperazine	$PhCH_2$	benzyl
Py	pyridine	Pyr	pyrrolo	spO	spirooxetano
spPHO	spirophenyldioxino	tb	*tert*-butyl	tDec	tetradecyl
Th	thiophene	THF	tetrahydrofurano	Ts	*p*-toluene sulfonyl
uDec	undecyl	V	vinyl		

The melting points or boiling points and yields are also listed in the table, along with literature references.

TABLE I

Crown Compound Syntheses

Compound	Remarks	M.P. (B.P.) (°C)	Yield (%)	Reference[a]
	9-Membered Rings			
	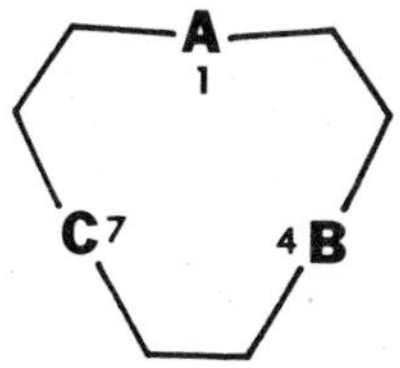			
9-O_3	A-C=O	0	–	*1*
9-O_2S	A,B=O; C = S	(40-48/0.15)	5	*2*
9-OS_2	A=O; B,C=S	(62-64/0.1)	6	*2*
9-S_3	A-C=S	113	Low	*3*
B-9-O_3	A-C=O; 2,3-benzo	67-69	5	*4,5*
B-9-O_2N	A,B=O; C=NH; 2,3-benzo	39-40	50	*6*
BTsO-9-O_2N	A,B=O; C=*N*-tosyl; 2,3-benzo	174-175	35	*6*
$(DMD)_4$-9-O_3	A-C=O; 2,3-bis-4′-(2′,2′-dimethyl-1′,3′-dioxolyl)	Liquid	6	*8a*
$(HBr)_3$-9-N_3	A-C=NH; tris-HBr	274-275	88	*7*
$(PhCh_2OCH_2)_2$-9-O_3	A-C=O; 2,3-bis-benzyl-oxymethyl	Liquid	4	*8a*
Ts_3-9-N_3	A-C=*N*-tosyl	222-223	71	*8*
	10-Membered Rings			
	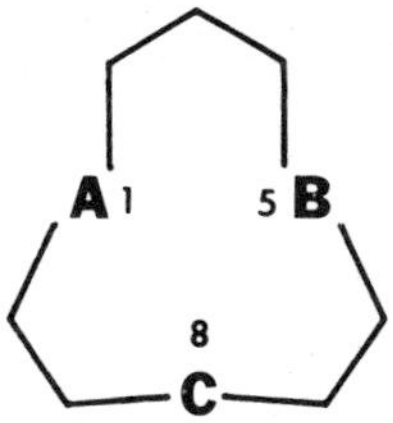			
BzNO_2-10-S_2	B,C=S; A=benzene C-nitro	145-148	4	*9*
$(HBr)_3$-10-N_3	A-C=NH; tris-HBr	242-243	82	*7*
Ts_3-10-N_3	A-C=*N*-tosyl	234-236	84	*8*
	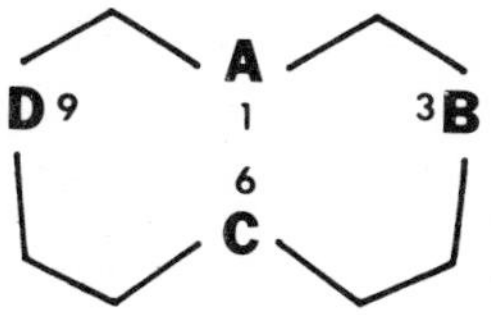			

(cont.)

Compound	Remarks	M.P. (B.P.) (°C)	Yield (%)	Reference[a]
Py-10-ONS_2	C=O; A=pyridineN; B,D=S	92-94	13	*10*
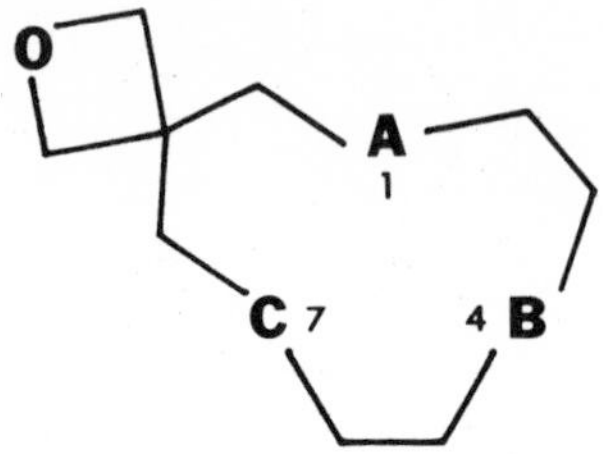				
spO-10-O_3	A-C=O	(100-120/0.3)	0.2	*11*
spO-10-OS_2	A,C=S; B=O	104-105	16	*11*
	11-Membered Rings			
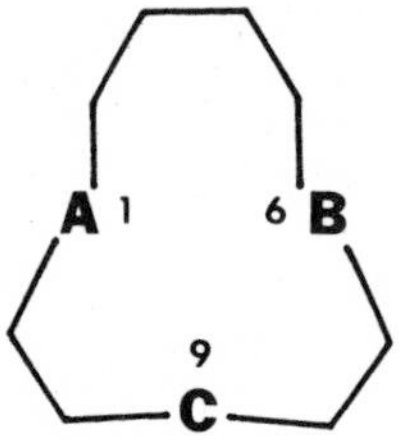				
BzMeS-11-S_2	A,B=S; C=benzeneC-SCH_3	100-103	3	*9*
Bz$MeSO_2$-11-S_2	A,B=S; C=benzene C-SO_2CH_3	144-147	5	*9*
BzNO_2-11-S_2	A,B=S; C=benzeneC-NO_2	102	45	*9*
BzPh-11-S_2	A,B=S; C=benzeneC-phenyl	133-134	22	*9*
Np_2-11-O_3	A-C=O; 2,3- and 4,5-dinaphtho	230-231 (Opt. pure)	5	*12*
	12-Membered Rings			
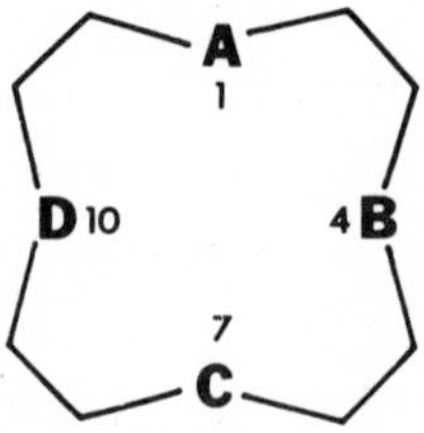				
12-O_4	A-D=O	(67-70/0.5)	13	*13*
12-O_3S	A,B,C,=O; D=S	(80-87/1)	14	*2*

(cont.)

Compound	Remarks	M.P. (B.P.) (°C)	Yield (%)	Reference[a]
12-O_2N_2	A,C=O; B,D=NH	83-84	55	*14*
12-O_2S_2[7,10]	A,B=O; C,D=S	(134-136/0.1)	19	*2*
12-O_2S_2[4,10]	A,C=O; B,D=S	(118-120/0.1)	12	*2*
12-ONS_2	A=O; B,D=S; C=NH	64	18	*15-17*
12-OS_3	A=O; B,C,D=S	89-91	26	*2*
12-S_4	A-D=S	224-225	6	*18*
		215-217	4	*19,20*
B-12-O_4	A-D=O; 2,3-benzo	44-44.5	4	*5*
B_2-12-O_4	A-D=O; 2,3- and 8,9-dibenzo	208-209	11-15	*5*
B-12-O_3N	A-C=O; D=NH; 8,9-benzo	80	–	*21*
B-12-O_2N_2	A,B=O; C,D=NH; 8,9-benzo	92-94	–	*21*
Bz-12-O_3	B-D=O; A=benzeneCH	–	2	*22*
Bz-12-S_3	B-D=S; A=benzeneCH	143-144	37	*10*
BzCl-12-S_3	B-D=S; A=benzeneC-Cl	130-132	46	*10*
BzMeS-12-S_3	B-D=S; A=benzeneC-SCH_3	94-97	15	*10*
$BzMe_2$-12-ON_2	C=O; B,D=*N*-CH_3; A=benzeneCH		40	*23*
BzPh-12-S_3	B-D=S; A=benzeneC-phenyl	157-159	17	*10,24*
C_2-12-O_4	A-D=O; 2,3- and 8,9-dicyclohexo	Liquid	48	*5*
$(HCl)_4$-12-N_4	A-D=NH•HCl	–	90	*8*
k_2-12-O_2N_2	A,C=O; B,D=NH; 5,9-diketo	182-183	65	*14*
k_2Py-12-ON_3	D=O; A,C=NH; B=pyridineN; 2,6-diketo	228-230	14	*10*
Ph-12-O_3N	A-C=O; D=*N*-phenyl	90	–	*21*
Py-12-ONS_2	A=O; B,D=S; C=pyridineN	–	–	*25*
Py_2-12-O_2N_2	A,C=O; B,D=pyridineN	172-175	6	*26*
Py-12-NS_3	A=pyridineN; B-D=S	162-163	50	*10*
Ts_2-12-O_2N_2	A,C=O, B,D=*N*-tosyl	203-204	80	*8*
Ts_4-12-N_4	A-D=*N*-tosyl	–	80	*8*

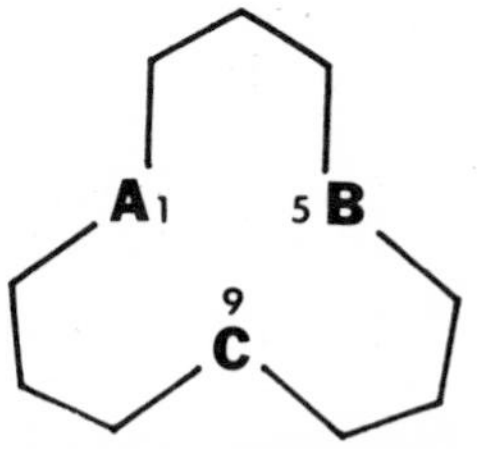

(cont.)

Compound	Remarks	M.P. (B.P.) (°C)	Yield (%)	Reference[a]
12-S_3	A-C=S	87-88	3	*19,20*

13-Membered Rings

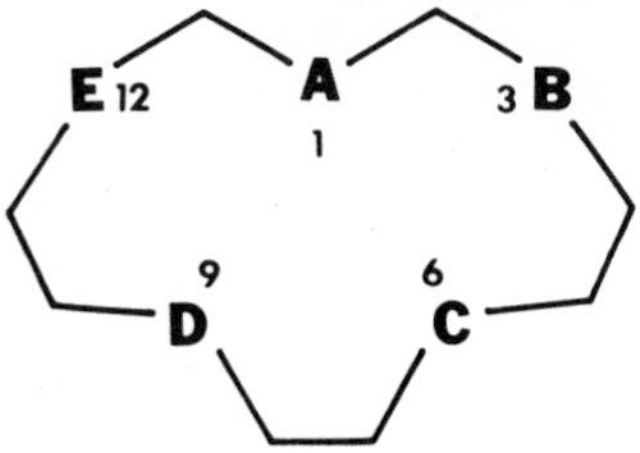

Py-13-O_4N	B-E=O; A=pyridineN	83-84	16	*27*
Py-13-O_2NS_2	C,D=O; A=pyridineN; B,E=S	75-77	18	*10*

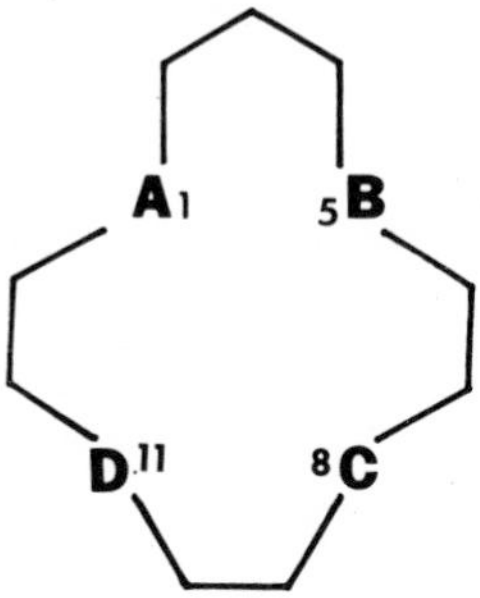

13-N_4	A-D=NH	40-41	58	*28*
13-S_4	A-D=S	134-135	16	*19,20*
k_2-13-O_4	A-D=O; 2,4-diketo	51.5-52.5	29	*29*

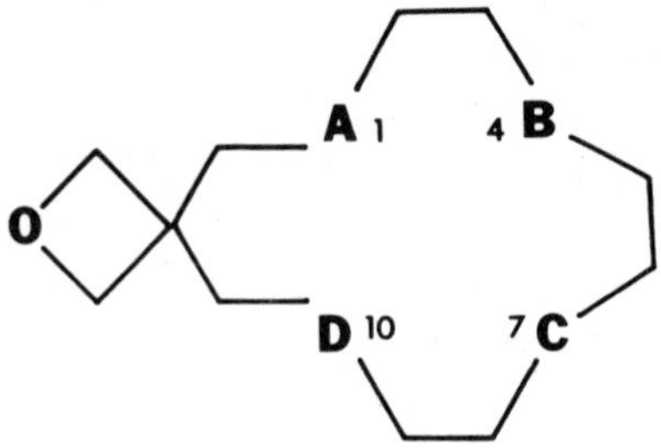

spO-13-O_4	A-D=O	(97-98/0.6)	30-35	*11*

(cont.)

Compound	Remarks	M.P. (B.P.) (°C)	Yield (%)	Reference[a]

14-Membered Rings

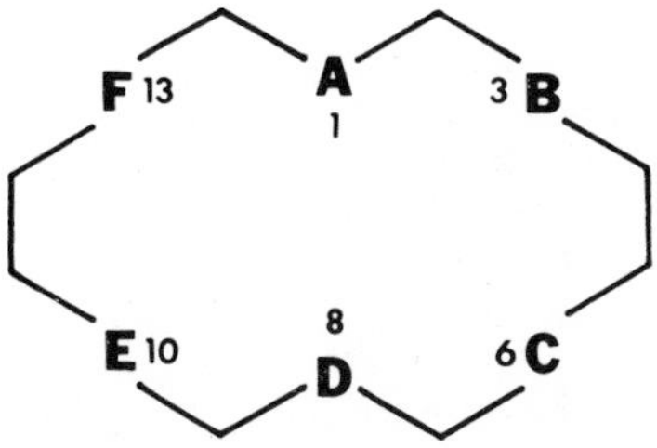

Compound	Remarks	M.P. (B.P.) (°C)	Yield (%)	Reference[a]
Py_2-14-O_4N_2	B,C,E,F=O A,D=pyridineN	215-216	16	*27*

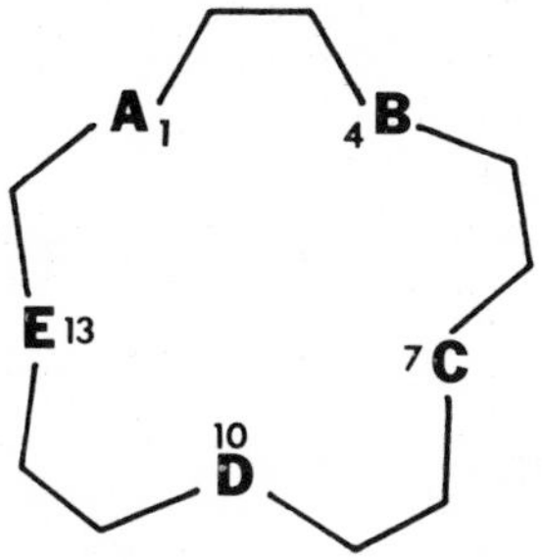

Compound	Remarks	M.P. (B.P.) (°C)	Yield (%)	Reference[a]
B_2-14-O_3S_2	B,C,D=O; A,E=S 2,3- and 11,12-dibenzo	150-153	3	*30*
C_2-14-O_5	A-E=O; 2,3- and 11,12-dicyclohexo	Oil (Isomer)	57	*31*
C_2-14-O_5	A-E=O; 2,3- and 11,12-dicyclohexo	62-63 (Isomer)	21	*31*

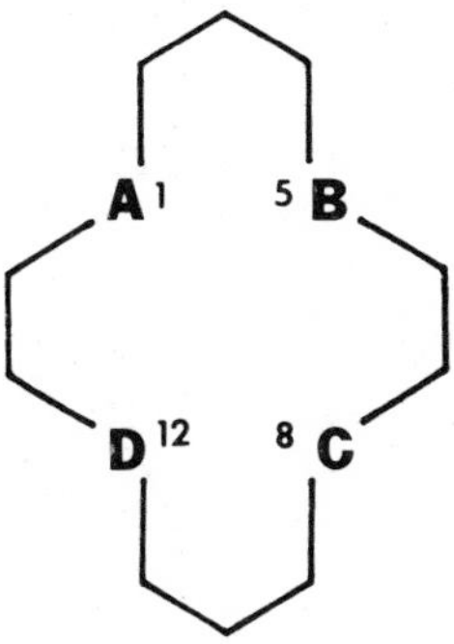

(cont.)

Compound	Remarks	M.P. (B.P.) (°C)	Yield (%)	Reference[a]
14-S_4	A-D=S	121-122.5 119-120	5-22 7.5	*18* *3,32* *33,34*
B_2-14-O_4	A-D=O; 6,7- and 13,14-dibenzo	150-152	27	*5*
B_2-14-O_2N_2	A,C=O; B,D=N; 2,3- and 9,10-dibenzo; 4,11-diene	73-77	87	*35*
C_2-14-O_4	A-D=O; 6,7- and 13,14-dicyclohexo	153.5-155.5	55	*5*
Me-14-N_4	B-D=NH; A=*N*-CH_3	220d	35	*36*
Me_2-14-N_4	C,D=NH; A,B=*N*-CH_3	201d	20	*36*
Me_4-14-N_4	A,B,C,D=*N*-CH_3	230d	40	*36*
$MeTs_2$-14-N_4	A=NH; B=*N*-CH_3; C,D=*N*-tosyl	207-209	60	*36*
$(tbB)_2$-14-O_4	A-D=O; 6,7- and 13,14-bis(3′-*tert*-butylbenzo)	149-152	11	*5*
$(tbC)_2$-14-O_4	A-D=O; 6,7- and 13,14-bis(3′-*tert*-butylcyclohexo)	Liquid	41	*5*

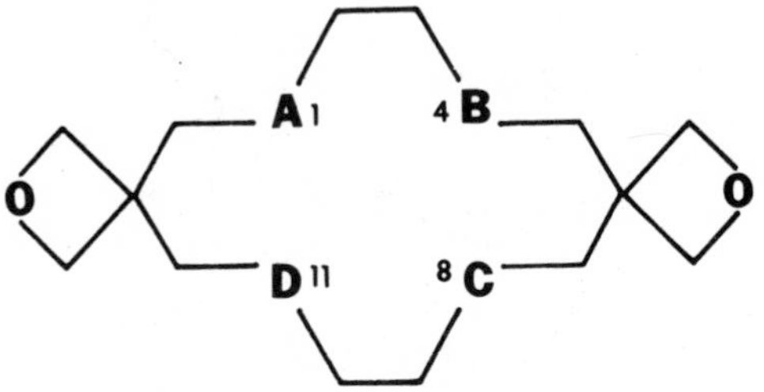

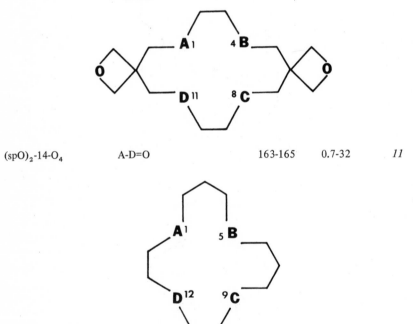

Compound	Remarks	M.P. (B.P.) (°C)	Yield (%)	Reference[a]
$(spO)_2$-14-O_4	A-D=O	163-165	0.7-32	*11*

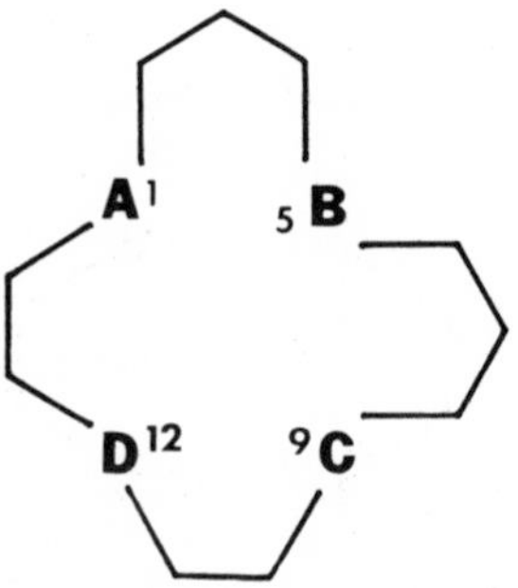

(cont.)

Compound	Remarks	M.P. (B.P.) (°C)	Yield (%)	Reference[a]
Bk_2Me-14-N_3	A,C=NH; B=*N*-CH_3; D=benzeneCH;10, 14-diketo	189-191	26	*10*
k_2MePy-14-N_4	A,C=NH; B=*N*-CH_3,D= pyridineN; 10,14-diketo	164-166	59	*10*
Ts_4-14-N_4	A-D=N-tosyl	234-236	58	*8*

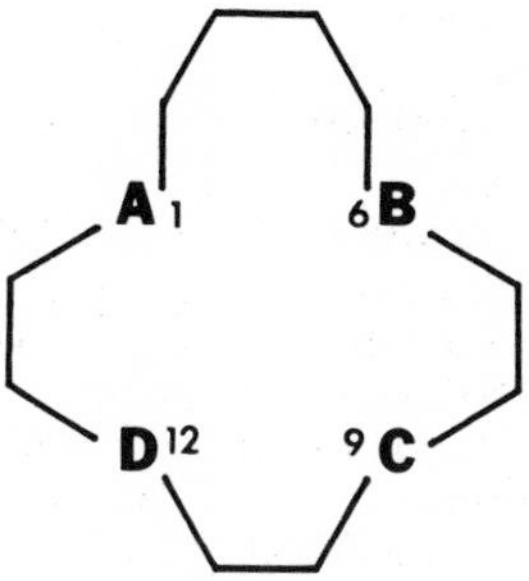

$BNorPh_4$-14-S_4	A-D=S; 3,4-benzo; 10,11-norborneno; 7,8,13,14-tetraphenyl	258	–	*37*

15-Membered Rings

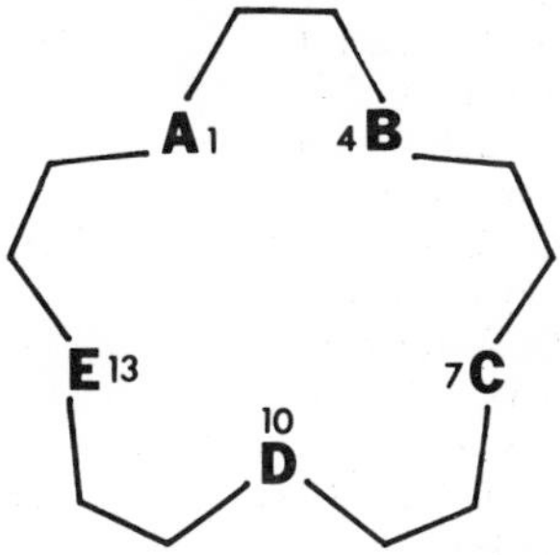

15-O_5	A-E=O	(100-135/12)	14	*13*
15-O_4S	A-D=O; E=S	(123-124/0.1)	29	*2*
15-O_3N_2	A,B,D=O; C,E=NH	89-90	85	*14*
15-O_3S_2^{7,13}	A,B,D=O; C,E=S	(150-151/0.1)	27	*2*
15-O_3S_2^{10,13}	A,B,C=O; D,E=S	51-52	20	*2*
15-O_2S_3^{4,10,13}	A,C=O; B,D,E=S	(180/1)	5	*38*
15-O_2S_3^{7,10,13}	A,B=O; C,D,E=S	43-44.5	41	*38*

(cont.)

Compound	Remarks	M.P. (B.P.) (°C)	Yield (%)	Reference[a]
$15\text{-}ON_2S_2^{4,13}$	A=O; C,D=NH; B,E=S	Liquid	–	*39*
$15\text{-}ON_2S_2^{7,10}$	A=O; B,E=NH; C,D=S	Liquid	–	*39*
$15\text{-}OS_4$	A=O; B-E=S	93-95	13	*38*
$15\text{-}N_2S_3$	A,C=NH; B,D,E=S	31	–	*39*
$15\text{-}S_5$	A-E=S	97.5-99	11	*18*
$AcB\text{-}15\text{-}O_5$	A-E=O; 2,3-(4′-acetobenzo)	96-97	45	*40*
$B\text{-}15\text{-}O_5$	A-E=O; 2,3-benzo	79-79.5	62	*5*
$B_2\text{-}15\text{-}O_5$	A-E=O; 2,3- and 8,9-dibenzo	113.5-115	43	*5*
$B\text{-}15\text{-}O_4N^{10}$	A,B,C,E=O; D=NH; 2,3-benzo	142-143	5	*6*
$B\text{-}15\text{-}O_4N^{13}$	A-D=O; E=NH; 11,12-benzo	101.5	–	*21*
$B\text{-}15\text{-}O_3N_2$	A-C=O; D,E=NH; 11,12-benzo	110	–	*21*
$BTs\text{-}15\text{-}O_4N$	A,B,C,E=O; D=*N*-tosyl; 2,3-benzo	136-137	36	*6*
$Bz\text{-}15\text{-}O_4$	B-E=O; A=benzeneCH	–	16	*12*
$BzMe_2\text{-}15\text{-}O_2N_2$	C,D=O; B,E=*N*-CH_3; A=benzeneCH	–	~40	*23*
$C\text{-}15\text{-}O_5$	A-E=O; 2,3-cyclohexo	Liquid	58	*5*
$C_2\text{-}15\text{-}O_5$	A-E=O; 2,3- and 8,9-dicyclohexo	62-63	28	*41*
$CarbB\text{-}15\text{-}O_5$	A-E=O; 2,3-(4′-carboxy-benzo)	180	92	*42*
$CarbBz\text{-}15\text{-}O_4$	B-E=O; A=benzeneC-carboxyl	106-112	98	*43*
$CarbMeB\text{-}15\text{-}O_5$	A-E=O; 2,3-(4′-carbo-methoxybenzo)	82	81	*42*
$dNp\text{-}15\text{-}O_5$	A-E=O; 2,3-(2′,3′-decahydronaphtho)	Liquid	28	*5*
$Fur\text{-}15\text{-}O_5$	B-E=O; A=furanO	–	–	*12*
$k_2\text{-}15\text{-}O_3N_2$	A,B,D=O; C,E=NH; 8,12-diketo	149-150	75	*14*
$k_2\text{-}15\text{-}ON_2S_2^{4,13}$	A=O; C,D=NH; B,E=S; 6,11-diketo	161	54	*39*
$k_2\text{-}15\text{-}ON_2S_2^{7,10}$	A=O; B,E=NH; C,D=S; 3,14-diketo	158	45	*39*
$k_2\text{-}15\text{-}N_2S_3$	B,E=NH; A,C,D=S; 3,14-diketo	190	57	*39*
$k_2Me_4\text{-}15\text{-}ON_4$	A=O; B-E=*N*-CH_3; 3,14-diketo	146	59	*39*
$k_2Py\text{-}15\text{-}O_3N_2$	D,E=O; A,C=NH; B=pvridineN; 2,6-diketo	200-201	60	*10*

(cont).

Compound	Remarks	M.P. (B.P.) (°C)	Yield (%)	Reference[a]
MeB-15-O_3S_2	A,B,C=O; D,E=S; 11,12-(3′-methylbenzo)	Liquid	30	*30*
MeB-15-O_5	A-E=O; 2,3-(4′-methylbenzo)	51.2-52.0	57	*44,45*
MeCarbBz-15-O_4	B-E=O; A=benzeneC-methylcarboxyl	Liquid	34	*43*
Me_4-15-ON_4	A=O; B-E=*N*-CH_3	Liquid	–	*39*
Np-15-O_5	A-E=O; 2,3-(2′,3′-naphtho)	117-119	20	*5*
Py-15-O_2NS_2	A,D=O; B,C=S; E=pyridineN	90-91	–	*25*
Py-15-N_5	A=pyridineN; B,E=N; C, D=NH; 3,14-dimethyl; 3,13-diene	–	–	*46*
Py-15-NS_4	A=pyridineN; B-E=S	131-133	38	*10*
tbB-15-O_5	A-E=O; 2,3-(3′-*tert*-butylbenzo)	43.5-44.5	61	*5*
tbC-15-O_5	A-E=O; 2,3-(3′-tert-butylcyclohexo)	Liquid	17	*5*
Ts_5-15-N_5	A-E=*N*-tosyl	278-280	83	*8*
VB-15-O_5	A-F=O; 2,3-(4′-vinylbenzo)	43-44	60	*40*

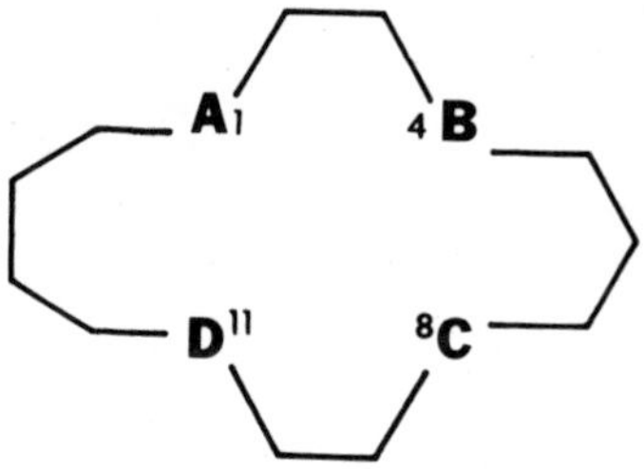

Compound	Remarks	M.P. (B.P.) (°C)	Yield (%)	Reference[a]
B-15-S_4	A-D=S; 13,14-benzo	84-86	38	*19*

16-Membered Rings

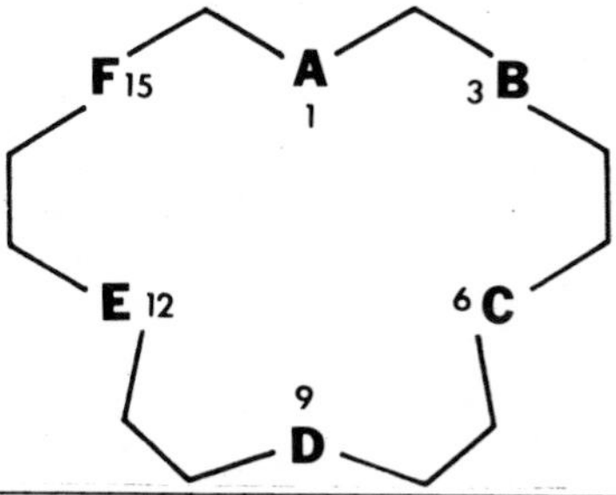

(cont).

Compound	Remarks	M.P. (B.P.) (°C)	Yield (%)	Reference[a]
Py-16-O_5N	B-F=O; A=pyridineN	76-78	4	*27*
Py-16-O_3NS_2	C-E=O; A=pyridineN; B-F=S	74-76	28	*10*

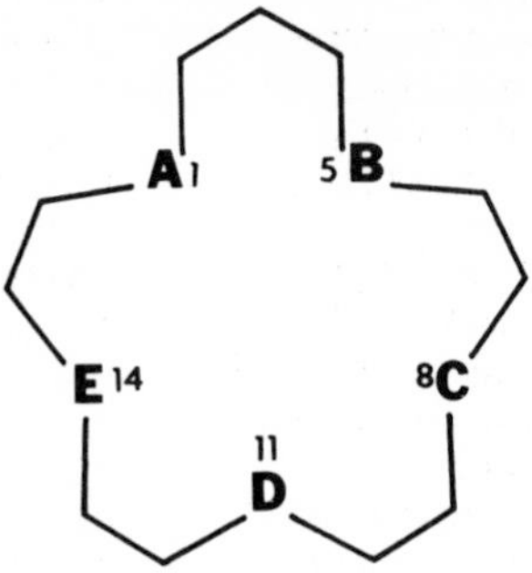

Compound	Remarks	M.P. (B.P.) (°C)	Yield (%)	Reference[a]
B_2-16-O_5	A-E=O; 6,7- and 15,16-dibenzo	117-118	18	*5*
C_2-16-O_5	A-E=O; 6,7- and 15,16-dicyclohexo	Liquid	66	*41*
k_2-16-O_5	A-E=O; 2,4-diketo	66.5-68	37	*29*
Np-16-O_5	A-E=O; 2,3,4,-(1′,8′-naphtho)	112-115	28	*41*

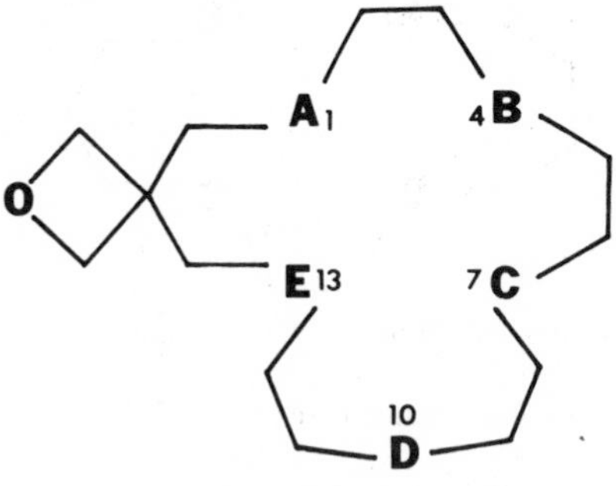

Compound	Remarks	M.P. (B.P.) (°C)	Yield (%)	Reference[a]
spO-16-O_5	A-E=O	28-30	53	*11*
spO-16-O_4N	A,B,D,E=O; C=NH	80-81	67	*47*
spO-16-O_3N_2	B,C,D=O; A,E=NH	–	8.5	*47*

(cont.)

Compound	Remarks	M.P. (B.P.) (°C)	Yield (%)	Reference[a]
	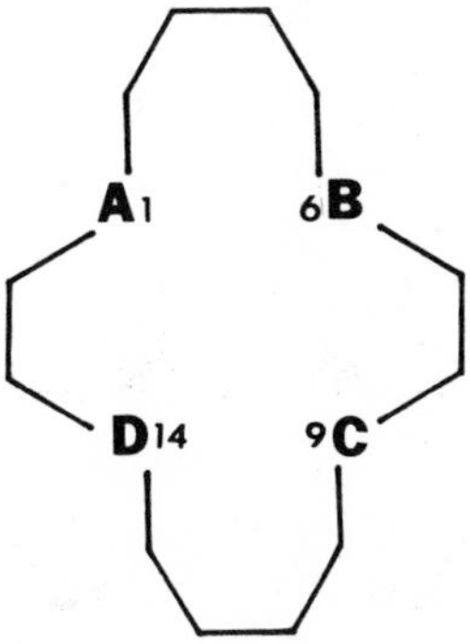			
Np_4-16-O_4	A-D=O; 2,3-, 4,5-, 10,11- and 12,13-tetranaphtho	355	1	*48*
	17-Membered Rings			
	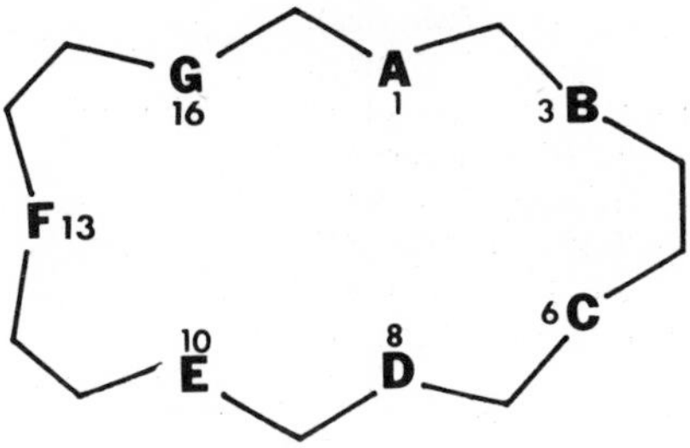			
Py_2-17-O_5N_2	B,C,E,F,G=O; A,D= pyridineN	94.5-95.5	5	*27*
	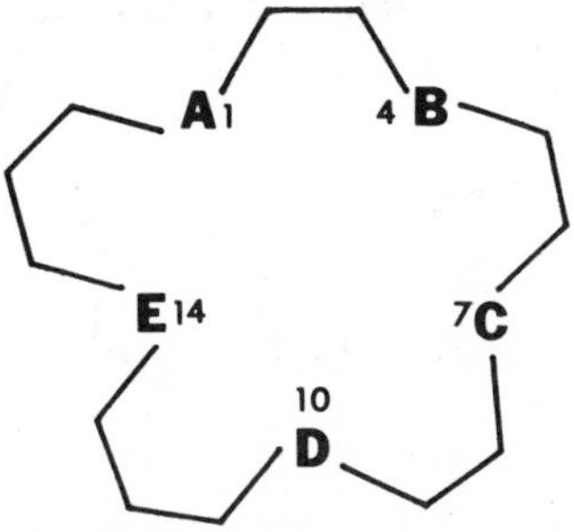			

(cont.)

Compound	Remarks	M.P. (B.P.) (°C)	Yield (%)	Reference[a]
k_2Me-17-N_3S_2	A,D=NH; E=*N*-CH_3; B,C=S; 2,9-diketo	160	51	*39*
Me-17-N_3S_2	A,D=NH; E=*N*-CH_3, B,C=S	Liquid	2	*39*

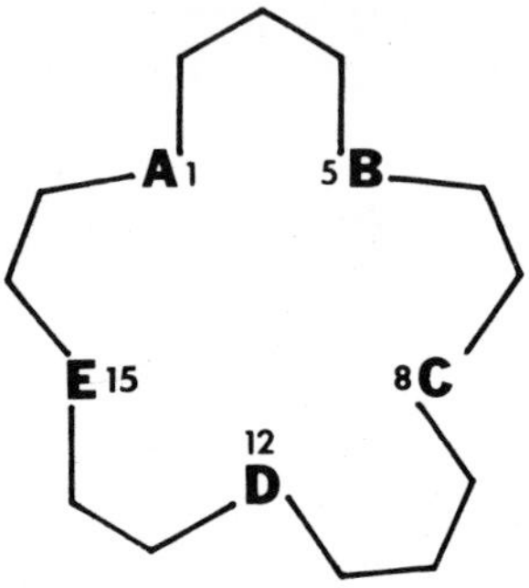

Bk_2pip-17-N_4	A,D=NH; E=benzeneCH; B,C=(1,4-piperazine)nitrogens; 13,17-diketo	253-256	31	*10*
$k_2$pipPy-17-N_5	A,D=NH; E=pyridineN; B,C=(1,4-piperazine)nitrogens; 13,17-diketo	266-268	37	*10*
$(nB)_2$-17-N_2S_3	A,D,E=S; B,C=N; 2,3- and 10,11-bis(4′-nitrobenzo); 4,8-diene	–	–	*49*

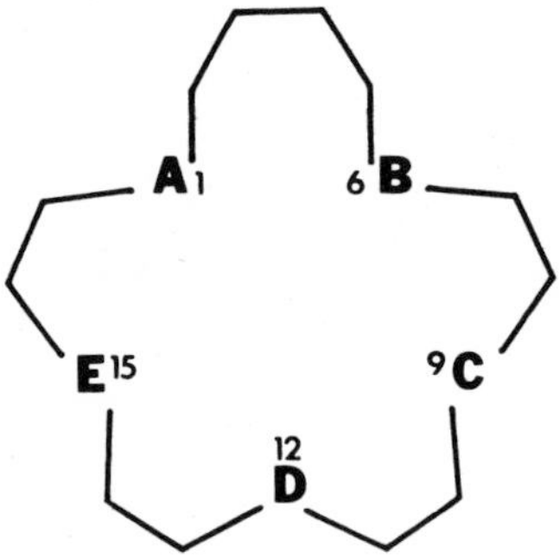

B-17-O_3S_2	C-E=O; A,B=S; 3,4-benzo	47-49	–	*25*

(cont.)

Compound	Remarks	M.P. (B.P.) (°C)	Yield (%)	Reference[a]
$(AcetMeNp)_2$-17-O_5	A-E=O; 2,3- and 4,5-bis(3′-α-acetic acid methylnaphtho)	Liquid	85	*50,51*
$(HyMeNp)_2$-17-O_5	A-E=O; 2,3- and 4,5-bis(3′-hydroxymethyl-naphtho)	Liquid	10	*51*
$(MeAcMeNp)_2$-17-O_5	A-E=O; 2,3- and 4,5-bis(3′-methyl-α-acetyl-oxymethylnaphtho)	Liquid	55	*51*
Np_2-17-O_4S	A,B,C,E=O; D=S; 2,3- and 4,5-dinaphtho	127-128	53	*52*

18-Membered Rings

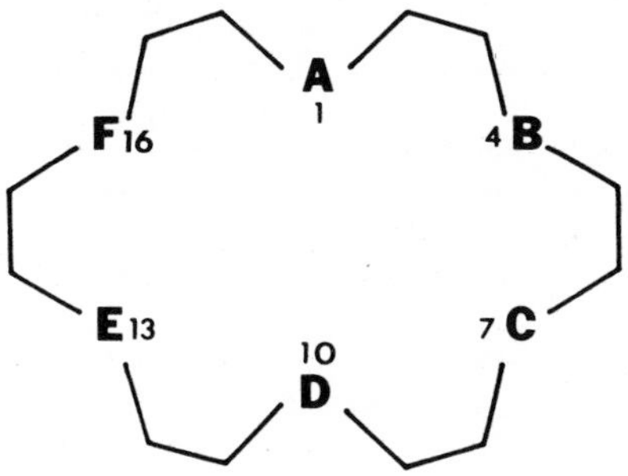

Compound	Remarks	M.P. (B.P.) (°C)	Yield (%)	Reference[a]
18-O_6	A-F=O	39-40	2	*5*
		36.5-38	~40	*53*
		39.5-40.5	30-93	*54*
18-O_5N	A-E=O,F=NH	–	43	*54*
18-O_5S	A-E=O; F=S	(164-170/0.1)	36	*38*
18-O_4N_2	A,B,D,E=O; C,F=NH	114-116	75	*14,55-57*
18-O_4S_2^{7,16}	A,B,D,E=O; C,F=S	88-89	12	*2,17,58*
18-O_4S_2^{10,16}	A,B,C,E=O; D,F=S	(174-179/1)	29	*38*
18-O_4S_2^{13,16}	A-D=O; E,F=S	54-56	28	*38*
18-O_3S_3	A,B,C=O; D,E,F=S	(200-206/2)	11	*38*
18-$O_2N_2S_2$	A,B=O; C,F=NH; D,E=S	47	55	*56,59*
18-O_2S_4	A,D=O; B,C,E,F=S	125-126	4	*2,60*
18-N_2S_4	A,D=NH; B,C,E,F=S	125	45	*15-17,56,59*
18-S_6	A-F=S	89-90	35	*17*
		89.6-90	1.7	*60*
		91-93	8	*18*

(cont.)

Compound	Remarks	M.P. (B.P.) (°C)	Yield (%)	Reference[a]
$AcB\text{-}18\text{-}O_6$	A-F=O; 2,3-(4′-acetobenzo)	77.5-78.5	32	*40*
$AcB_2\text{-}18\text{-}O_6$	A-F=O; 2,3-(4′-acetobenzo); 11,12-benzo	169-171	55	*40*
$(AcOMe)_4\text{-}18\text{-}O_6$	A-F=O; 2,3,11,12-tetrakis-acetyloxymethyl	69-74	–	*8a*
$[(AcO)_2Et]_4\text{-}18\text{-}O_6$	A-F=O; 2,3,11,12-tetrakis(1′,2′-acetyloxyethyl)	–	–	*8a*
$B\text{-}18\text{-}O_6$	A-F=O; 2,3-benzo	43-44	60	*5*
$B\text{-}18\text{-}O_4S_2{}^{7,16}$	A,B,D,E=O; C,F=S; 2,3-benzo	91	1	*30*
$B_2\text{-}18\text{-}O_6a$	A-F=O; 2,3- and 11,12-dibenzo	162-164	45-80	*4,5,61,62*
$B_2\text{-}18\text{-}O_6b$	A-F=O; 2,3- and 8,9-dibenzo	117-118	25	*5*
$B_2\text{-}18\text{-}O_5$	A-E=O; F=CH_2; 2,3- and 11,12-dibenzo	157-158	28	*5*
$B_2\text{-}18\text{-}O_5N$	A,B,D,E,F=O; C=NH; 2,3- and 11,12-dibenzo	150-152	35	*6*
$B_2\text{-}18\text{-}O_4N_2{}^{4,10}$	A,C,E,F=O; B,D=NH; 2,3- and 11,12-dibenzo	203-204	71	*6*
$B_2\text{-}18\text{-}O_4N_2{}^{4,13}$	A,C,D,F=O; B,E=NH; 2,3- and 11,12-dibenzo	175-177	45	*6*
$B_2\text{-}18\text{-}O_4S_2{}^{7,16}$	A,B,D,E=O; C,F=S; 5,6- and 14,15-dibenzo	114-115	5	*30*
$B_2\text{-}18\text{-}O_4S_2{}^{10,16}$	A,B,C,E=O; D,F=S; 8,9- and 17,18-dibenzo	143-144	15	*30*
$B_2\text{-}18\text{-}O_3N_3$	A,C,E=O; B,D,F=NH 2,3- and 11,12-dibenzo	198-200	82	*6*
$B_2\text{-}18\text{-}O_2N_4$	A,E=O; B,C,D,F=NH 2,3- and 11,12-dibenzo	182-183	30	*6*
$B_3\text{-}18\text{-}O_6$	A-F=O; 2,3-, 8,9- and 14, 15-tribenzo	190-192	28	*5*
$BC\text{-}18\text{-}O_6$	A-F=O; 2,3-benzo; 11, 12-cyclohexo	Liquid	8	*41*
$B_2Me_4\text{-}18\text{-}O_6$	A-F=O; 2,3- and 11,12-dibenzo; 5,9,14,18-tetramethyl (5 isomers)	A109-110 B137 C136-137 D121-122 E 84- 86	6.5 5.6 3 – –	*63*
$B_2Me_4\text{-}18\text{-}O_6$	A-F=O; 2,3- and 11,12-	F199-200	6.7	*63*

(cont.)

Compound	Remarks	M.P. (B.P.) (°C)	Yield (%)	Reference[a]
	dibenzo; 6,8,15,17-tetramethyl (5 isomers	G134	3.8	
		H 92	1.9	
		I160	8.4	
		J120-122	31	
BMs-18-O_5N	A,C-F=O; B=*N*-mesyl; 2,3-benzo	91-92	32	*6*
B_2Ms_2-18-O_4N_2	A,C,E,F=O; B,D=*N*-mesyl; 2,3- and 11,12-dibenzo	200-202	40	*6*
BMs_2-18-O_4N_2	C-F=O; A,B=*N*-mesyl; 2,3-benzo	191-192	5	*6*
B_2Py_2-18-O_4N_2	B,C,E,F=O; A,D=pyridineN; 5,6- and 14,15-dibenzo	184-186	9	*26*
BrBz-18-O_5	B-F=O; A=benzeneC-Br	Liquid	7	*43*
B_2Ts-18-O_5N	A,B,D,E,F=O; C=*N*-tosyl; 2,3- and 11,12-dibenzo	158-160	34	*6*
B_2Ts_2-18-O_4N_2	A,B,D,E=O; C,F=*N*-tosyl; 2,3- and 11,12-dibenzo	215-216	10	*6*
B_2Ts_3-18-O_3N_3	A,C,E=O; B,D,F=*N*-tosyl; 2,3- and 11,12-dibenzo	150-153	54	*6*
B_2Ts_4-18-O_2N_4	A,E=O; B,C,D,F=*N*-tosyl; 2,3- and 11,12-dibenzo	Glass	20	*6*
Bz-18-O_5	B-F=O; A=benzeneCH	44-46	67	*22,43*
Bz-18-O_3S_2	C,D,E=O; B,F=S; A=benzeneCH	56-57	–	*25*
$BzMe_2$-18-O_3N_2	C-E=O; B,F=*N*-CH_3; A=benzeneCH	–	~40	*23*
$Bz_2(NO_2)_2$-18-S_4	B,C,E,F=S; A,D=bis(benzeneC-NO_2)	179-184	12	*9*
Bz_2Ph_2-18-S_4	B,C,E,F=S; A,D=bis-(benzeneC-phenyl)	217-218	5	*9*
$Bz_2(SMe)_2$-18-S_4	B,C,E,F=S; A,D=bis-(benzeneC-SCH_3)	170-173	5	*9*
C-18-O_6	A-F=O; 2,3-cyclohexo	Liquid	46	*5*
C_2-18-O_6	A-F=O; 2,3- and 11,12-dicyclohexo	68.5-69.5	45-67	*5,62*

(cont.)

Compound	Remarks	M.P. (B.P.) (°C)	Yield (%)	Reference[a]
C_2-18-O_6	A-F=O; 2,3- and 11,12-dicyclohexo (2 isomers)	77-78 120-121	25 30	*31* *31*
C_2-18-O_2S_4	A,D=O; B,C,E,F=S; 5,6- and 14,15-dicyclohexo	Liquid	24	*30*
CarbBz-18-O_5	B-F=O; A=benzeneC-carboxyl	100-101	98	*43*
ClBz-18-O_5	B-F=O; A=benzeneC-Cl	Liquid	53	*43*
CynBz-18-O_5	B-F=O; A=benzeneC-cyano	Liquid	10	*43*
$(DMD)_4$-18-O_6	A-F=O; 2,3,11,12-tetrakis-4′-(2′,2′-dimethyl-1′,3′-dioxolyl)	Liquid	14	*8a*
Fur-18-O_6	B-F=O; A=furanO	Liquid	10-30	*22,64*
Fur_2-18-O_6a	A,C=furanO; B,D,E,F=O	68-70	35	*64*
Fur_2-18-O_6b	A,D=furanO; B,C,E,F=O	109-111	11	*64*
Fur_3-18-O_6	A,C,E=furanO; B,D,F=O	124-126	10	*64*
$(HyMe)_4$-18-O_6	A-F=O; 2,3,11,12-tetrakis-hydroxymethyl	Liquid	75	*8a*
HyMeBz-18-O_5	B-F=O; A=benzeneC-hydroxymethyl	–	80	*43*
$(Hy_2\ Et)_4$-18-O_6	A-F=O; 2,3,11,12-tetrakis(1′,2′-dihydroxyethyl)	58-60	–	*8a*
k_2-18-O_4N_2	A,B,D,E=O; C,F=NH; 8,15-diketo	110-111	75	*14,55*
k_2Py-18-O_3N_3	D-F=O; A,C=NH; B=pyridineN; 2,6-diketo	126-129	72	*10*
k_4Py_2-18-N_6	A,C,D,F=NH; B,E=pyridineN; 2,6,11,15-tetraketo	>250	36	*10*
MeB-18-O_6	A-F=O; 2,3-(4′-methylbenzo)	54-55	52	*44,45*
MeB-18-O_4S_2	A-D=O; E,F=S; 14,15-(3′-methylbenzo)	Liquid	56	*30*
$(MeB)_2$-18-O_2S_4	A,D=O; B,C,E,F=S; 5,6- and 14,15-bis-(3′-methylbenzo)	147	6	*30*
MeCarbBz-18-O_5	B-F=O; A=benzeneC-methylcarboxyl	Liquid	82	*43*
$(Me_2NCO)_4$-18-O_6	A-F=O; 2,3,11,12-tetrakis(*N,N*-dimethylcarboxamido)	186	20	*64a*
MeOBz-18-O_3S_2	C,D,E=O; B,F=S; A=benzeneC-methoxy	Liquid	–	*25*

(cont.)

Compound	Remarks	M.P. (B.P.) (°C)	Yield (5)	Reference[a]
$[(MeO)_2Et]_4$-18-O_6	A-F=O; 2,3,11,12-tetrakis(1′,2′-methoxyethyl)	–	–	*8a*
MeOMeBz-18-O_5	B-F=O; A=benzeneC-methoxymethyl	70-71	50	*43*
MoMeB-18-O_6	A-E=O; 2,3-(3′-methyl-6′-*N*-morpholinyl-methylbenzo)	Liquid	35	*65*
Np-18-O_6	A-F=O; 2,3-(2′,3′-naphtho)	110-111.5	25	*5*
Np_2-18-O_6	A-F=O; 2,3- and 11,12-bis(2′,3′-naphtho)	224-226	25	*5*
PC-18-O_6	A-E=O; 2,3-(4′-12′-paracyclopano)	73-74	18	*65*
$(PhCH_2OCH_2)_4$-18-O_6	A-F=O; 2,3,11,12-tetrakis-benzyloxy-methyl	Liquid	12	*8a*
$(Ph_3COCH_2)_4$-18-O_6	A-F=O; 2,3,11,12-tetrakis-triphenyl-methoxymethyl	–	–	*8a*
Py-18-O_5N	B-F=O; A=pyridineN	40-41	29	*26*
Py-18-O_3NS_2	C,D,E=O; A=pyridineN; B,F=S	58-59	–	*25*
$Py_2(Me_2NCO)_4$-18-O_4N_2	B,C,E,F=O; A,C=pyridineN; 5,6,14,15-tetrakis(*N,N*-dimethylcarboxamido)	224	15	*64a*
Py_2-18-O_4N_2 [1,7]	B,D,E,F=O; A,C=pyridineN	–	–	*26*
Py_2-18-O_4N_2 [1,10]	B,C,E,F=O; A,D=pyridineN	147-148	18	*26*
Py_3-18-O_3N_3	B,D,F=O; A,C,E=pyridineN	125-128	32	*26*
$PyMe_2$-18-N_6	A=pyridineN; B,F=N C,D,E=NH; 3,17-dimethyl; 3,16-diene	–	–	*46*
$PyTs_2$-18-O_3N_3	D-F=O; A,C=*N*-tosyl; B=pyridineN	163-165	27	*10*
tbB-18-O_6	A-F=O; 2,3-(3′-*tert*-butylbenzo)	35-37	62	*5*
$(tbB)_2$-18-O_6	A-F=O; 2,3- and 11,12-bis(3′-*tert*-butylbenzo)	135-137	17	*5*
tbC-18-O_6	A-F=O; 2,3-(3′-*tert*-butylcyclohexo)	Liquid	12	*5*
$(tbC)_2$-18-O_6	A-F=O; 2,3- and 11,12-bis-(3′-*tert*-butylcyclo-hexo)	Liquid	52	*5*
Th-18-O_3S_3	C,D,E=O; B,F=S; A=thiopheneS	Liquid	–	*25*

(cont.)

Compound	Remarks	M.P. (B.P.) (°C)	Yield (%)	Reference[a]
THF-18-O_6	B-F=O; A=tetrahydro-furanO	Liquid	–	*64*
Ts_2-18-O_4N_2	A,B,D,E=O; C,F=*N*-tosyl	163.5-164.5	80	*8*
Ts_6-18-N_6	A-F=*N*-tosyl	311-313	60-75	*8*
VB-18-O_6	A-F=O; 2,3-(4′-vinyl-benzo)	61-62	71	*40*
VB_2-18-O_6	A-F=O; 2,3-(4′-vinyl-benzo); 11,12-benzo	148-149	66.5	*40*

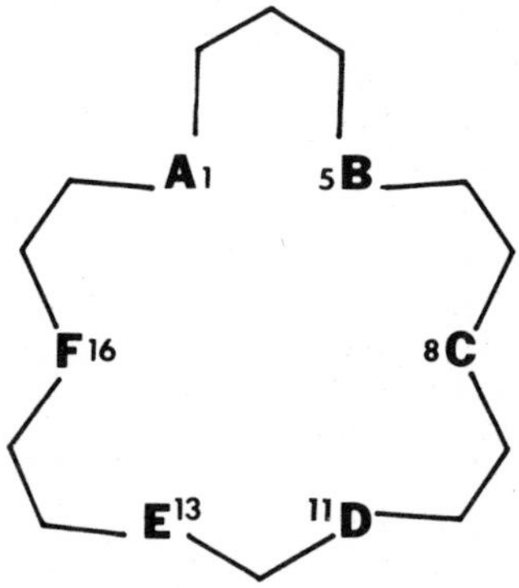

B_2-18-O_6	A-F=O; 6,7- and 17,18-dibenzo	118	31	*41*

19-Membered Rings

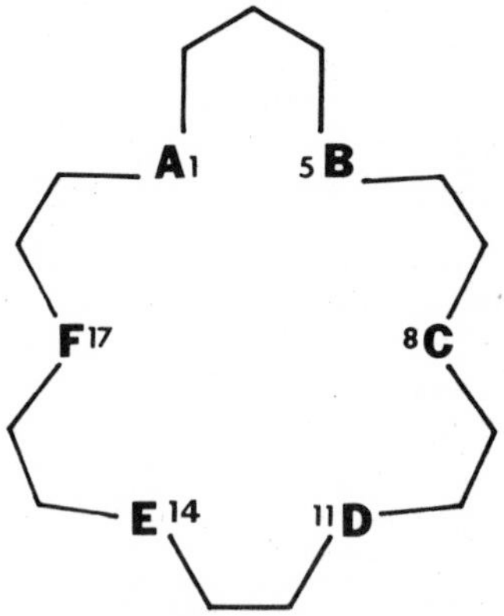

(cont.)

Compound	Remarks	M.P. (B.P. (°C)	Yield (%)	Reference[a]
$19\text{-}O_4S_2$	C-F=O; A,B=S	Liquid	20	*66*
$B_2\text{-}19\text{-}O_6$	A-F=O; 6,7- and 18,19-dibenzo	84.5-86	16	*5*
$B_3\text{-}19\text{-}O_6$	A-F=O; 6,7-, 12,13-, 18,19-tribenzo	147-149	16	*5*
$Hy\text{-}19\text{-}O_4S_2$	C-F=O; A,B=S; 3-hydroxy	Liquid	8	*66*
$k_2\text{-}19\text{-}O_6$	A-F=O; 2,4-diketo	68-69	38	*29*

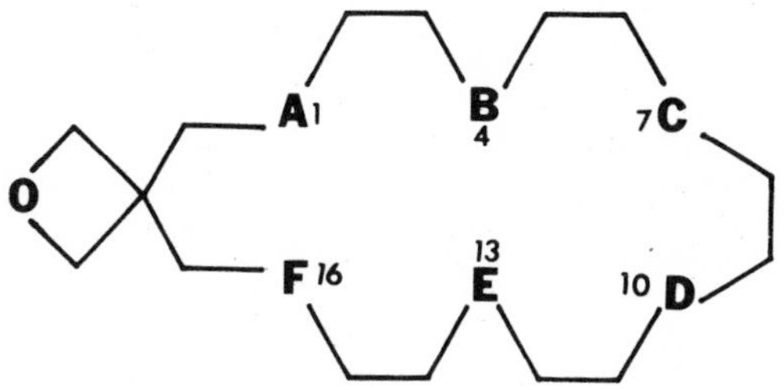

$spO\text{-}19\text{-}O_6$	A-F=O	(136-137/1)	60	*11*

20-Membered Rings

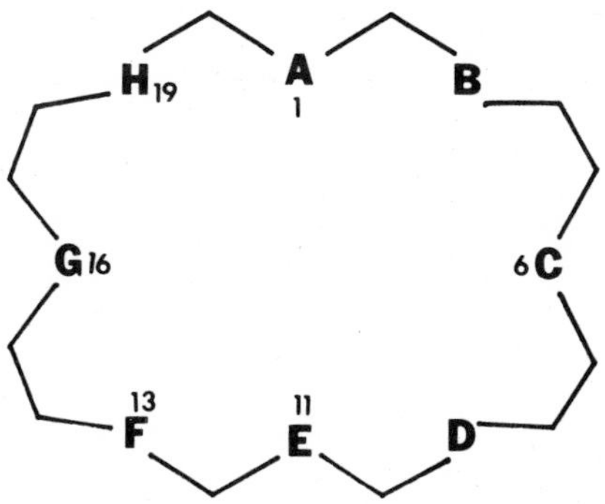

$Py_2\text{-}20\text{-}O_6N_2$	B,C,D,F,G,H=O; A,E= pyridineN	111-112	18	*27*

(cont.)

Compound	Remarks	M.P. (B.P.) (°C)	Yield (%)	Reference[a]

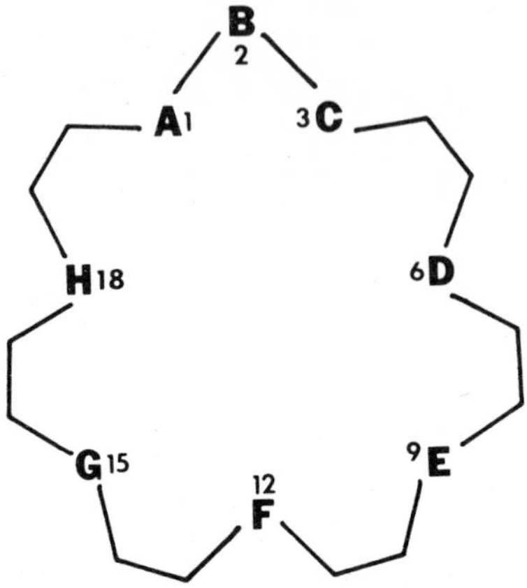

Compound	Remarks	M.P. (B.P.) (°C)	Yield (%)	Reference[a]
B_2-20-O_7	A,C-H=O; B=CH_2; 7,8- and 16,17-dibenzo	151-152	62	*41*
B_2k-20-O_7	A,C-H=O; B=CO; 7,8- and 16,17-dibenzo	122-125	48	*41*
B_2SO-20-O_7	A,C-H=O; B=SO; 7,8- and 16,17-dibenzo	133	33	*30*

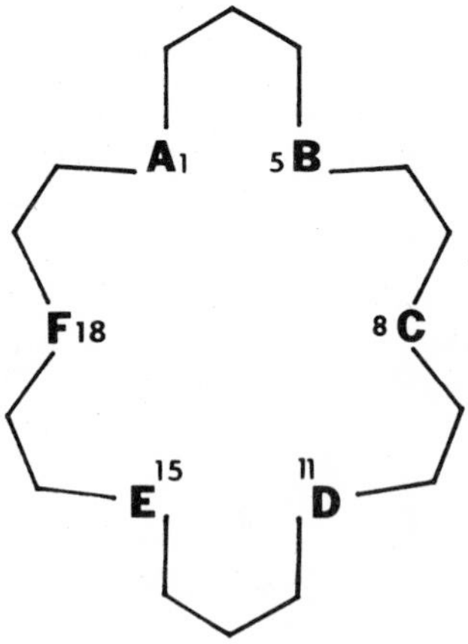

Compound	Remarks	M.P. (B.P.) (°C)	Yield (%)	Reference[a]
B_2-20-O_4	A,B,D,E=O; C,F=CH_2; 6,7- and 16,17-dibenzo	139-141	4	*5*

(cont.)

Compound	Remarks	M.P. (B.P.) (°C)	Yield (%)	Reference[a]
B_4-20-N_2S_4	A,D=NH; B,C,E,F=S; 2,3-, 6,7-, 12,13- and 16,17-tetrabenzo; 4,14-diene	Complexes Only	–	*67*
Bz_2-20-O_6	A-F=O; 3,13-benzeneCH	164	–	*68*
$Bz_2(NO_2)_2$-20-S_4	A,B,D,E=S; C,F=bis-(benzeneC-NO_2)	124	18	*9*
Bz_2Ph_2-20-S_4	A,B,D,E=S; C,F=bis-(benzeneC-phenyl)	236-237	7	*9*
$(HyMe)_2(PhCH_2O$ $Me)_2$-20-O_6	A-F=O; 3,13-bis-hydroxymethyl; 3,13-bis-benzyloxymethyl	Oil	–	*69*

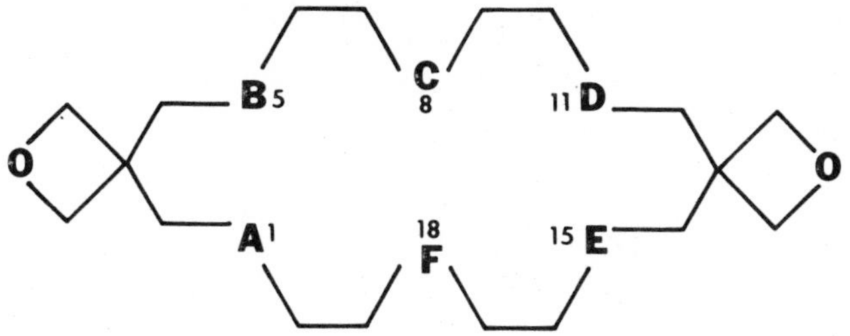

Compound	Remarks	M.P. (B.P.) (°C)	Yield (%)	Reference[a]
$(spO)_2$-20-O_6	A-F=O; 3,13-bis-spirooxetano	86-87	47	*11*
$(spO)_2$-20-O_4N_2	A,B,D,E=O; C,F=NH; 3,13-bis-spirooxetano	118.5-119	17	*47*
$(spO)_2$-20-O_2S_4	C,F=O; A,B,D,E=S; 3, 13-bis-spirooxetano	100-101	15	*11*
$(spO)_2$-20-S_6	A-F=S; 3,13-bis-spiro-oxetano	129-131	11	*11*

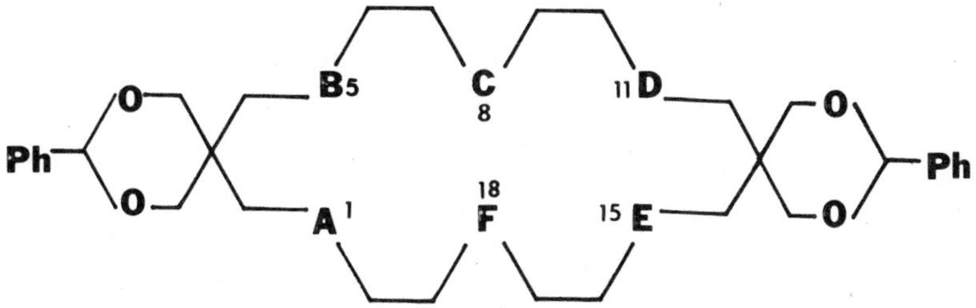

Compound	Remarks	M.P. (B.P.) (°C)	Yield (%)	Reference[a]
$(spPhO)_2$-20-O_6	A-F=O; 3,13-bis(spiro-2′-phenyl-1′,3′-dioxino)	99	70	*69*

(cont.)

Compound	Remarks	M.P. (B.P.) (°C)	Yield (%)	Reference[a]

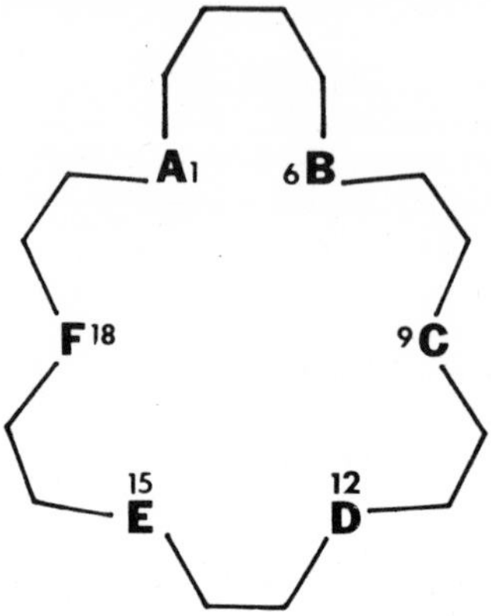

Compound	Remarks	M.P. (B.P.) (°C)	Yield (%)	Reference[a]
(acetOMeNP)Np-20-O_6	A-F=O; 2,3-naphtho; 4,5-(3′-α-acetic acid oxymethylnaphtho)	Oil	70	*50,51*
$(acetOMeNp)_2$-20-O_6	A-F=O; 2,3- and 4,5-bis-(3′-α-acetic acid oxy-methylnaphtho)	Oil	80	*50,51*
(acetOMeNp)(Mor-MeNp)-20-O_6	A-F=O; 2,3-(3′-α-acetic acid oxymethyl-naphtho); 4,5-(3′-morpholinylmethyl-naphtho)	Oil	65	*51*
$amMeBNp_2$-20-O_6	A-F=O; 13,14-acetomido-methylbenzo; 2,3- and 4,5-dinaphtho	193-194	36	*52*
B_2-20-O_6	A-F=O; 2,3- and 4,5-dibenzo	64-65	12	*52*
BNp_2-20-O_6	A-F=O; 13,14-benzo; 2,3- and 4,5-dinaph-tho	147-148	41	*52*
BNp_2-20-O_5S	A,B,C,D,F=O; E=S; 13,14-benzo; 2,3- and 4,5-dinaphtho	111-113	38	*52*
BNp_2-20-O_4S_2	A,B,C,F=O; D,E=S; 13,14-benzo; 2,3- and 4,5-dinaphtho	149-150	72	*52*

(cont.)

Compound	Remarks	M.P. (B.P.) (°C)	Yield (%)	Reference[a]
$B_2Me_4Ph_2$-20-O_4N_2	A,B,D,E=O; C,F=*N*-CH_3; 8,19-dimethyl; 7,20-diphenyl; 3,4- and 13,14-dibenzo	94-95	40	*70*
B_2Pyr_2-20-O_4N_2	A,B,D,E=O; C,F=N; 3,4- and 13,14-dibenzo; 8,9- and 18,19-dipyrrolo	98-99	–	*70*
$CarbPBNp_2$-20-O_6	A-F=O; 13,14-(4′-γ-carboxypropylbenzo); 2,3- and 4,5-dinaphtho	Oil	70	*51*
$HyPBNp_2$-20-O_6	A-F=O; 13,14-(4′-γ-hydroxypropylbenzo); 2,3- and 4,5-dinaphtho	Liquid	16	*52*
$(HyMeNp)_2$-20-O_6	A-F=O; 2,3- and 4,5-bis-(3′-hydroxymethylnaphtho)	132-134	60	*51*
(HyMeNp)(MorMeNp)-20-O_6	A-F=O; 2,3-(3′-hydroxymethylnaphtho); 4,5-(3′-morpholinylmethylnaphtho)	Oil	55	*51*
(HyMeNp)Np-20-O_6	A-F=O; 2,3-naphtho; 4,5-(3′-hydroxymethylnaphtho)	136-137	50	*51*
$(MeAcMeNp)_2$-20-O_6	A-F=O; 2,3- and 4,5-bis-(3′-methyl-α-acetyloxymethylnaphtho)	Oil	60	*50,51*
(MeAcMeNp)(MorMeNp)-20-O_6	A-F=O; 2,3-(3′-methyl-α-acetyloxymethylnaphtho); 4,5-(3′-morpholinylmethylnaphtho)	Oil	35	*51*
(MeAcMeNp)Np-20-O_6	A-F=O; 2,3-naphtho; 4,5-(3′-methyl-α-acetyloxymethylnaphtho)	Oil	70	*51*
$MeVBNp_2$-20-O_6	A-F=O; 13,14-(3′-β-methylvinylbenzo); 2,3- and 4,5-dinaphtho	Liquid	41	*52*
$(MorMeNp)_2$-20-O_6	A-F=O; 2,3- and 4,5-bis-(3′-morpholinylmethylnaphtho)	Oil	65	*51*
Np_2-20-O_6	A-F=O; 2,3- and 4,5-dinaphtho	130-130.5	60	*51*
Np_2-20-O_6	A-F=O; 2,3- and 4,5-dinaphtho	Liquid (Opt. pure)	64	*12*

(cont.)

Compound	Remarks	M.P. (B.P.) (°C)	Yield (%)	Reference[a]
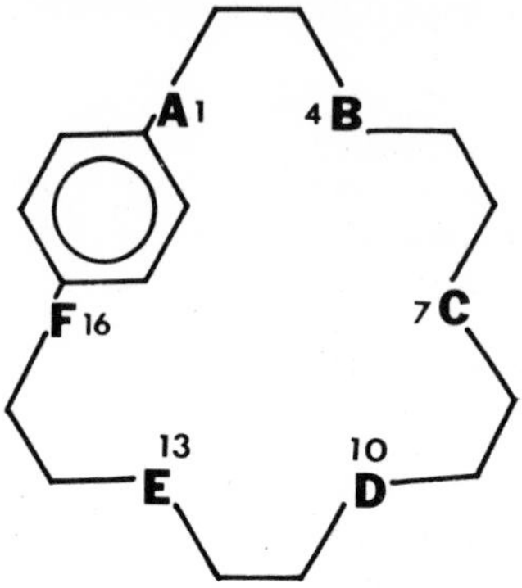				
p-Cycloph-20-O_6	A-F=O	Liquid	–	*65*
	21-Membered Rings			
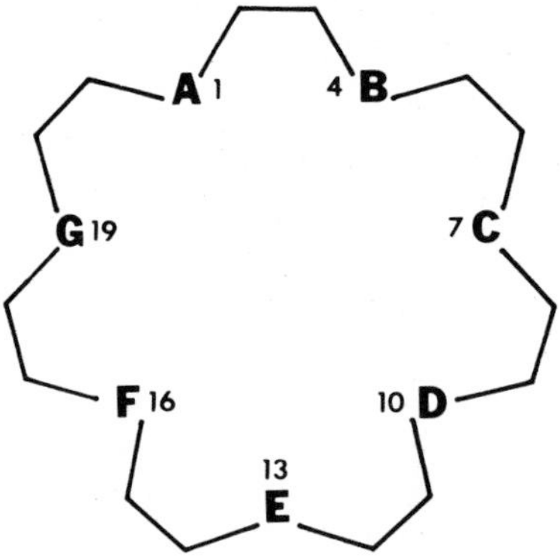				
21-O_7	A-G=O	–	18	*54*
21-O_5N_2	A,B,C,E,F=O; D,G=NH	Liquid	85	*14,55*
21-O_5S_2[13,19]	A,B,C,D,F=O; E,G=S	(180/1)	25	*38*
21-O_5S_2[16,19]	A-E=O; F,G=S	Liquid	10	*66*
21-O_4S_3	A-D=O; E,F,G=S	Liquid	11	*38*
B_2-21-O_7	A-G=O; 2,3- and 11,12-dibenzo	106.5-107.5	36	*5*
C_2-21-O_7	A-G=O; 2,3- and 11,12-dicyclohexo	Liquid	50	*5*
B_3-21-O_7	A-G=O; 2,3-, 8.9-, and 14,15-tribenzo	98.5-100	19	*5*
B_2k_2-21-O_5N_2	A-E=O; F,G=NH; 2,3- and 11,12-dibenzo; 15, 20-diketo	168-170	15	*61*
Bz-21-O_6	B-G=O; A=benzeneCH	–	49	*22*

(cont.)

Compound	Remarks	M.P. (B.P.) (°C)	Yield (%)	Reference[a]
CarbBz-21-O_6	B-G=O; A=benzeneC-carboxyl	86-95	98	*43*
k_2-21-O_5N_2	A,B,C,E,F=O; D,G=NH; 11,18-diketo	90-91	75	*14,55*
MeCarbBz-21-O_6	B-G=O; A=benzeneC-methylcarboxyl	Liquid	68	*43*
Py-21-O_4NS_2	C-E=O; A=pyridineN; B,G=S	–	–	*25*
Ts_7-21-N_7	A-G=*N*-tosyl	183-184	45	*8*

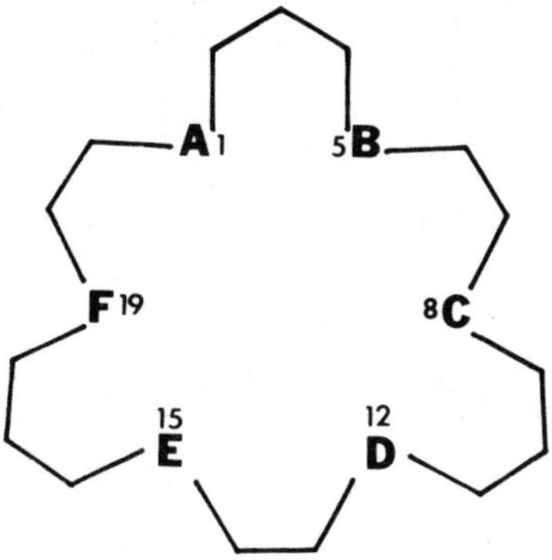

21-S_6	A-F=S	64-65	9.7	*18*

22-Membered Rings

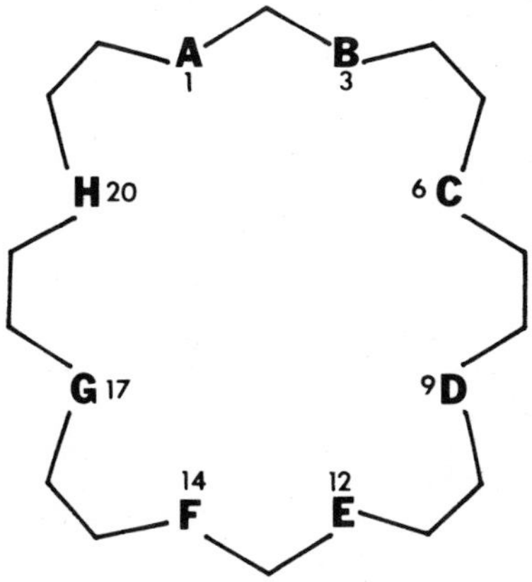

B_2-22-O_8	A-H=O; 7,8- and 18, 19-dibenzo	166-167	8	*41*

(cont.)

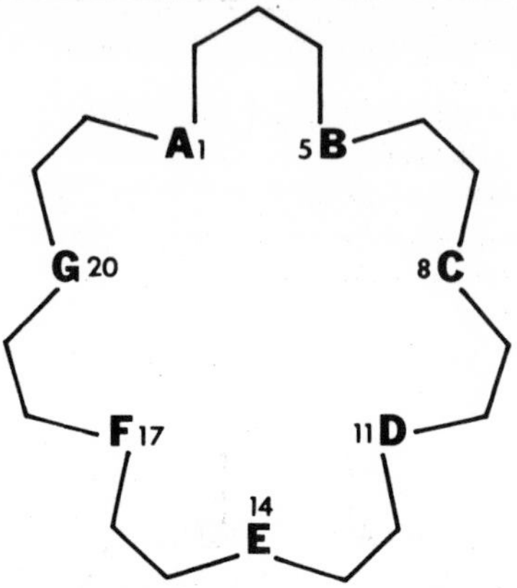

Compound	Remarks	M.P. (B.P.) (°C)	Yield (%)	Reference[a]
Hy-22-O_5S_2	C-G=O; A,B=S; 3-hydroxy	Liquid	15	*66*

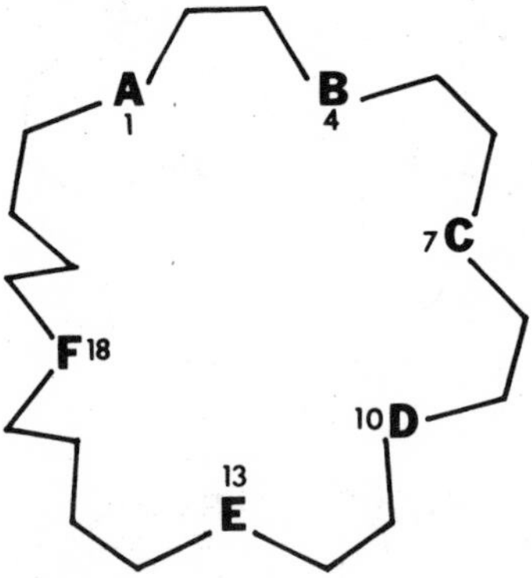

Compound	Remarks	M.P. (B.P.) (°C)	Yield (%)	Reference[a]
B_2-22-O_6	A-F=O; 2,3- and 11, 12-dibenzo	82-83	9	*5*

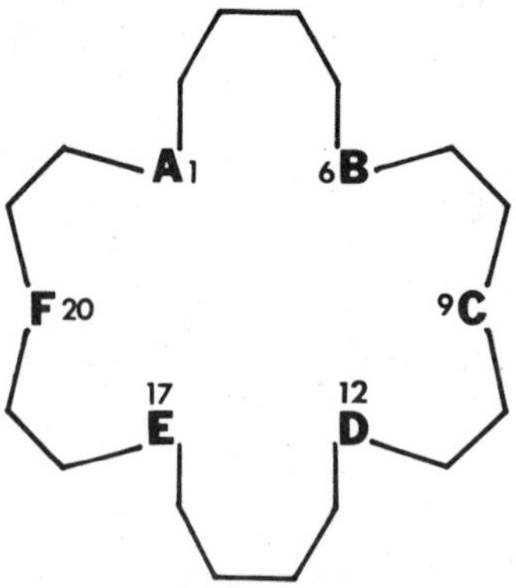

(cont.)

Compound	Remarks	M.P. (B.P.) (°C)	Yield (%)	Reference[a]
B_2Py_2-22-O_4N_2	A,B,D,E=O; C,F=pyridineN; 3,4- and 14,15-dibenzo	142-143	40	*71*
$(BrNp)_4$-22-O_6	A-F=O; 2,3-, 4,5-, 13,14- 15,16-tetrakis(6′-bromo-naphtho)	299-300 189-191 334-335	80 91 90	*72*
$BzNp_4$-22-O_5	A-E=O; F=benzeneCH; 2,3-, 4,5-, 13,14-, and 15,16-tetranaphtho	–	13	*73*
$BzNp_4Py$-22-O_4N	A,B,D,E=O; C=benzeneCH; F=pyridineN; 2,3-, 4,5-, 13,14-, and 15,16-tetra-naphtho	–	43	*73*
$(CMeNp)_2Np_2$-22-O_6	A-F=O; 2,3- and 4,5-di-naphtho; 13,14- and 15, 16-bis(3′-chloromethyl-naphtho)	Oil	76	*74*
$(DMCSNp)_4$-22-O_6	A-F=O; 2,3-, 4,5-, 13,14-, and 15,16-tetrakis(6′-dimethylchlorosilyl-naphtho)	95-96	84	*72*
$(HyMeNp)_2Np_2$-22-O_6	A-F=O; 2,3- and 4,5-di-naphtho; 13,14- and 15, 16-bis(3′-hydroxymethyl-naphtho)	Oil	28	*74*
$(MeNp)_2Np_2$-22-O_6	A-F=O; 2,3- and 4,5-di-naphtho; 13,14- and 15, 16-bis(3′-methylnaphtho)	Oil	80	*74*
Np_4-22-O_6	A-F=O; 2,3-, 4,5-, 13,14-, and 15,16-tetranaphtho	123-126	31	*12,74*
Np_4-22-O_5	A-E=O; F=CH_2; 2,3-, 4,5-, 13,14-, and 15,16-tetra-naphtho	–	41	*73*
Np_4Py-22-O_5N	A-E=O; F=pyridineN; 2,3-, 4,5-, 13,14-, and 15,16-tetranaphtho	–	43	*73*
Np_4Py_2-22-O_4N_2	A,B,D,E=O; C,F=pyridineN; 2,3-, 4,5-, 13,14-, and 15,16-tetranaphtho	–	26	*26*
Np_4Py-22-O_4N	A,B,D,E=O; C=CH_2; F= pyridineN; 2,3-, 4,5-, 13,14-, and 15,16-te-tranaphtho	–	29	*73*

(cont.)

Compound	Remarks	M.P. (B.P.) (°C)	Yield (%)	Reference[a]
	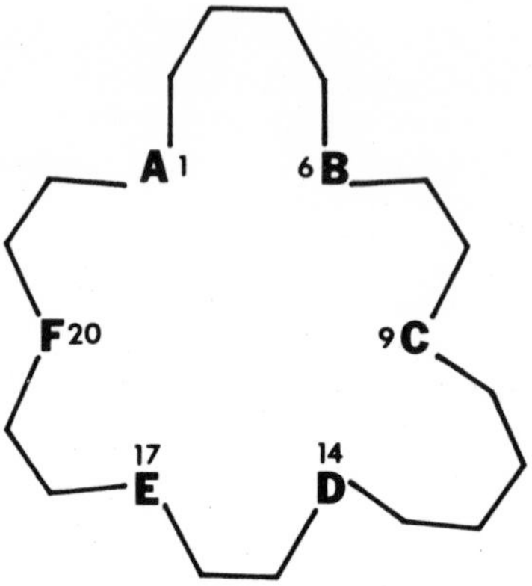			
Np_4-22-O_6	A-F=O; 2,3-, 4,5-, 10,11-, and 12,13-tetranaphtho	–	50	*75*
	23-Membered Rings			
	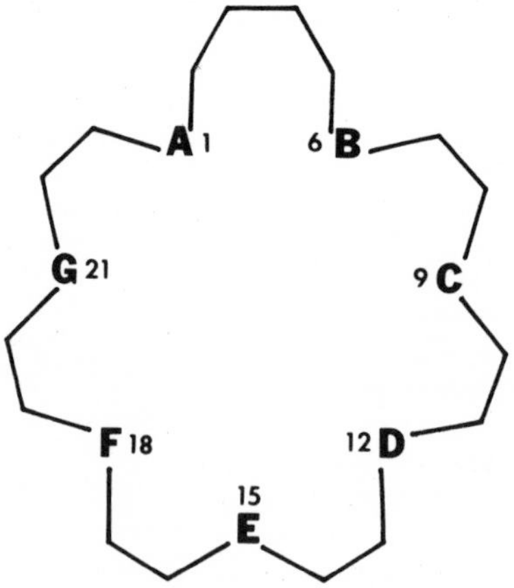			
$(AcetOMeNp)_2$-23-O_7	A-G=O; 2,3- and 4,5-bis(3′-acetic acid oxymethylnaphtho)	Oil	75	*50,51*

(cont.)

Compound	Remarks	M.P. (B.P.) (°C)	Yield (%)	Reference[a]
$(HyMeNp)_2$-23-O_7	A-G=O; 2,3- and 4,5-bis(3′-hydroxymethyl-naphtho)	Oil	50	*51*
$(MeAcMeNp)_2$-23-O_7	A-G=O; 2,3- and 4,5-bis(3′-methyl-α-acetyloxy-methylnaphtho)	Oil	50	*51*
Np_2-23-O_7	A-G=O; 2,3- and 4,5-dinaphtho	130-130.5	60	*52*

24-Membered Rings

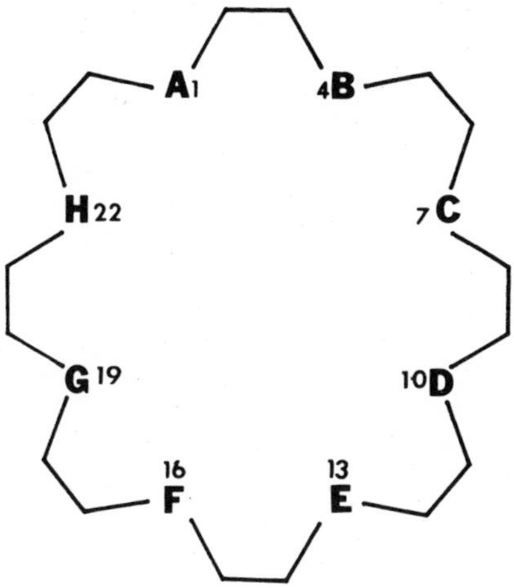

Compound	Remarks	M.P. (B.P.) (°C)	Yield (%)	Reference[a]
24-O_8	A-H=O	–	15	*54*
24-O_6N_2	A,B,C,E,F,G=O; D,H=NH	Liquid	90	*14,55*
24-$O_6S_2^{19,22}$	A-F=O; G,H=S	Liquid	5	*66*
24-$O_6S_2^{10,22}$	A,B,C,E,F,G=O; D,H=S	Liquid	1	*58,66*
B_2-24-O_8	A-H=O; 2,3- and 14,15-dibenzo	113-114	38	*5*
B_4-24-O_8	A-H=O; 2,3-, 8,9-, 14,15-, and 20,21-tetrabenzo	150-152	18	*5*
B_2k_2-24-O_6N_2	A-F=O; G,H=NH; 2,3- and 14,15-dibenzo; 18,23-diketo	176-177	16	*61*
Bz-24-O_7	B-H=O; A=benzeneCH	–	18	*22*
Bz_2-24-O_6	B,C,D,F,G,H=O; A,E=	–	30	*22*

(cont.)

Compound	Remarks	M.P. (B.P.) (°C)	Yield (%)	Reference[a]
	benzeneCH			
C_2-24-O_8	A-H=O; 2,3- and 14, 15-dicyclohexo	Liquid	63	*5*
Fur-24-O_8	B-H=O; A=furanO	–	–	*22*
k_2-24-O_6N_2	A,B,C,E,F,G=O; D,H= NH; 11,21-diketo	49-50	68	*14,55*
k_4Py_2-24-O_2N_6	D,H=O; A,C,E,G=NH; B,F=pyridineN; 2, 6,14,18-tetraketo	338-340	3	*10*
Np_2-24-O_8	A-H=O; 2,3- and 14,15-bis(2′,3′-naphtho)	190-191.5	63	*5*
Py_4-24-O_4N_4	A,C,E,G=O; B,D,F,H= pyridineN	173-176	20	*26*

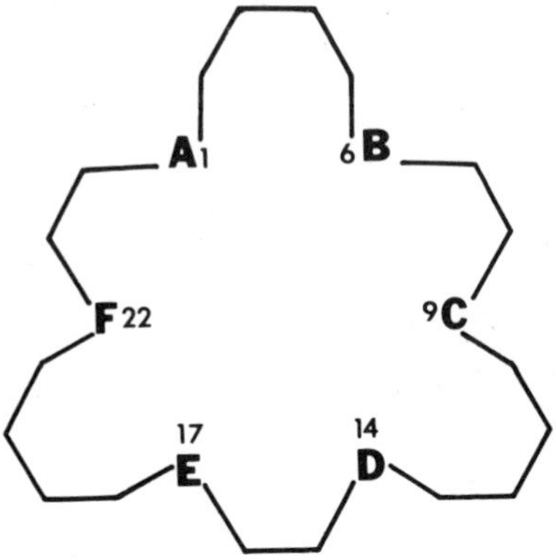

Np_6-24-O_6	A-F=O; 2,3-, 4,5-, 10,11-, 12,13-, 18,19- and 20, 21-hexanaphtho	338-342 188-190	1 1	*48* (Opt. active)

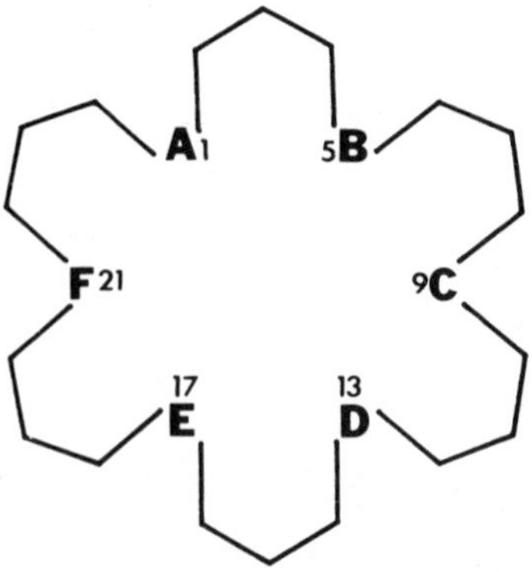

(cont.)

Compound	Remarks	M.P. (B.P.) (°C)	Yield (%)	Reference[a]
24-S_6	A-F=S	29-30	15	*18*
	25-Membered Rings			
	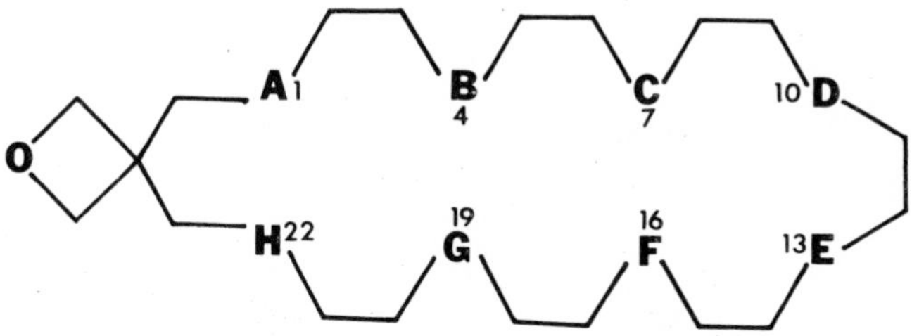			
spO-25-O_8	A-H=O	(182-183/0.3)	35	*11*
	26-Membered Rings			
	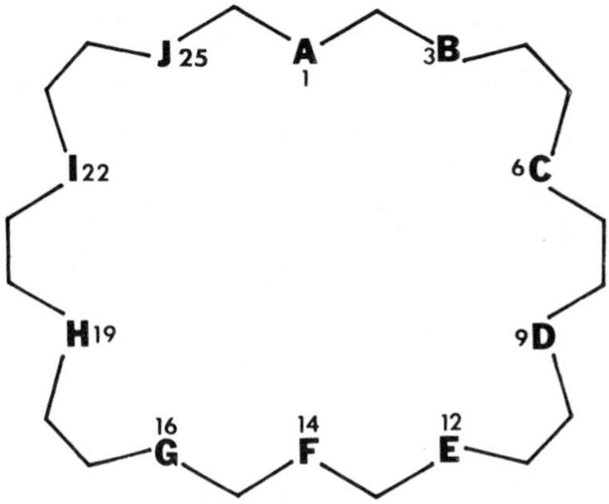			
Py_2-26-O_8N_2	B-E,G-J=O; A,F= pyridineN	117-120	2	*27*

(cont.)

Compound	Remarks	M.P. (B.P.) (°C)	Yield (%)	Reference[a]

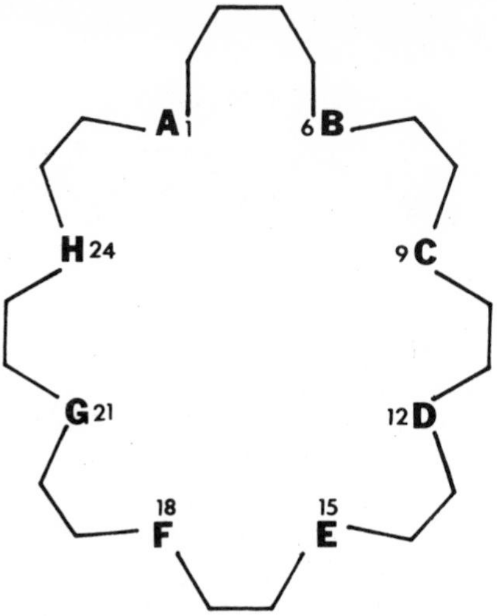

Np_2-26-O_8	A-H=O; 2,3- and 4,5-dinaphtho	Liquid	48	*52*

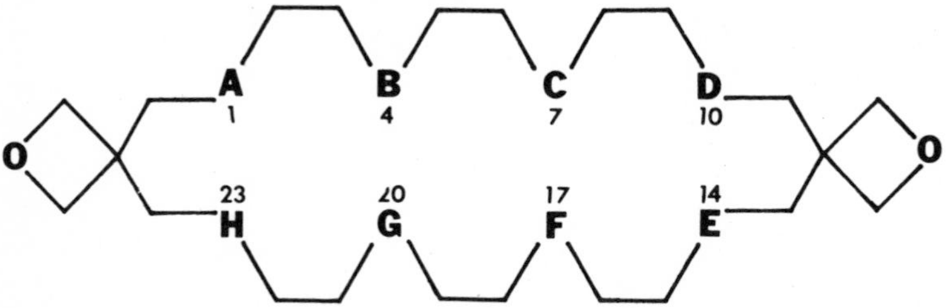

$(spO)_2$-26-O_8	A-H=O	52-53	25	*11*

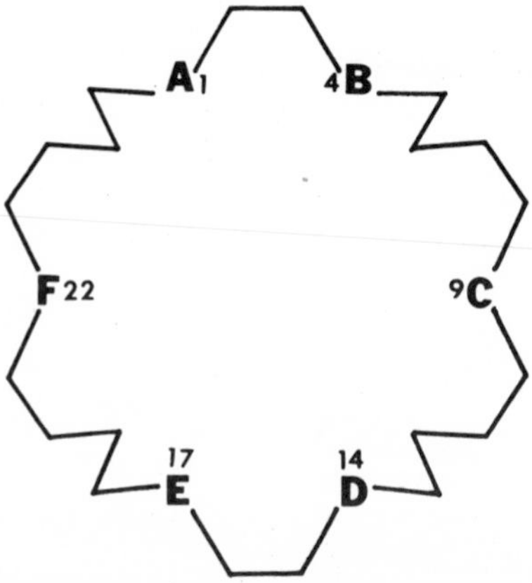

(cont.)

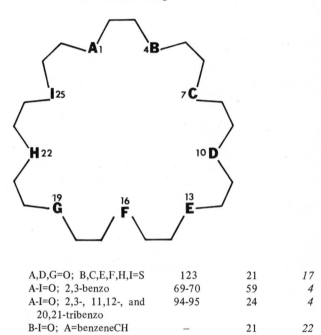

Compound	Remarks	M.P. (B.P.) (°C)	Yield (%)	Reference[a]
B_2-26-O_6	A-F=O; 2,3- and 15,16-dibenzo	125-127	17	*5*
	27-Membered Rings			
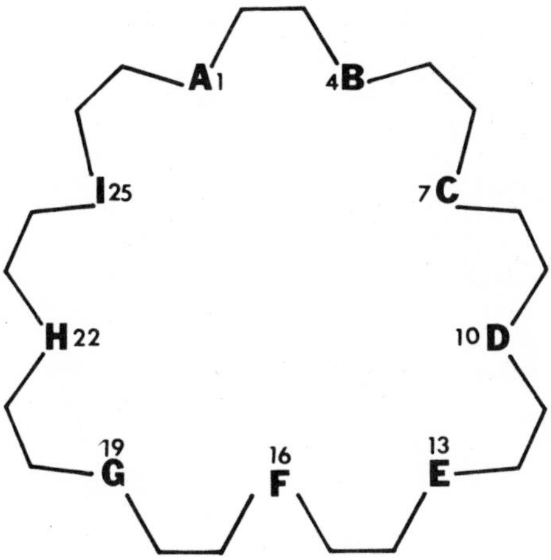				
27-O_3S_6	A,D,G=O; B,C,E,F,H,I=S	123	21	*17*
B-27-O_9	A-I=O; 2,3-benzo	69-70	59	*4*
B_3-27-O_9	A-I=O; 2,3-, 11,12-, and 20,21-tribenzo	94-95	24	*4*
Bz-27-O_8	B-I=O; A=benzeneCH	–	21	*22*
	28-Membered Rings			
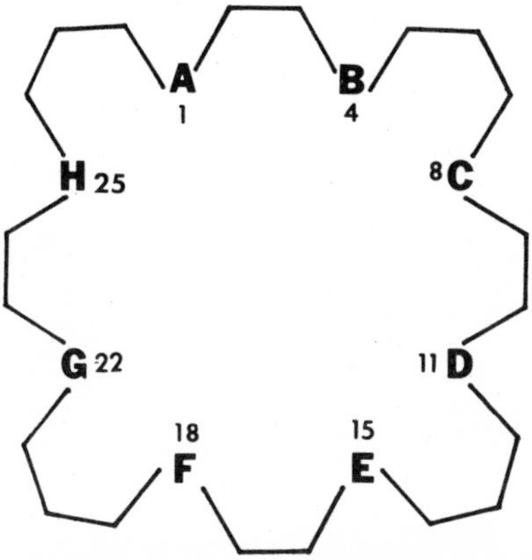				

(cont.)

Compound	Remarks	M.P. (B.P.) (°C)	Yield (%)	Reference[a]
28-S_8	A-H=S	66-67	25	*33*
B_2-28-O_4	A,B,E,F=O; C,D,G,H= CH_2; 2,3- and 16,17-dibenzo	137-138.5	3	*5*

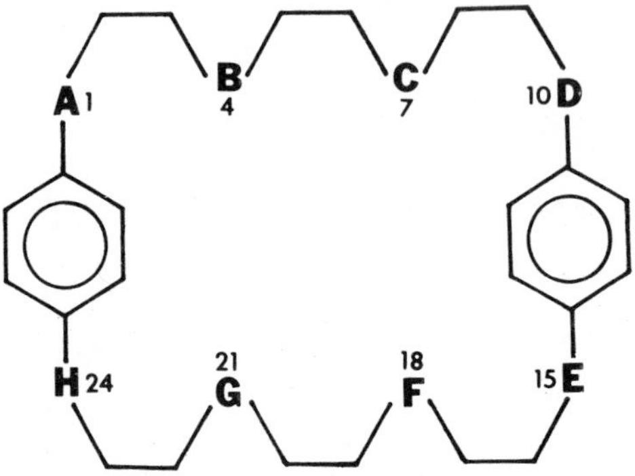

(p-Cycloph)$_2$-28-O_8	A-H=O	96-97	2	*65*

30-Membered Rings

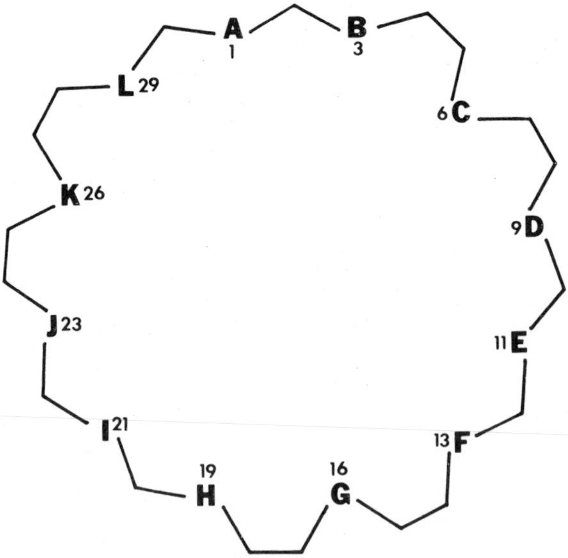

(cont.)

Compound	Remarks	M.P. (B.P. (°C)	Yield (%)	Reference[a]
Py_3-30-O_9N_3	B,C,D,F,G,H,J,K,L=O; A,E,I=pyridineN	120.5-121.5	3	*27*

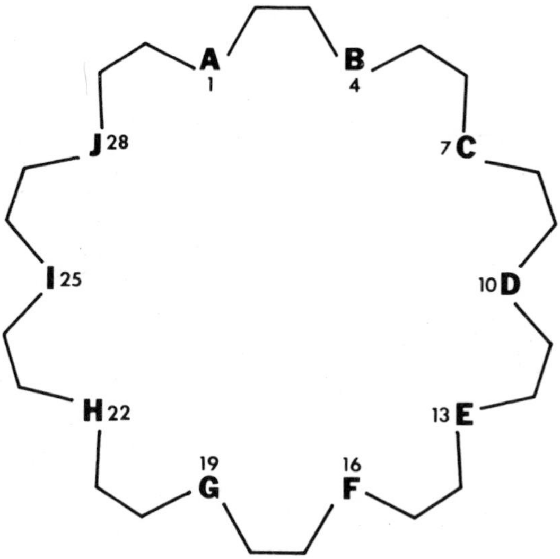

B_2-30-O_{10}	A-J=O; 2,3- and 17,18-dibenzo	106-107.5	6	*5*
Bz-30-O_9	B-J=O; A=benzeneCH	–	21	*22*
$(Bz)_2$-30-O_8	B,C,D,E,G,H,I,J=O; A,F=benzeneCH	–	9	*22*
C_2-30-O_{10}	A-J=O; 2,3- and 17,18-dicyclohexo	Liquid	61	*5*
CarbBz-30-O_9	B-J=O; A=benzeneC-carboxyl	Oil	98	*43*
Fur_2-30-O_{10}	B,C,D,E,G,H,I,J=O; A,F=furanO	–	10	*22*
MeCarbBz-30-O_9	B-J=O; A=benzeneC-methylcarboxyl	Oil	34	*43*

(cont.)

Compound	Remarks	M.P. (B.P.) (°C)	Yield (%)	Reference[a]

Larger than 30-Membered Rings

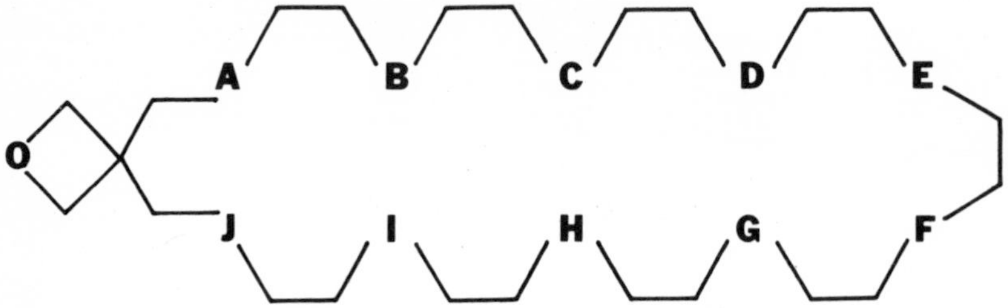

Compound	Remarks	M.P. (B.P.) (°C)	Yield (%)	Reference[a]
spO-31-O_{10}	A-J=O	(250-260/1)	11	*11*

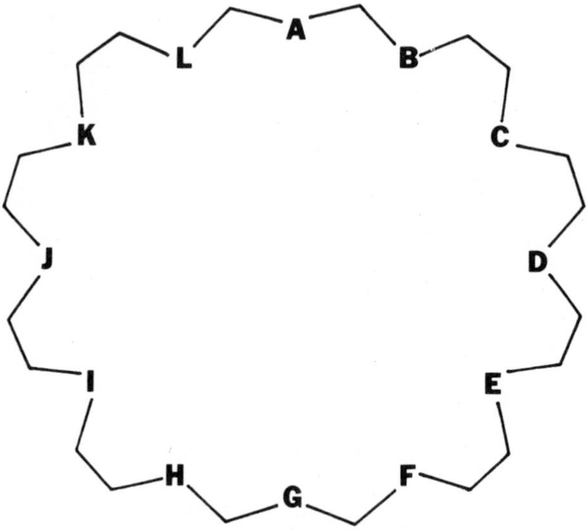

Compound	Remarks	M.P. (B.P.) (°C)	Yield (%)	Reference[a]
Py_2-32-$O_{10}N_2$	B-F,H-L=O; A,G= pyridineN	–	–	*27*

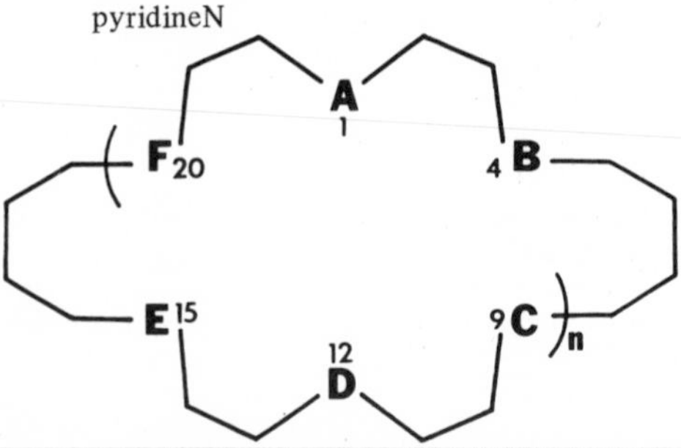

(cont.)

Compound	Remarks	M.P. (B.P.) (°C)	Yield (%)	Reference[a]
B_3Py_3-33-O_6N_3	*n*=2; B,C,E,F,=O; A,D= pyridineN; 6,7- and 17,18-dibenzo	129-130	9	*71*
B_4Py_4-44-O_8N_4	*n*=3; B,C,E,F=O; A,D= pyridineN; 6,7- and 17,18-dibenzo	108-109	–	*71*
B_5Py_5-55-$O_{10}N_5$	*n*=4; B,C,E,F=O; A,D= pyridineN; 6,7- and 17,18-dibenzo	104-105	–	*71*

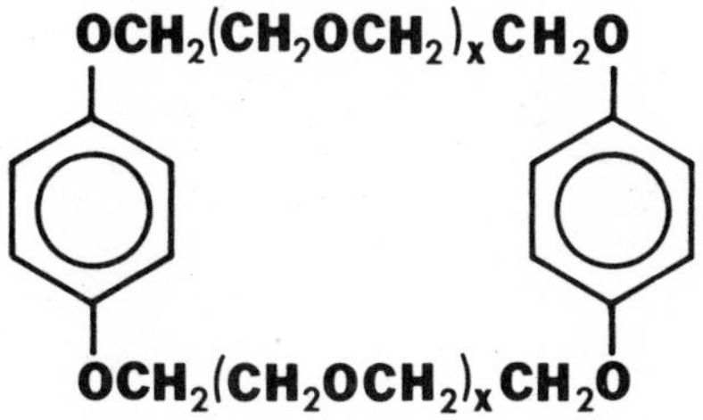

(*p*-Cycloph)$_2$-34-O_{10}	*x*=3	3-94	3	*65*
(*p*-Cycloph)$_2$-40-O_{12}	*x*=4	67-69	4	*65*

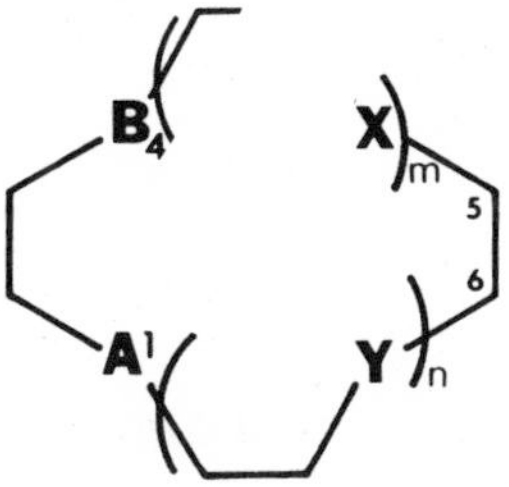

B_2-48-O_{16}	A,B,X,Y=O; *m*=5,*n*=9; 2,3- and 5,6-dibenzo	Liquid	32	*5*
C_2-48-O_{16}	A,B,X,Y=O; *m*=5, *n*=9; 2,3- and 5,6-dicyclo-hexo	Liquid	58	*5*
B_2-60-O_{20}	A,B,X,Y=O; *m,n*=9; 2,3- and 5,6-dibenzo	Liquid	41	*5*
C_2-60-O_{20}	A,B,X,Y=O; *m,n*=9 2,3- and 5,6-dicyclo-hexo	Liquid	45	*5*

(cont.)

Compound	Remarks	M.P. (B.P.) (°C)	Yield (%)	Reference[a]

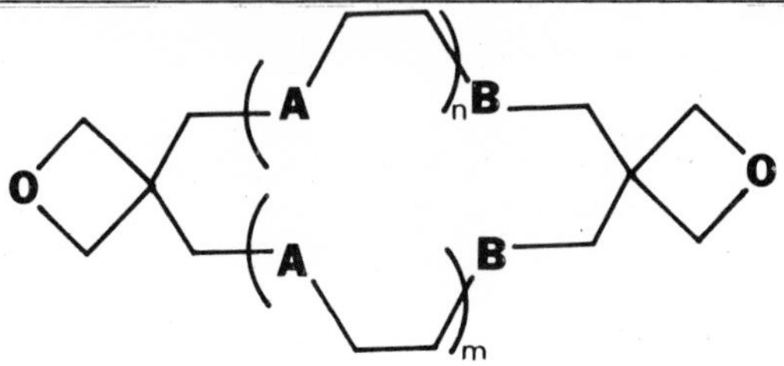

$(spO)_2$-50-O_{16}	A,B=O; n=m=7	42-43	3	*11*

Two-Ring Systems

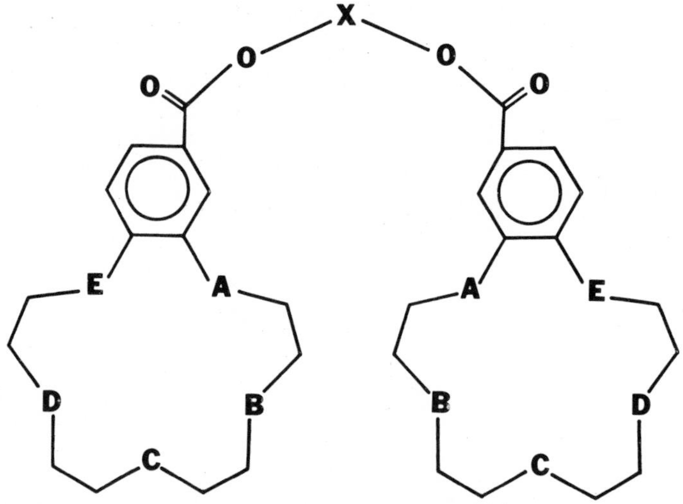

k_2B_2-$(15)_2$-O_{10}a	A-E=O; X=CH_2CH_2	124-125	87	*42*
k_2B_2-$(15)_2$-O_{10}b	A-E=O; X=$(CH_2)_5$	88-89	60	*42*
k_2B_2-$(15)_2$-O_{10}c	A-E=O; X=$(CH_2)_8$	82	56	*42*
k_2B_2-$(15)_2$-O_{10}d	A-E=O; X=$CH_2CH_2OCH_2CH_2$	89-91	64	*42*
k_2B_2-$(15)_2$-O_{10}e	A-E=O; X=$(CH_2CH_2O)_2$-CH_2CH_2	83-84	48	*42*

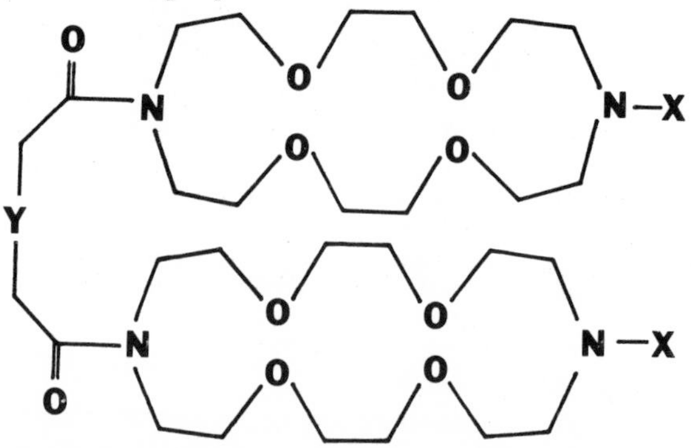

Compound	Remarks	M.P. (B.P.) (°C)	Yield (%)	Reference[a]
(18)$_2$-O$_8$N$_4$	Y=CH$_2$; X=H	Liquid	–	*76*
(18)$_2$-O$_9$N$_4$	Y=O; X=H	Liquid	–	*76*
Ts-(18)$_2$-O$_8$N$_5$	Y=*N*-tosyl; X=H	Liquid	–	*76*

Bicyclic Ring Systems

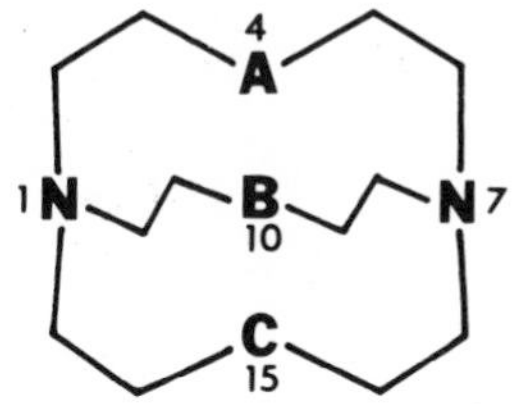

Compound	Remarks	M.P. (B.P.) (°C)	Yield (%)	Reference[a]
(12)$_3$-O$_3$N$_2$	A-C=O	81-83	33	*77*
k$_2$-(12)$_3$-O$_3$N$_2$	A-C=O; 2,6-diketo	178-180	10	*77*

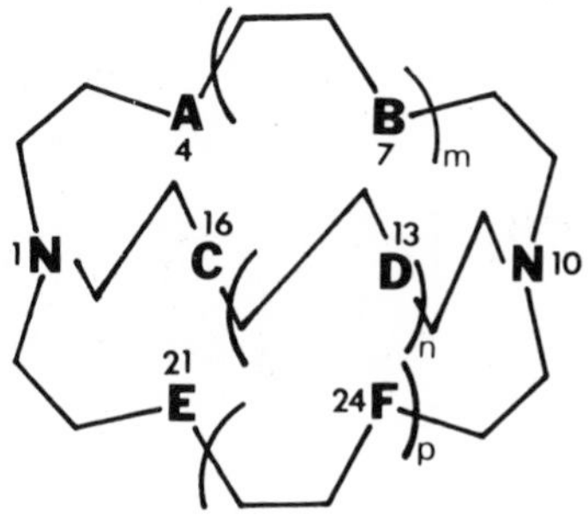

Compound	Remarks	M.P. (B.P.) (°C)	Yield (%)	Reference[a]
k$_2$-(12)$_3$-O$_3$N$_2$	A,C,F=O; *m,n,p*=0; 2,9-diketo	–	10	*78*
12(15)$_2$-O$_4$N$_2$	A-F=O; *n,p*=0; *m*=1	Liquid	–	*14*
k$_2$-12(15)$_2$-O$_4$N$_2$	A-F=O; *n,p*=0; *m*=1 2,9-diketo	151-152	45	*14*
15(18)$_2$-O$_5$N$_2$	A-F=O; *m.n*=1; *p*=0	Liquid	–	*14*
k$_2$-15(18)$_2$-O$_5$N$_2$	A-F=O; *m,n*=1; *p*=0; 2,9-diketo	114-115	40	*14*
18(21)$_2$-O$_7$N$_2$	A-F=O; *m,n*=1; *p*=2	Liquid	–	*14*
k$_2$-18(21)$_2$-O$_7$N$_2$	A-F=O; *m,n*=1; *p*=2 2,9-diketo	99-100	45	*14*
(21)$_2$24-O$_8$N$_2$	A-F=O; *m*=1; *n,p*=2	48-50	–	*14,55*
k$_2$-(21)$_2$24-O$_8$N$_2$	A-F=O; *m,n*=2; *p*=1; 2,9-diketo	114-115	45	*14*
(24)$_3$-O$_9$N$_2$	A-F=O; *m,n,p*=2	Liquid	–	*14,55*
k$_2$-(24)$_3$-O$_9$N$_2$	A-F=O; *m,n,p,*=2; 2,9-diketo	72-74	60	*14*

(cont.)

Compound	Remarks	M.P. (B:P.) (°C)	Yield (%)	Reference[a]
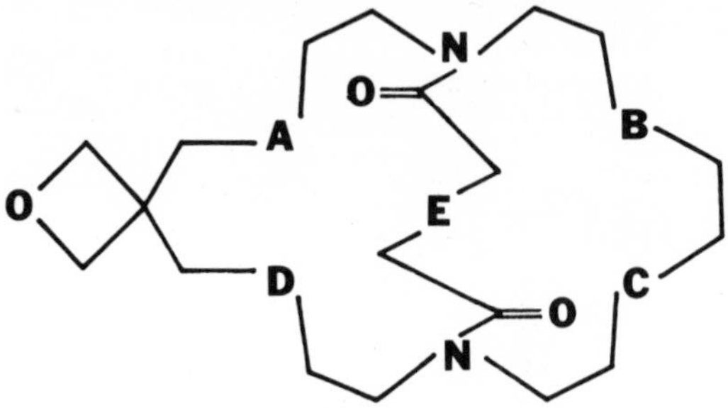				
$k_2$spO-15,16,19-O_5N_2	A-E=O	180.5-182	30	*47*
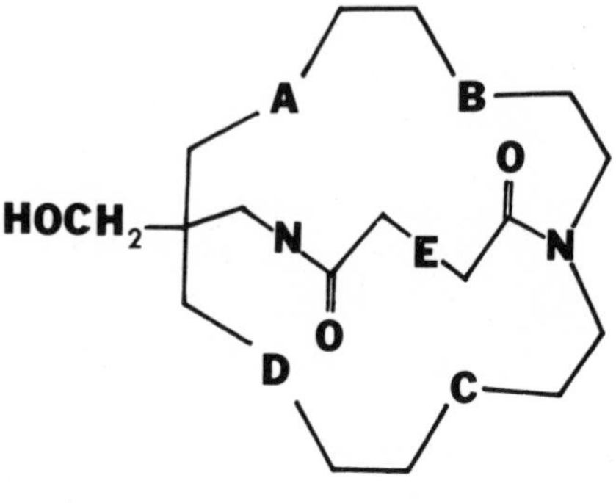				
HyMek_2-$(16)_3$-O_5N_2	A-E=O	186-187.5	35	*47*
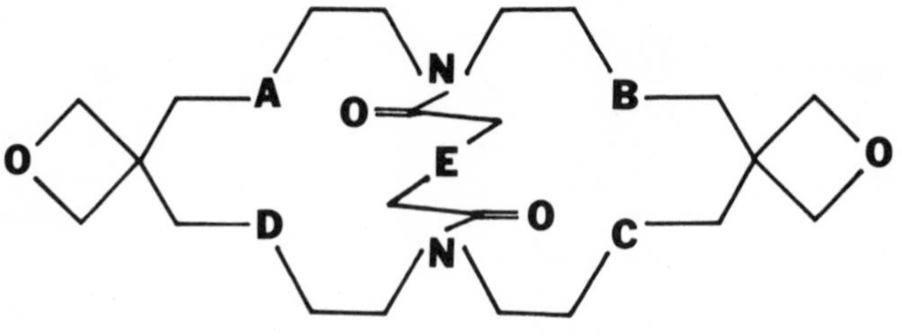				
$k_2(spO)_2$-$(16)_2$20-O_5N_2	A-E=O	190-192	43	*47*

(cont.)

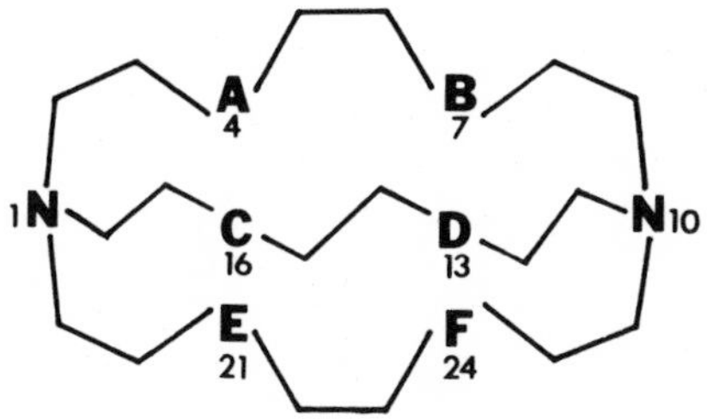

Compound	Remarks	M.P. (B.P.) (°C)	Yield (%)	Reference[a]
$(18)_3$-O_6N_2	A-F=O	66-69	95	*14,55,79*
$(18)_3$-O_4N_4	A-D=O; E,F=NH	69	85	*80*
$(18)_3$-$O_4N_2S_2$	A-D=O; E,F=S	78-80	25	*56*
$(18)_3$-O_2N_6	A,B=O; C-F=NH	75	75	*80*
$(18)_3$-$O_2N_2S_4$	E,F=O; A-D=S	86-87	20	*56*
$(18)_3$-N_2S_6	A-F=S	172	7	*56*
B-$(18)_3$-O_6N_2	A-F=O; 5,6-benzo	–	–	*81*
B_2-$(18)_3$-O_6N_2	A-F=O; 5,6- and 14,15-dibenzo	–	–	*81*
Ac_2-$(18)_3$-O_4N_4	A-D=O; E,F=*N*-acetyl	86	98	*80*
Ac_4-$(18)_3$-O_2N_6	A,B=O; C-F=*N*-acetyl	Liquid	90	*80*
k_2-$(18)_3$-O_6N_2	A-F=O; 2,9-diketo	113-115	45	*14,55*
k_2Ts_2-$(18)_3$-O_4N_4	A-D=O; E,F=*N*-tosyl; 20,24-diketo	208	55	*80*
k_2Ts_4-$(18)_3$-O_2N_6	A,B=O; C-F=*N*-tosyl; 20,24=diketo	188	55	*80*
Me_2-$(18)_3$-O_4N_4	A-D=O; E,F=*N*-CH_3	Liquid	90	*80*
Me_4-$(18)_3$-O_2N_6	A,B=O; C-F=*N*-CH_3	32-33	85	*80*
tDec-$(18)_3$-O_6N_2	A-F=O; 5-tetradecyl	–	–	*82*
Ts_2-$(18)_3$-O_4N_4	A-D=O; E,F=*N*-tosyl	83-84	93	*80*
Ts_4-$(18)_3$-O_2N_6	A,B=O; C-F=*N*-tosyl	104-107	90	*80*
uDec-$(18)_3$-O_6N_2	A-F=O; 5-undecyl	–	–	*82*

(cont.)

Compound	Remarks	M.P. (B.P.) (°C)	Yield (%)	Reference[a]

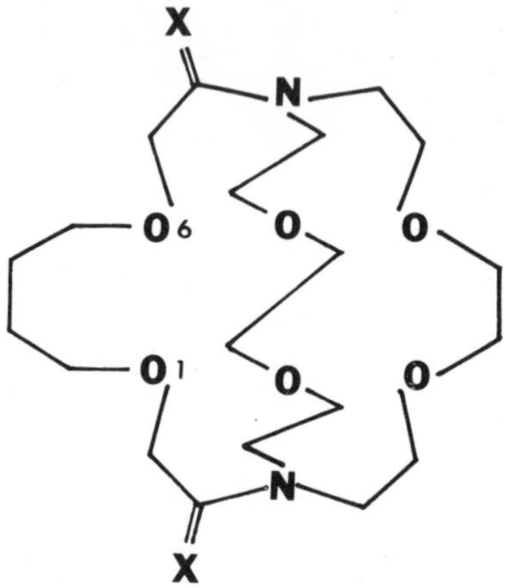

k_2Np_2-18(20)$_2$-O_6N_2	X=O; 2,3- and 4,5-dinaphtho	190-192	60	*83*
Np_2-18(20)$_2$-O_6N_2	X=H_2; 2,3- and 4,5-dinaphtho	Liquid	90	*83*

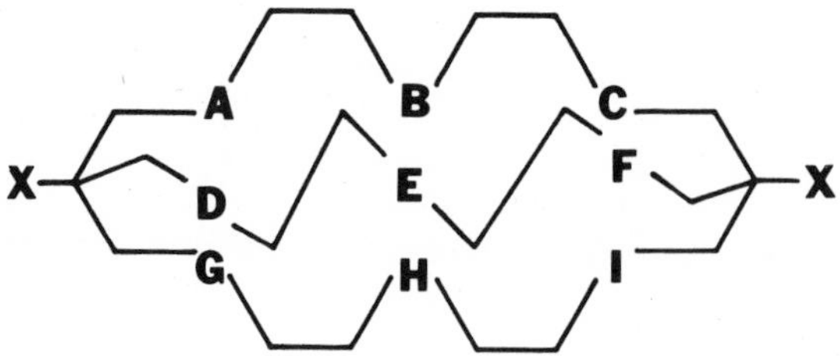

$(HyMe)_2$-(20)$_3$-O_9	A-I=O; X=hydroxymethyl	64-65	–	*69*
$(MesOMe)_2$-(20)$_3$-O_9	A-I=O; X=mesyloxymethyl	90-91	–	*69*
$(PhCH_2OMe)_2$-(20)$_3$-O_9	A-I=O; X=benzyloxymethyl	48-49	30	*69*

(cont.)

Compound	Remarks	M.P. (B.P.) (°C)	Yield (%)	Reference[a]
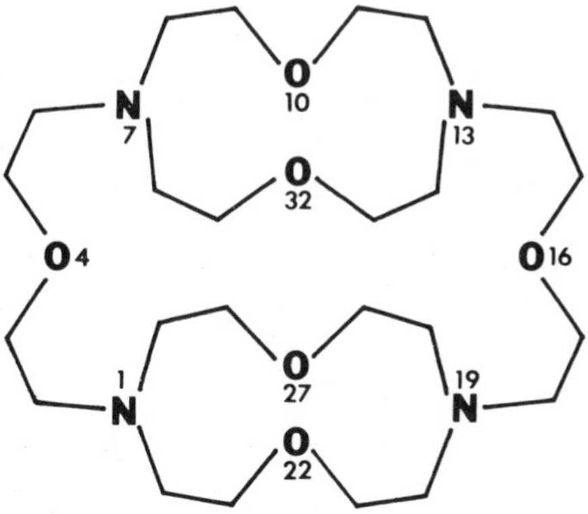				
$(12)_2(24)_2$-O_6N_4		54-55	–	*78*
k_4-$(12)_2(24)_2$-O_6N_4	2,6,14,18-tetraketo	243-246	30	*78*
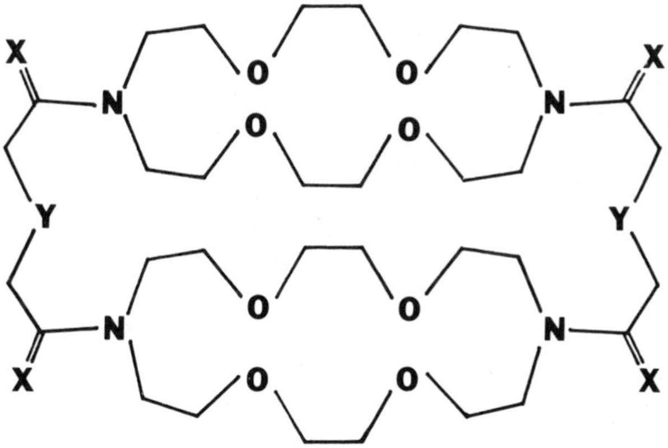				
$(18)_2(30)_2$-$O_{10}N_4$	Y=O; $X=H_2$	64	90	*76*
$(18)_2(30)_2$-O_8N_4	$Y=CH_2$; $X=H_2$	45-46	90	*76*
$(18)_2(30)_2$-O_8N_6	Y=NH; $X=H_2$	92-94	70	*76*
k_4-$(18)_2(30)_2$-O_{10}-N_4	Y=O; X=O	185	70	*76*
k_4-$(18)_2(30)_2$-O_8-N_4	$Y=CH_2$; X=O	185-186	75	*76*
k_4Ts_2-$(18)_2(30)_2$-O_8N_6	Y=*N*-tosyl; X=O	223	55	*76*
Ts_2-$(18)_2(30)_2$-O_8N_6	Y=*N*-tosyl; $X=H_2$	152	90	*76*

(cont.)

Compound	Remarks	M.P. (B.P.) (°C)	Yield (%)	Reference[a]
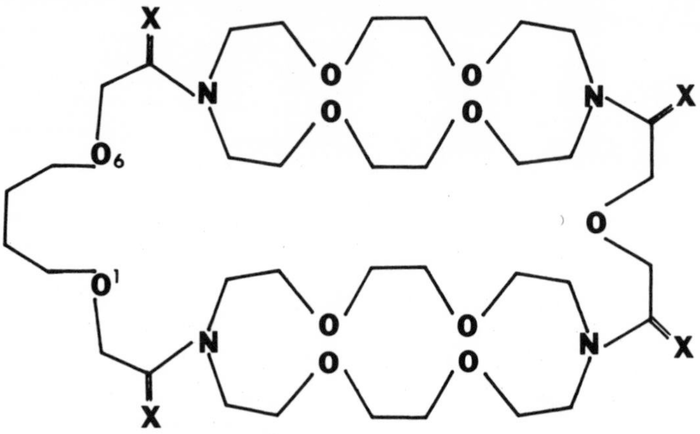				
Np_2-$(18)_2(35)_2$-O_{11}-N_4	X=H_2; 2,3- and 4,5-dinaphtho	Liquid	50	*83*
k_4Np_2-$(18)_2(35)_2$-$O_{11}N_4$	X=O; 2,3- and 4,5-dinaphtho	Liquid	90	*83*
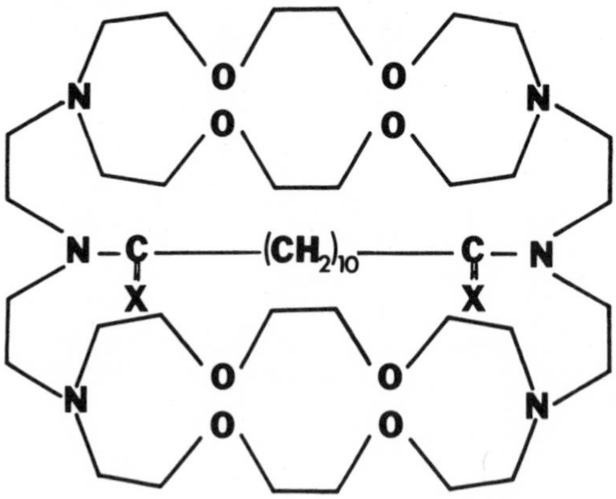				
$(18)_2(28)_2(30)_2$-O_8N_6	X=H_2	Liquid	95	*76*
k_2-$(18)_2(28)_2$ $(30)_2$-O_8N_6	X=O	Liquid	40	*76*

(cont.)

Compound	Remarks	M.P. (B.P.) (°C)	Yield (%)	Reference[a]

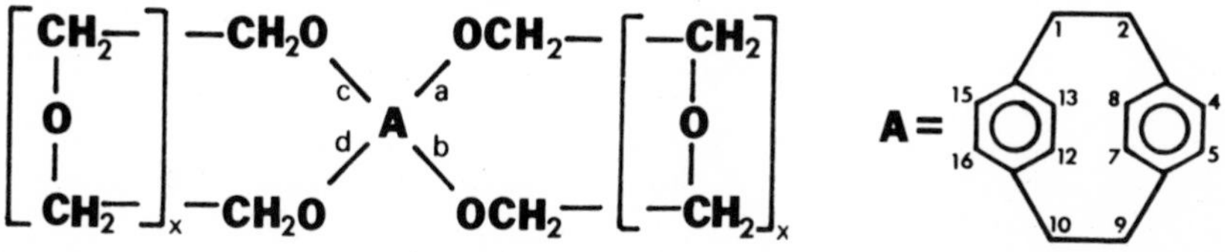

[2,2]paracyclophano Compounds

(a).	Attachments: *a*=4; *b*=5; *c*=15; *d*=16; *x*=3	192-193	6	*65*
(b).	Attachments: *a*=4; *b*=5; *c*=15; *d*=16; *x*=4	157-158	14	*65*
(c).	Attachments: *a*=4; *b*=12; *c*=7; *d*=15; *x*=3	118-119	3	*65*
(d).	Attachments: *a*=4; *b*=12; *c*=7; *d*=15; *x*=4	121-122	25	*65*
(e).	Attachments: *a*=4; *b*=13; *c*=7; *d*=16; *x*=4	133-134	23	*65*

Polymeric Systems

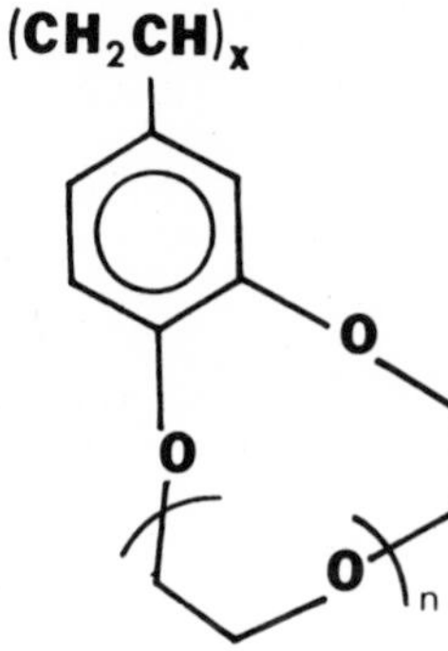

(a). *n*=3		–	–	*84*
(b). *n*=4		–	–	*84*
Silica gel-Np_4-22-O_6 (see 22-membered rings)		–	–	*72*

[a]References

1 Dale *et al.* (1974)
2 Bradshaw *et al.* (1973)
3 Ray (1920)
4 Madan and Cram (1975)
5 Pedersen (1967)
6 Hogberg and Cram (1975)
7 Koyama and Yoshino (1972)
8 Richman and Atkins (1974)
8a Curtis *et al.* (1975)
9 Vögtle *et al.* (1975)
10 Vögtle *et al.* (1974)
11 Krespan (1974)
12 Kyba *et al.* (1973b)
13 Cook *et al.* (1974)
14 Dietrich *et al.* (1973a)
15 Black and McLean (1968)
16 Black and McLean (1969)
17 Black and McLean (1971)

[a]References *(cont.)*

18 Ochrymowycz *et al.* (1974)
19 Rosen and Busch (1969b)
20 Rosen and Busch (1970)
21 Lockhart *et al.* (1973)
22 Reinhoudt and Gray (1975)
23 Leigh and Sutherland (1975)
24 Vögtle and Neuman (1973)
25 Vögtle and Weber (1974)
26 Newcomb *et al.* (1974)
27 Newkome *et al.* (1975)
28 Martin *et al.* (1974)
29 Bradshaw *et al.* (1975)
30 Pedersen (1971)
31 Stoddart and Wheatley (1974)
32 Rosen and Busch (1969a)
33 Travis and Busch (1970)
34 Travis and Busch (1974)
35 Kluiber and Sasso (1970)
36 Buxtorf and Kaden (1974)
37 Schrauzer *et al.* (1970)
38 Bradshaw *et al.* (1974)
39 Pelissard and Louis (1972)
40 Kopolow *et al.* (1973)
41 Pedersen (1970)
42 Bourgoin *et al.* (1975)
43 Newcomb and Cram (1975)
44 Takaki *et al.* (1971)
45 Takaki *et al.* (1972)
46 Curry and Busch (1964)
47 Krespan (1975)
48 deJong *et al.* (1975)
49 Black and McLean (1970)
50 Helgeson *et al.* (1973a)
51 Helgeson *et al.* (1973b)
52 Kyba *et al.* (1973a)
53 Gokel *et al.* (1974)
54 Greene (1972)
55 Dietrich *et al.* (1969)
56 Dietrich *et al.* (1970)
57 King and Krespan (1974)
58 Dann *et al.* (1961)
59 Lehn (1973)
60 Meadow and Reid (1934)
61 Ashby *et al.* (1974)
62 Pedersen (1973)
63 Parsons (1975)
64 Timko and Cram (1974)
64a Girodeau *et al.* (1975)
65 Helgeson *et al.* (1974b)
66 Bradshaw *et al.* (1976)
67 Lindoy and Busch (1968)
68 Lutringhaus (1937)
69 Coxon and Stoddart (1974)
70 Wudl and Gaeta (1972)
71 Newkome and Robinson (1973)
72 Dotsevi *et al.* (1975)
73 Gokel *et al.* (1975b)
74 Helgeson *et al.* (1974a)
75 Gokel *et al.* (1975a)
76 Lehn *et al.* (1973)
77 Cheney and Lehn (1972)
78 Cheney *et al.* (1972)
79 Dye *et al.* (1973)
80 Lehn and Montavan (1972)
81 Dietrich *et al.* (1973b)
82 Cinquini *et al.* (1975)
83 Dietrich *et al.* (1974)
84 Kopolow *et al.* (1971)

REFERENCES

Ashby, J., Hull, R., Cooper, M. J., and Ramage, E. M. (1974). *Synth. Commun.*, **4**, 113-117.

Black, D. S. C., and McLean, I. A. (1968). *Chem. Commun.* 1004.

Black, D. S. C., and McLean, I. A. (1969). *Tetrahedron Lett.* 3961-3964.

Black, D. S. C., and McLean, I. A. (1970). *Inorg. Nucl. Chem. Lett.* **6**, 675-678.

Black, D. S. C., and McLean, I. A. (1971). *Aust. J. Chem.* **24**, 1401-1411.

Black, D. S. C., and Hartshorn (1973). *Coord. Chem. Rev.* **9**, 219-230.

Bourgoin, M., Wong, K. W., Hui, J. Y., and Smid, J. (1975). *J. Am. Chem. Soc.*, **97**, 3462-3467.

Bradshaw, J. S., Hui, J. Y., Haymore, B. L., Christensen, J. J., and Izatt, R. M. (1973). *J. Heterocyclic Chem.* **10**, 1-4.

Bradshaw, J. S., Hui, J. Y., Chan, Y., Haymore, B. L., Izatt, R. M., and Christensen, J. J. (1974). *J. Heterocyclic Chem.* **11**, 45-49.

Bradshaw, J. S., and Hui, J. Y. (1974). *J. Heterocyclic Chem.* **11**, 649-673.

Bradshaw, J. S. *et al.* (1975). *J. Chem. Soc. Chem. Commun.* 874-875.

Bradshaw, J. S. *et al.* (1976). *J. Org. Chem.* **41**, 134-136.

Buxtorf, R., and Kaden, T. A. (1974). *Helv. Chim. Acta* **57**, 1035-1042.

Cheney, J., and Lehn, J. M. (1972). *J. Chem. Soc. Chem. Commun.* 487-489.

Cheney, J., Lehn, J. M., Sauvage, J. P., and Stubbs, M. E. (1972). *J. Chem. Soc. Chem. Commun.* 1100-1101.

Christensen, J. J., Hill, J. O., and Izatt, R. M. (1971). *Science* **174**, 459-467.

Christensen, J. J., Eatough, D. J., and Izatt, R. M. (1974). *Chem. Rev.* **74**, 351-384.

Cinquini, M., Montanari, F., and Tundo, P. (1975). *J. Chem. Soc. Chem. Commun.* 393-394.

Cook, F. L., Caruso, T. C., Byrne, M. P., Bowers, C. W., Speck, D. H., and Liotta, C. L. (1974). *Tetrahedron Lett.* 4029-4032.

Cooper, J., and Plesch, P. H. (1974). *J. Chem. Soc. Chem. Commun.* 1017-1018.

Coxon, A. C., and Stoddart, J. F. (1974). *J. Chem. Soc. Chem. Commun.*, 537.

Cram, D. J., and Cram, J. M. (1974). *Science* **183**, 803-809.

Curry, J. D., and Busch, D. H. (1964). *J. Am. Chem. Soc.* **86**, 592-597.

Curtis, W. D., Laidler, D. A., Stoddart, J. F., and Jones, G. H., (1975). *J. Chem. Soc. Chem. Commun.* 833-835.

Dale, J., Borgen, G., and Daasvatn, K. (1974). *Acta Chem. Scand. Ser. B* **28**, 378-379.

Dann, J. R., Chiesa, P. P., and Gates, J. W., Jr. (1961). *J. Org. Chem.* **26**, 1991-1995.

deJong, F., Siegel, M. G., and Cram, D. J. (1975). *J. Chem. Soc. Chem. Commun.* 551-553.

Dietrich, B., Lehn, J. M., and Sauvage, J. P. (1969). *Tetrahedron Lett.* 2885-2888.

Dietrich, B., Lehn, J. M., and Sauvage, J. P. (1970). *Chem. Commun.* 1055-1056.

Dietrich, B., Lehn, J. M., Sauvage, J. P., and Blanzat, J. (1973a). *Tetrahedron* **29**, 1629-1645.

Dietrich, B., Lehn, J. M., and Sauvage, J. P. (1973b). *J. Chem. Soc. Chem. Commun.* 15-16.

Dietrich, B., Lehn, J. M., and Simon, J. (1974). *Angew. Chem. Int. Ed. English* **13**, 406-407.

Dotsevi, G., Sogah, G. D. Y., and Cram, D. J. (1975). *J. Am. Chem. Soc.* **97**, 1259-1261.

Dye, J. L., Lok, M. T., Tehan, F. J., Ceraso, J. M., and Voorhees, K. J. (1973). *J. Org. Chem.* **38**, 1773-1775.

Frensdorff, H. K. (1971). *J. Am. Chem. Soc.* **93**, 4684-4686.

Girodeau, J. M., Lehn, J. M., and Sauvage, J. P. (1975). *Angew. Chem. Int. Ed. English* **14**, 764.

Gokel, G. W., Cram, D. J., Liotta, C. L., Harris, H. P., and Cook, F. L. (1974). *J. Org. Chem.* **39**, 2445-2446.

Gokel, G. W., Timko, J. M., and Cram, D. J., (1975a). *J. Chem. Soc. Chem. Commun.* 394-396.

Gokel, G. W., Timko, J. M., and Cram, D. J. (1975b). *J. Chem. Soc. Chem. Commun.* 444-446.

Graf, E., and Lehn, J. M. (1975). *J. Am. Chem. Soc.* **97**, 5022-5024.

Greene, R. N. (1972). *Tetrahedron Lett.* 1793-1796.

Helgeson, R. C., Koga, K., Timko, J. M., and Cram, D. J. (1973a). *J. Am. Chem. Soc.* **95**, 3021-3023.

Helgeson, R. C., Timko, J. M., and Cram, D. J. (1973b). *J. Am. Chem. Soc.* **95**, 3023-3025.

Helgeson, R. C., Timko, J. M., Moreau, P., Peacock, S. C., Moyer, J. M., and Cram, D. J. (1974a). *J. Am. Chem. Soc.* **96**, 6762-6763.

Helgeson, R. C., Timko, J. M., and Cram, D. J. (1974b). *J. Am. Chem. Soc.* **96**, 7380-7382.

Hogberg, S. A. G., and Cram, D. J. (1975). *J. Org. Chem.* **40**, 151-152.

Izatt, R. M., Eatough, D. J., and Christensen, J. J. (1973). *Structure and Bonding* **16**,161-192.

Izatt, R. M., Haymore, B. L., Bradshaw, J. S., and Christensen, J. J. (1975). *Inorg. Chem.* **14**, 3132.

Kapoor, P. N., and Mehrotra, R. C. (1974). *Coord. Chem. Rev.* **14**, 1-27.

Kappenstein, C. (1974). *Bull. Soc. Chim. Fr.* 89-109.

King, A. P., and Krespan, C. G. (1974). *J. Org. Chem.* **39**, 1315-1316.

Kluiber, R. W., and Sasso, G. (1970). *Inorg. Chim. Acta* **4**, 226-230.

Kopolow, S., Hogen Esch, T. E., and Smid, J. (1971). *Macromolecules* **4**, 359-360.

Kopolow, S., Hogen Esch, T. E., and Smid, J. (1973). *Macromolecules* **6**, 133-142.

Koyama, H., and Yoshino, T. (1972), *Bull. Chem. Soc. Jpn.* **45**, 481-484.

Krespan, C. G. (1974). *J. Org. Chem.* **39**, 2351-2355.

Krespan, C. G. (1975). *J. Org. Chem.* **40**, 1205-1209.

Kyba, E. P., Siegel, M. G., Soussa, L. R., Sogah, G. D. Y., and Cram, D. J. (1973a). *J. Am. Chem. Soc.* **95**, 2691-2692.

Kyba, E. P., Koga, K., Sousa, L. R., Siegel, M. G., and Cram, D. J. (1973b). *J. Am. Chem. Soc.* **95**, 2692-2693.

Leigh, S. J., and Sutherland, I. O. (1975). *J. Chem. Soc. Chem. Commun.* 414-415.

Lehn, J. M., and Montavan, F. (1972). *Tetrahedron Lett.* 4557-4560.

Lehn, J. M. (1973). *Structure and Bonding* **16**, 1-65.

Lehn, J. M., Simon, J., and Wagner, J. (1973). *Angew. Chem. Int. Ed. English* **12**, 578-579.

Lindoy, L. F. (1975). *Chem. Soc. Rev.* **4**, 421-441.

Lindoy, L. F., and Busch, D. H. (1968). *Chem. Commun.* 1589-1590.

Lockhart, J. C., Robson, A. C., Thomspon, M. E., Furtado, D., Kaura C. K., and Allan, A. R. (1973). *J. Chem. Soc. Perkin Trans.* **1**, 577-581.

Luttringhaus, A. (1937). *Ann. Chem.* **528**, 181-209.

Madan, K., and Cram, D. J. (1975). *J. Chem. Soc. Chem. Commun.* 427-428.

Martin, M. Y., DeHayes, L. J., Zompa, L. J., and Busch, D. H. (1974). *J. Am. Chem. Soc.* **96**, 4046-4048.

Meadow, J. R., and Reid, E. E. (1934). *J. Am. Chem. Soc.* **56**, 2177-2180.

Newcomb, M., Gokel, G. W., and Cram, D. J. (1974). *J. Am. Chem. Soc.* **96**, 6810-6811.

Newcomb, M., and Cram, D. J. (1975). *J. Am. Chem. Soc.* **97**, 1257-1259.

Newkome, G. R., and Robinson, J. M. (1973). *J. Chem. Soc. Chem. Commun.* 831-832.

Newkome, G. R., McClure, G. L., Broussard, J. B., and Danesh-Khoshboo, F. (1975). *J. Am. Chem. Soc.* **97**, 3232-3234.

Ochrymowycz, L. A., Mak, C. P., and Michna, J. D. (1974). *J. Org. Chem.* **39**, 2079-2084.

Parsons, D. J. (1975). *J. Chem. Soc. Perkin Trans.* **1**, 245-250.
Pedersen, C. J. (1967). *J. Am. Chem. Soc.* **89**, 7017-7036.
Pedersen, C. J. (1970). *J. Am. Chem. Soc.* **92**, 391-394.
Pedersen, C. J. (1971). *J. Org. Chem.* **36**, 254-257.
Pedersen, C. J., and Frensdorff, H. K. (1972). *Angew. Chem. Int. Ed. English* **11**, 16-25.
Pedersen, C. J. (1973). *Organ. Synthesis* **52**, 66-74.
Pelissard, D., and Louis, R. (1972). *Tetrahedron Lett.* 4589-4592.
Ray, P. C. (1920). *J. Chem. Soc.* 1090-1094.
Reinhoudt, D. N., and Gray, R. T. (1975). *Tetrahedron Lett.* 2105-2108.
Richman, J. E., and Atkins, T. J. (1974). *J. Am. Chem. Soc.* **96**, 2268-2270.
Rosen, W., and Busch, D. H. (1969a). *Chem. Commun.* 148-149.
Rosen, W., and Busch, D. H. (1969b). *J. Am. Chem. Soc.* **91**, 4696-4697.
Rosen, W., and Busch, D. H. (1970). *Inorg. Chem.* **9**, 262-265.
Schrauzer, G. N., Ho, R. K. Y., and Murillo, R. P. (1970). *J. Am. Chem. Soc.* **92**, 3508-3509.
Stoddart, J. F., and Wheatley, C. M. (1974). *J. Chem. Soc. Chem. Commun.* 390-391.
Takaki, U., Hogen Esch, T. E., and Smid, J. (1971). *J. Am. Chem. Soc.* **93**, 6760-6766.
Takaki, U., Hogen Esch, T. E., and Smid, J. (1972). *J. Phys. Chem.* **76**, 2152-2155
Timko, J. M., and Cram, D. J. (1974). *J. Am. Chem. Soc.* **96**, 7159-7160.
Travis, K., and Busch, D. H. (1970). *Chem. Commun.* 1041-1042.
Travis, K., and Busch, D. H. (1974). *Inorg. Chem.* **13**, 2591-2598.
Truter, M. R. (1973). *Structure and Bonding* **16**, 71-111.
Vögtle, F., and Neumann, P. (1973). *Chem. -Z.* **97**, 600-610.
Vögtle, F., and Weber, E. (1974). *Angew. Chem.* **86**, 126-127.
Vögtle, F., and Weber, E. (1975). *Angew. Chem. Int. Ed. English* **13**, 814-816.
Vögtle, F., Weber, E., Wehner, W., Nätscher, R., and Grütze, J. (1974). *Chem. -Z* **98**, 562-563.
Vögtle, F., Grütze, J., Nütscher, R., Wieder, W., Weber, E., and Grün, R. (1975). *Chem. Ber.* **108**, 1694-1711.
Wudl, F., and Gueta, F. (1972). *J. Chem. Soc. Chem. Commun.* 107.

3 APPLICATION OF MACROCYCLIC POLYDENTATE LIGANDS TO SYNTHETIC TRANSFORMATIONS

Charles L. Liotta
School of Chemistry
Georgia Institute of Technology
Atlanta, Georgia

I. Introduction 111
II. Solvation, Solubility, and Nucleophilicity 113
III. Anion Activation Promoted by Macrocyclic Polydentate Ligands: Survey of Reactions of Naked Anions . . 118
A. Halide Ions 118
B. Carboxylate Ions 125
C. Cyanide Ion 131
D. Nitrite Ion 141
E. Nitrate Ion 142
F. Oxygen Nucleophiles and Bases 143
G. Superoxide 163
H. Hydride Reductions 171
I. Oxidation Reactions 175
J. Bromine Addition Reactions 179
K. Sulfur Nucleophiles 180
L. Carbenes 180
M. Carbanions and Anion Rearrangements . . . 183
N. Alkali Metals 201
References 201

I. INTRODUCTION

In 1967, Charles Pedersen of the Elastomers Chemical Department at DuPont reported the synthesis of a variety of new macrocyclic polydentate ligands (see Chapter 1). He demonstrated their ability to complex metal ions and solubilize inorganic and organic salts in polar and nonpolar solvents with the production of chemically active anionic species (Pedersen, 1967a, b, 1968; 1970a, b, 1971a, b, 1972; Pedersen and Frensdorff, 1972.) This classic and elegant work could be termed "the shot heard 'round the [chemical] world," since it ushered in a revolution in many areas of chemistry (Christensen *et al.*, 1971, 1974; Cram and Cram, 1974; Vögtle and Neumann, 1973; Lehn, 1973; Gokel and Durst, 1976a, b), especially in synthetic organic and physical organ-

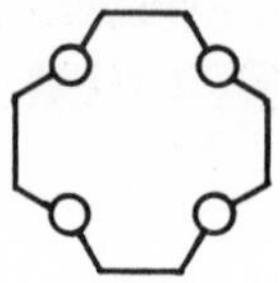

I. 12–Crown–4

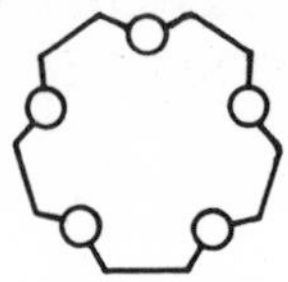

II. 15–Crown–5

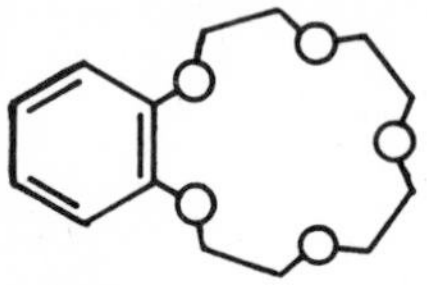

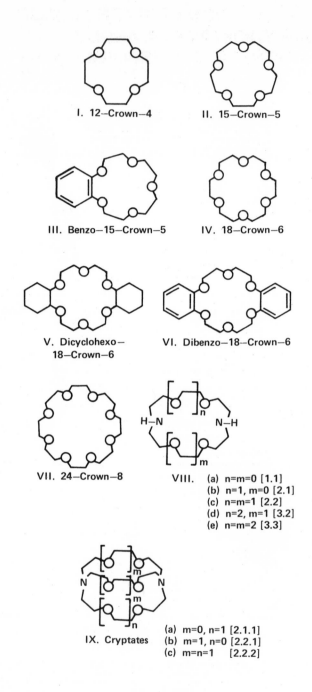

III. Benzo–15–Crown–5

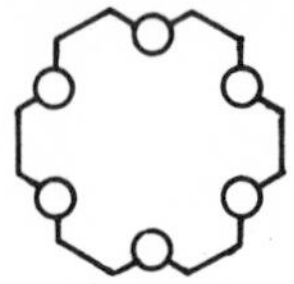

IV. 18–Crown–6

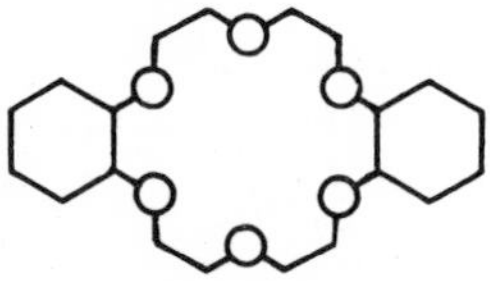

V. Dicyclohexo–18–Crown–6

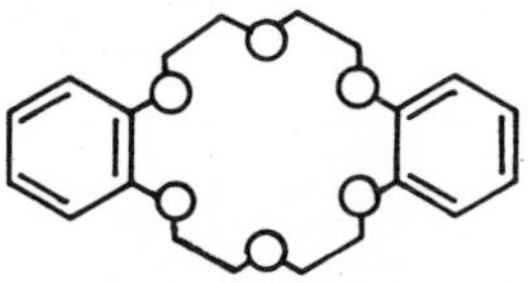

VI. Dibenzo–18–Crown–6

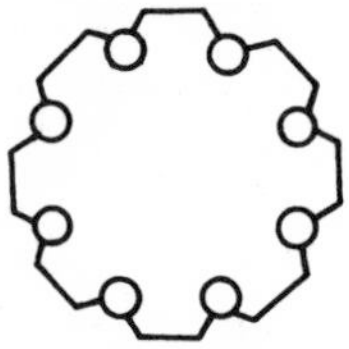

VII. 24–Crown–8

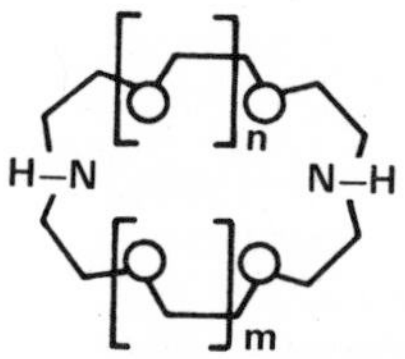

VIII. (a) n=m=0 [1.1]
(b) n=1, m=0 [2.1]
(c) n=m=1 [2.2]
(d) n=2, m=1 [3.2]
(e) n=m=2 [3.3]

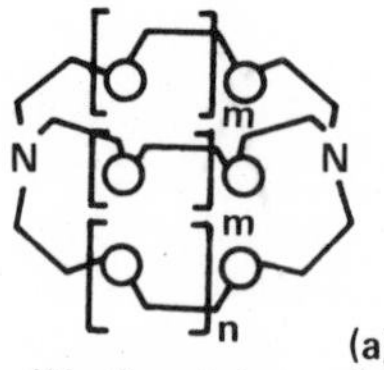

IX. Cryptates

(a) m=0, n=1 [2.1.1]
(b) m=1, n=0 [2.2.1]
(c) m=n=1 [2.2.2]

Fig. 1. Representative macrocyclic polydentate ligands.

ic chemistry. During the past nine years an enormous amount of work has been reported concerning the use of these novel ligands in catalyzing synthetic organic reactions and in probing reaction mechanisms. It is these areas which will be the main concern of this chapter. Reactions carried out under homogeneous conditions as well as those carried out under liquid-solid and liquid-liquid phase transfer catalytic (heterogeneous) conditions will be surveyed. No attempt will be made to include the enormous amount of work reported on the use of quaternary ammonium and phosphonium salts in phase transfer reactions nor to include details of the theory of these types of reactions, since excellent discussions already exist elsewhere in the literature (Starks, 1971, Dockx, 1973; Dehmlow, 1974a, b, 1975; Makosza, 1966; Brandstrom *et al.*, 1969).

It should be noted that there is a certain degree of frustration which accompanies the writing of a survey of this kind. It seems that every new issue of the journals contains some novel and interesting application of macrocyclic polydentate ligands. As a result, this review will cover only the work reported up to the early part of 1976. What will be attempted here is to give the reader a broad survey of the kinds of applications for which these ligands have been used with enough details so as not to miss any of the subtleties. By using this approach it is hoped that imaginative chemists will come up with further uses for these interesting species.

In order to conserve space in the body of this review, the structures and names of the macrocyclic polydentate ligands most commonly used in organic reactions are illustrated in Fig. 1. Structures I-VIII represent two-dimensional ligands belonging to the "crown ether" class, while structures IXa-IXc represent three-dimensional ligands belonging to the "cryptate" class. The crown and cryptate nomenclature will be used throughout this review (Christensen *et al.*, (1974). (See also Chapter 1, B.)

II. Solvation, Solubility, and Nucleophilicity

Anions, unencumbered by strong solvation forces, should prove to be potent nucleophiles and potent bases and should provide the basis for the development of new and valuable reagents for organic synthesis. Ideally, this condition is best achieved with anions in the gas phase where solvation forces are completely absent (Young *et al.*, 1973; Bohme *et al.*, 1974; Brauman *et al.*, 1974). Indeed, some of the most significant steps forward in this area have been the relatively recent development of ion-cyclotron resonance, high-pressure mass spectrometric, and flowing afterglow techniques to investigate gas phase rate and equilibrium processes (Grunwald and Leffler, 1963; Bowers *et al.*, 1971; McIver, 1970; Taft, 1975; Kebarle, 1972). Unfortunately, the high concentrations of gas phase anions necessary to make synthetic transformations practical cannot be easily and conveniently attained at this time.

With the advent of crown ethers and related macrocyclic polydentate ligands, simple and efficient means have become available for solubilizing simple metal salts in nonpolar and dipolar aprotic solvents where solvation of the anionic portion of the salt should be minimal. Figure 2 illustrates the structures of

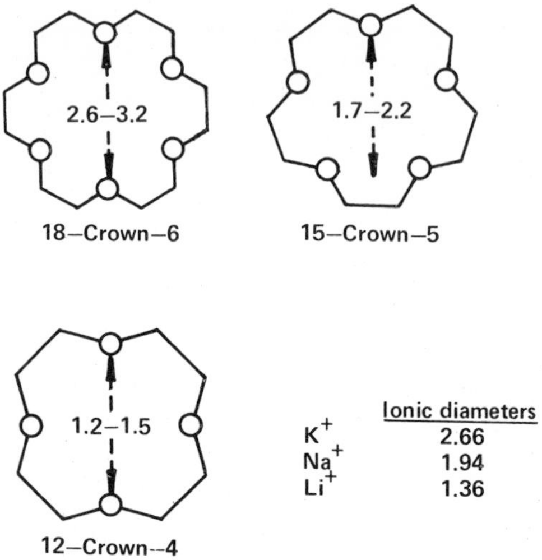

Fig. 2. Several cyclic polyether cavity diameters (Pedersen, Chapter 1, Table VII) and cation ionic diameters (Pedersen, Chapter 1, Table VI).

some of the simple but most synthetically useful crown ethers, presenting an estimate of each of their cavity diameters and the ionic diameters of some alkali metal ions. Using the simple "lock and key" approach, it is evident that 18-crown-6 has cavity dimensions of the same magnitude as the diameter of the potassium ion, while 15-crown-5 and 12-crown-4 have cavity sizes suited for the ionic diameters of sodium ion and lithium ion, respectively. Therefore, in principle, particular crowns should be more specific for particular metal ions than for others. It should be emphasized that this does not mean that 18-crown-6 cannot solubilize sodium salts. It only means that it is more specific for potassium ion than for sodium ion. It will be shown that in many cases 18-crown-6 is used for both sodium and potassium salts in organic reactions, and exact correspondence between cavity size and ionic diameter is not always a critical factor.

The solubilization process of a metal ion salt in crown-solvent system in the hypothetical one molal infinite dilution state may be viewed in terms of a complete thermodynamic cycle (Fig. 3.) Examination of the cycle shows that the solubilization depends on a combination of factors: (1) the crystal lattice free energy of the salt (ΔG_1); (2) the free energy of solution of the crown (ΔG_2); the free energy of complexation of the metal ion by the crown in the

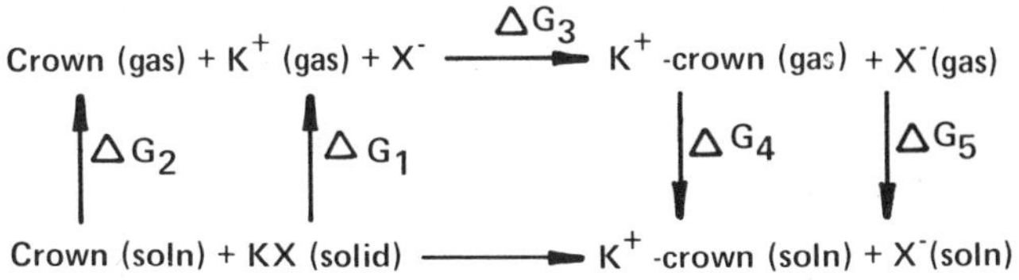

Fig. 3. Thermodynamic cycle.

gas phase (ΔG_3); the free energy of solution of the metal-crown complex (ΔG_4); and the free energy of solution of the anion (ΔG_5). In nonpolar aprotic solvents, additional terms due to ion-pairing and aggregation effects must also be considered. Nevertheless it is clear that three of the major driving forces for solubilization in nonpolar and dipolar aprotic solvents are ΔG_1, ΔG_3, and ΔG_5.

Table I summarizes the unpublished results of Liotta and Dabdoub concerning the solubilities of a variety of common potassium salts in acetonitrile at 25°C in

TABLE I

Solubilities[a] of Potassium Salts in Acetonitrile at 25°C in the Presence and Absence of 18-Crown-6 (IV)

Potassium salt	Sol. in 0.15 *M* (IV) in acetonitrile	Sol. in acetonitrile	Solubility enhancement
KF	4.3×10^{-3}	3.18×10^{-4}	14
KCl	5.55×10^{-2}	2.43×10^{-4}	228
KBr	1.35×10^{-1}	2.08×10^{-3}	65
KI	2.02×10^{-1}	1.05×10^{-1}	2
KCN	1.29×10^{-1}	1.19×10^{-3}	108
KOAc	1.02×10^{-1}	5.00×10^{-4}	204
KN_3	1.38×10^{-1}	2.41×10^{-3}	57
KSCN	8.50×10^{-1}	7.55×10^{-1}	1.13

[a] Solubility was measured using a Coleman model 21 flame photometer.

the presence and in the absence of 18-crown-6 (0.15 *M*). The solubility enhancement is quite dramatic for KCl, KCN, and KOAc (greater than 10^2), only moderate for KBr and KN_3, and marginal for KF, KI, and KSCN. It should be noted that KI and KSCN already have high inherent solubilities in pure acetonitrile.

Along related lines, Knochel *et al.* (1975) have reported the solubility of potassium acetate in acetonitrile in the presence of a variety of macrocyclic polydentate ligands. The solubilization effectiveness was found to be as follows: (VIII)d > (VIII)e > (VIII)b > (VIII)c > (VI) > (IV) > (IX)c > (IX)a >> (X) > (V) > nonactin > (XI). Arguments related to cavity diameter, lipophilicity, and rigidity were advanced to explain, at least partially, the observed structure-solubilization order.

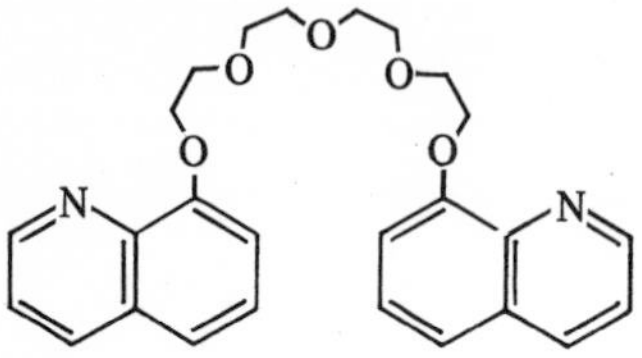

(X)

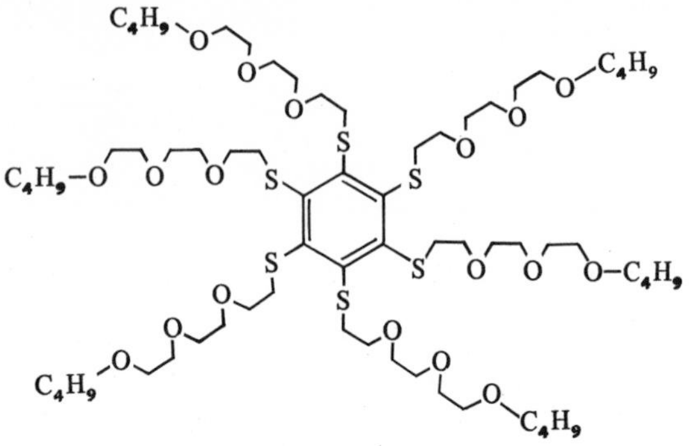

(XI)

A simplistic view of the solubilization process is shown in Fig. 4. The dimensions of the 18-crown-6 are such that it can effectively coordinate with a potassium ion. Since the complex has a hydrophobic ("greasy") exterior, it is readily solubilized by the nonpolar or dipolar aprotic solvent. In order to preserve electrical neutrality, the anion must accompany the potassium-crown complex into solution. Since a nonpolar aprotic solvent such as benzene or a dipolar aprotic solvent such as acetonitrile should not have a great affinity for the anion, as compared to protic media, the anion is not expected to be highly solvated. As a result the anion should be a potent nucleophile as well as a potent base whether it is in solution as an ion pair or a free anion. These anionic species have been termed "naked" anions (Liotta *et al.*, 1975).

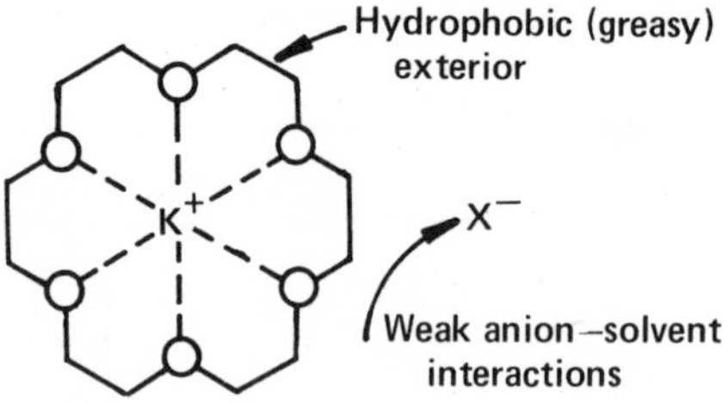

Fig. 4. Solubility of potassium salts in nonpolar and dipolar aprotic solvents.

Studies related to the relative nucleophilicities of naked anions toward benzyl tosylate in acetonitrile at 30°C [Reaction (1)] have recently been reported by Liotta *et al.* (1975). The results are summarized in Table II. It was shown by conductance techniques that each of the naked anions existed as free ions under

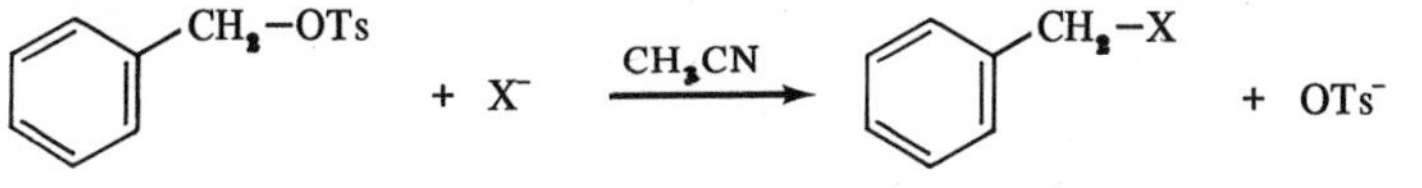

(1)

the conditions of the kinetic studies. This finding suggested that acetonitrile was indeed an excellent solvent for these species.

TABLE II

Second-Order Rate Constants at 30°C in Acetonitrile for the Reaction with Benzyl Tosylate[a,b]

Nucleophile	K (liters mole^{-1} sec^{-1})	Relative rates: Liotta *et al.* (1975)	Relative rates: Swain and Scott (1953)
N_3^-	1.02	10.0	20.0
CH_3COO^-	0.95	9.6	1.0
CN^-	0.23	2.4	250
Br^-	0.12	1.3	16.0
Cl^-	0.12	1.3	2.0
I^-	0.09	1.0	200.0
F^-	0.14	1.4	0.2
SCN^-	0.02	0.3	125

[a]The second-order rate coefficients of Cl^-, Br^-, and SCN^- toward *n*-Bu-I have been determined by Grimsrud and Kratcchvil (1973) in CH_3CN, taking into account ion-pairing effects, to be 4.0, 3.4, and 0.58 ($\times 10^{-3}$ M^{-1} sec^{-1}), respectively, which gives approximately the same order of nucleophilicity reported here.

[b]The second-order rate coefficients of SCN^-, Br^-, N_3^-, Cl^-, and AcO^- toward CH_3-OTs have been determined by Engemyr and Songstad (1972) in CH_3CN by spectrophotometric procedures to be 2.3, 50, 25, 230, and 2600 ($\times 10^{-4}$ M^{-1} sec^{-1}), respectively. The relative nucleophilicities determined from these rate constants are considerably different from the values reported here.

The rate constants shown in Table II have a total variation of less than a factor of 10 if SCN^- is deleted. If a statistical factor of 2 is applied to N_3^- and CH_3COO^-, these rates vary by less than a factor of 5. These results were in direct contrast to the previously observed relative nucleophilicities in protic solvents (see also Table II) (Grunwald and Leffler, 1963; Streitweiser, 1962; Kosower, 1968; Edwards, 1954, 1968; Swain and Scott, 1953; Ritchie and Virtanen, 1972; Pearson and Songstad, 1967; Parker, 1969; Ritchie 1974). Furthermore, in these studies the halides appeared to be nucleophiles with virtually identical reactivities, whereas the reactivities varied by a factor of 200 in water (Streitweiser, 1962) and a factor of 20 in dimethylformamide (Kosower, 1968; Parker 1969). There appeared to be a general leveling of the nucleophilicities of anions in acetonitrile. Several reversals of the usual order of nucleophilicities were also seen in Table II. Quite surprisingly, the "best" nucleophile was CH_3COO^- (within experimental error), which was normally considered a very poor nucleophile, whereas SCN^-, one of the more potent nucleophiles in aqueous solution, was approximately 30 times less nucleophilic than CH_3COO^-. Apparently, the reactivities of anions dissolved in acetonitrile did not vary appreciably, a situation reminiscent of most anions studied in the gas phase. This seemed to indicate that variations in anion solvation in acetonitrile were not important factors in determining the relative reactivities. *The results tended to support the notion that naked anions in acetonitrile were solvated by much weaker forces than in protic solvents.*

Whereas the relative nucleophilicities in acetonitrile were similar to those found in the gas phase, the absolute gas phase rates were approximately 11 orders of magnitude greater than those found for the anions in acetonitrile. This large difference in absolute rates was indicative of the moderating influence of the solvent on all the reactivities. In fact, Yamdagni and Kabarle (1972) demonstrated that acetonitrile forms a stable adduct in the gas phase with halide ions, and Coetzee and Sharpe (1972) had shown that several anions in acetonitrile caused the C—H stretching frequencies of acetonitrile to shift to lower wave-number values. Consequently, it was concluded that the anions are solvated but the differences in solvation in the series studied did not appreciably affect the relative kinetic properties of the anions.

III. ANION ACTIVATION PROMOTED BY MACROCYCLIC POLYDENTATE LIGANDS: SURVEY OF REACTIONS OF NAKED ANIONS

A. Halide Ions

The solubilization of alkali metal halides in dipolar and nonpolar aprotic solvents containing crown ether has been reported. These naked halides have been shown to be effective reagents in a wide variety of substitution and elimination reactions. Some data on the reactions of naked fluoride reported by Liotta and Harris (1974) are summarized in Table III. The reactions were carried out in acetonitrile and benzene in the presence of (IV) under relatively mild conditions. The conversions were essentially quantitative. Less than 5% reaction took place in the absence of crown (IV) under identical conditions covering the same periods of time.

The products of reaction were either fluorides, alkenes, or mixtures of these, indicating that naked fluoride acted as either nucleophile or base. Benzyl bromide (Run 1, Table III) reacted rapidly to produce benzyl fluoride. Primary halides (Run 2, Table III) gave predominantly primary fluorides with only small amounts of alkene, whereas secondary halides (Runs 3 and 4, Table III) gave exclusively or predominantly alkene products. An interesting reaction which

TABLE III

Reactions of Naked Fluoride with Organic Substrates

Run	Substrate	Solvent	Products[a]	Concentrations (M) (IV)	Substrate	Temp. (°C)	$t_{1/2}$[b] (hr)
1	Benzyl bromide	CH_3CN	Benzyl fluoride	0.19	2.0	83	11.5
2	1-Bromooctane	CH_3CN	1-Fluorooctane (92%) 1-Octene (8%)	0.19	1.16	83	115
		C_6H_6	1-Fluorooctane (92%) 1-Octene (8%)	0.68	2.9	90	128
3	2-Bromooctane	C_6H_6	2-Fluorooctane (32%) 1- and 2-octenes (68%)	0.50	2.8	90	240
4	Bromocyclohexane	CH_3CN	Cyclohexene	0.15	3.61	83	104
5	2-Chloro-2-methyl-cyclohexanone	CH_3CN	2-Fluoro-2-methyl-cyclohexanone (31%) 2-Methyl-2-cyclo-hexenone (69%)	0.15	3.3	83	20
6	2,4-Dinitrochloro-benzene	CH_3CN	2,4-Dinitrofluoro-benzene	0.15	3.5	25 83	5 0.12
7	Acetyl chloride	CH_3CN	Acetyl fluoride	0.14	7.0	25	5.5

[a] In all cases conversion to products was quantitative.

[b] The time for one-half conversion of starting materials to products is tabulated as an approximate indication of the relative rates of reaction.

illustrated the competition between displacement and elimination processes was the reaction of naked fluoride with 2-chloro-2-methylcyclohexanone, producing 2-fluoro-2-methyl-cyclohexanone and 2-methyl-2-cyclohexenone [Run 5, Table III; Reaction (2)]. It was found that alkyl chlorides reacted slowly with naked

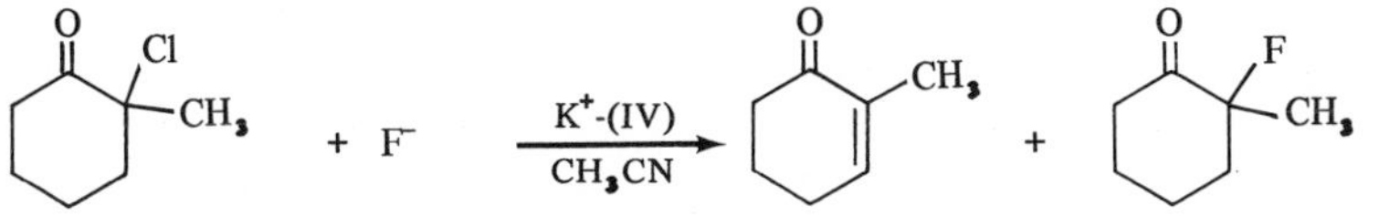

(2)

fluoride, while the corresponding tosylates had reactivities comparable to that of bromides. These observations regarding leaving group abilities were consistent with the tabulation reported by Streitweiser (1962) for homogeneous reactions. In addition, reaction appeared to be faster in acetonitrile than in benzene (Run 2, Table III). Displacement at sp^2 hybridized carbon, as illustrated by the reactions of 2,4-dinitrochlorobenzene and acetyl chloride, occurred smoothly at room temperature and rapidly at reflux, giving 100% conversion to the corresponding fluorides (Runs 6 and 7, Table III).

The reagent was prepared by dissolving (IV) in dry acetonitrile or dry benzene and then adding dry potassium fluoride. After the heterogeneous system was stirred for 30 minutes, the organic substrate was added and the resulting mixture stirred until reaction was complete. It was emphasized that efficient stirring is important for complete reaction to be attained. This, like the solubilization of $KMnO_4$ in benzene reported by Sam and Simmons (1972), is an example of the dissolution of an insoluble salt directly in a solvent, such as acetonitrile or benzene, simply by adding crown ether. A solvent exchange procedure has been reported as an alternate method to solubilize salts (Pedersen, 1967a, b, 1968; 1970a, b; 1971). In all cases reported in Table III, the crown is present in catalytic concentrations.

The concentration of naked fluoride in both benzene and acetonitrile at 25°C has been determined from analysis of the potassium ion concentration by flame photometry. The results are shown in Table IV. It is interesting to note that a

TABLE IV

Solubility of Potassium Fluoride in Benzene and Acetonitrile Containing 18-Crown-6 (IV)

Solvent	(IV) (*M*)	KF (*M*)
Benzene	1.01	5.2×10^{-2}
	0.34	1.4×10^{-2}
Acetonitrile	0.16	3.5×10^{-3}

plot of the solubility of potassium-IV-fluoride versus the concentration of crown (IV) in solution produced a reasonable straight line passing near the origin. It appeared, therefore, that the concentration of solubilized KF is independent of the dielectric constant of the medium.

Terrier *et al.* (1975) have detected the formation of sigma anionic complex (Meisenheimer complex) (XII) [Reaction (3)] using ^1H and ^{19}F nmr spectro-

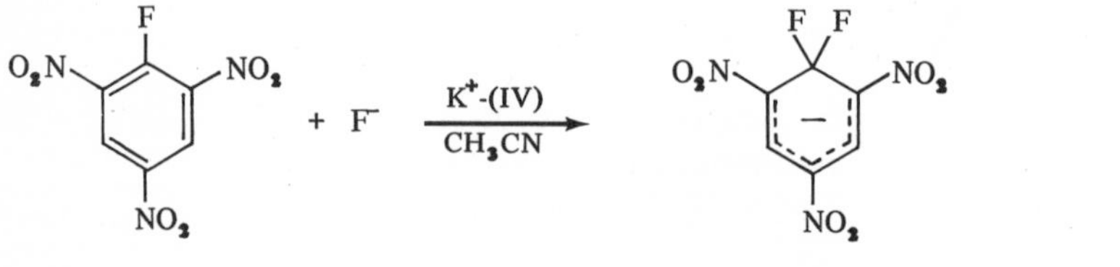

(XII)

(3)

scopy. Complex (XII) was formed by adding picryl fluoride to a heterogeneous mixture of (IV) and potassium fluoride in dry acetonitrile.

Naso and Rozine (1974) have reported that the KF-promoted elimination reaction of *cis*-β-bromo-*p*-nitrostyrene produced *p*-nitrophenylacetylene [Reaction (4)]. The reactions were carried out in the solvents acetonitrile, dimethyl-

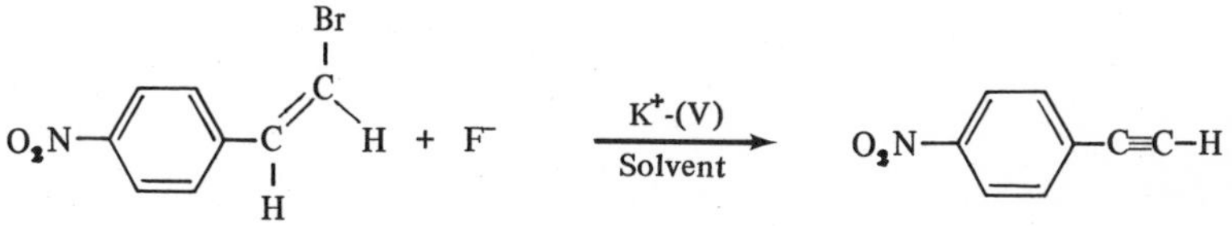

(4)

formamide, and *n*-butyl cellosolve in the absence and in the presence of (V). The effect of crown on the rate of reaction is summarized in Table V. It was clear that crown accelerated the reaction in all cases but the effect appeared to be quite

TABLE V

The Effect of the Addition of Dicyclohexo-18-Crown-6 (V) on the Reactions of *cis*-β-Bromo-*p*-Nitrostyrene with Potassium Fluoride[a]

Run	Solvent	Temp. (°C)	Base		Time (min)	Conversion[b] (%)
1	CH_3CN	80	KF		60	~0
2	CH_3CN	80	KF	(V)	60	53
3	CH_3CN	80	KF		90	~0
4	CH_3CN	80	KF	(V)	90	71
5	DMF	70	KF		30	28
6	DMF	70	KF	(V)	30	58
7	DMF	70	KF		45	35
8	DMF	70	KF	(V)	45	68
9	Bu-cellosolve	150	KF		10	15
10	Bu-cellosolve	150	KF	(V)	10	43
11	Bu-cellosolve	150	KF		20	22
12	Bu-cellosolve	150	KF	(V)	20	80

[a]The concentrations of substrate and (V) were 3×10^{-2} *M*; undissolved KF was present during the runs.

[b]The percentages of conversion were obtained from glc analysis.

dramatic in acetonitrile (Runs 1-4, Table V). The authors reported that tetraethylammonium fluoride in acetonitrile at 25° was also effective in carrying out this transformation.

Ykman and Hall (1975) have reported that reaction of dimethyl 2-chloroethylene-1,1-dicarboxylate with KF in the presence of (V) produced dimethyl

2-fluoroethylene-1,1-dicarboxylate in unspecified yield [Reaction (5)]. Under a similar set of conditions, methyl 1,2,2-trichloroethylene-1-carboxylate was

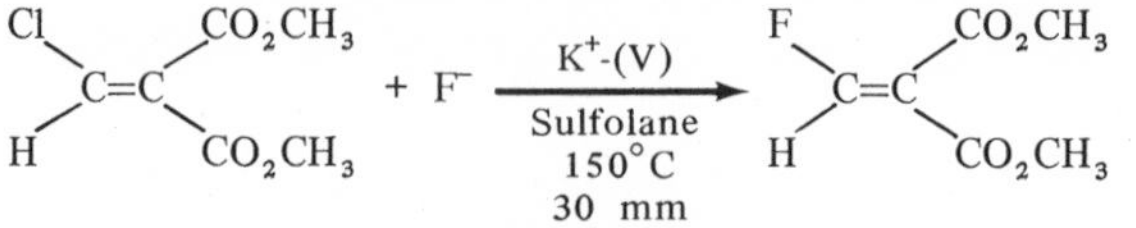

$$
\underset{H}{\overset{Cl}{\diagdown}}C=C\underset{CO_2CH_3}{\overset{CO_2CH_3}{\diagup}} + F^- \xrightarrow[\substack{\text{Sulfolane} \\ 150^\circ C \\ 30 \text{ mm}}]{K^+\text{-(V)}} \underset{H}{\overset{F}{\diagdown}}C=C\underset{CO_2CH_3}{\overset{CO_2CH_3}{\diagup}} \qquad (5)
$$

produced in 70% yield from dimethyl 2,2-dichloro-1,1-dicarboxylate [Reaction (6)]. This yield was calculated assuming that 2 moles of starting material gave 1 mole of product. Similar results were obtained with potassium chloride solubilized

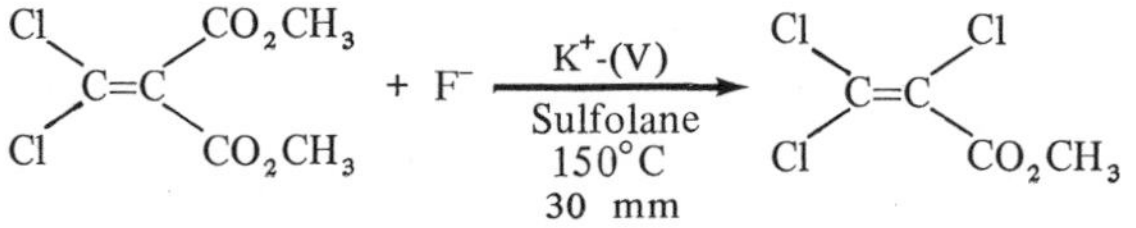

$$
\underset{Cl}{\overset{Cl}{\diagdown}}C=C\underset{CO_2CH_3}{\overset{CO_2CH_3}{\diagup}} + F^- \xrightarrow[\substack{\text{Sulfolane} \\ 150^\circ C \\ 30 \text{ mm}}]{K^+\text{-(V)}} \underset{Cl}{\overset{Cl}{\diagdown}}C=C\underset{CO_2CH_3}{\overset{Cl}{\diagup}} \qquad (6)
$$

with crown. Reaction (6) was believed to proceed via an initial decarboalkoxylation followed by abstraction of Cl^+ from another reactant. Other transformations promoted by naked chloride are shown in Reactions (7) and (8).

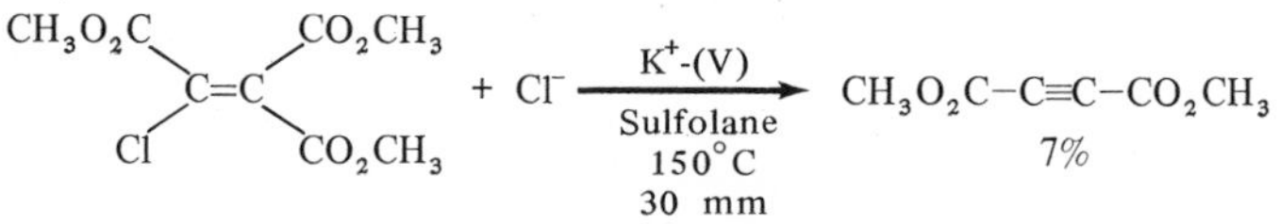

$$
\underset{Cl}{\overset{CH_3O_2C}{\diagdown}}C=C\underset{CO_2CH_3}{\overset{CO_2CH_3}{\diagup}} + Cl^- \xrightarrow[\substack{\text{Sulfolane} \\ 150^\circ C \\ 30 \text{ mm}}]{K^+\text{-(V)}} CH_3O_2C-C\equiv C-CO_2CH_3 \atop 7\% \qquad (7)
$$

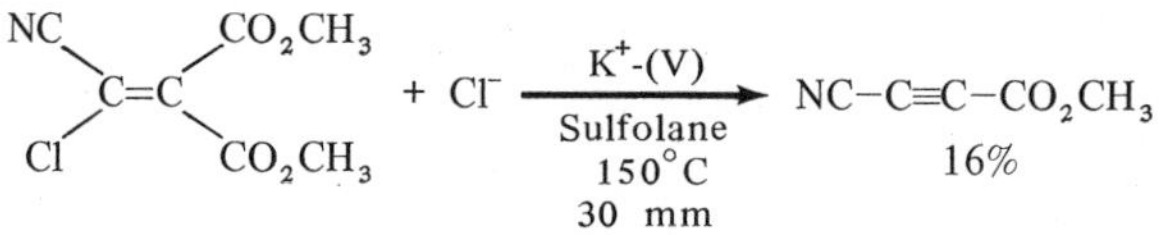

$$
\underset{Cl}{\overset{NC}{\diagdown}}C=C\underset{CO_2CH_3}{\overset{CO_2CH_3}{\diagup}} + Cl^- \xrightarrow[\substack{\text{Sulfolane} \\ 150^\circ C \\ 30 \text{ mm}}]{K^+\text{-(V)}} NC-C\equiv C-CO_2CH_3 \atop 16\% \qquad (8)
$$

Sam and Simmons (1974) have reported the observation of enhanced reactivity of naked bromide and naked iodide in nucleophilic substitution and elimination reactions. Reaction of n-butyl brosylate with each of the naked halides in acetone at $25^\circ C$ gave clean second-order kinetics. The corresponding n-butyl halides were produced in quantitative yield [Reaction (9)]. It was also reported

$$
CH_3-CH_2-CH_2-CH_2-OBs + X^- \xrightarrow[\text{Acetone}]{K^+\text{-(V)}} CH_3-CH_2-CH_2-CH_2-X + OBs^- \qquad (9)
$$

$$
X = Br, I
$$

that the KI-(V) complex acts as a base in aprotic solvents. Reaction with 2-bromooctane in dimethylformamide at $100^\circ C$ for 6 hr produced only 2-octene in 75-80% yield. Under identical conditions, tetra-n-butylammonium bromide produced 60-65% 2-octene. In acetone, the crown complex of potassium iodide gave low yields of 2-octene, the major product being mesityl oxide from the condensation of acetone. In these experiments, no cis/trans ratios of the 2-octene products were reported.

Landini *et al.* (1974) have reported the reactions of a series of octyl derivatives with a variety of potassium and sodium halides under liquid-liquid phase transfer catalytic conditions using (V) as the phase transfer catalyst. The reactions were carried out by stirring a mixture of the organic substrate and a saturated aqueous solution of alkali metal halide at 80-115°C in the presence of 0.05 molar equivalent of crown. The results are summarized in Table VI. Several conclusions

TABLE VI

Nucleophilic Substitutions in Octyl Derivatives Catalyzed by Dicyclohexo-18-Crown-6 (V)[a] under Liquid-Liquid Phase Transfer Conditions

Run	Substrate	Reagent[b]	t (°C)	Reaction Time (hr)	Reaction Yield (%)	Product[c] n-$C_8H_{17}X$, X=
1	n-$C_8H_{17}OSO_2Me$	KI	100	0.11	100	I
2	n-$C_8H_{17}OSO_2Me$	NaI	100	0.16	100	I
3	n-$C_8H_{17}OSO_2Me$	KBr	100	0.5	96	Br
4	n-$C_8H_{17}OSO_2Me$	NaBr	100	0.5	88	Br
5	n-$C_8H_{17}OSO_2Me$	KCl	100	3.0	89	Cl
6	n-$C_8H_{17}OSO_2Me$	NaCl	100	3.0	75	Cl
7	n-$C_8H_{17}OSO_2Me$	KF	115	42.0	65	F
8	n-$C_8H_{17}Br$	KI	100	1.5	92	I
9	n-$C_8H_{17}Br$	KI	80	3.0	100	I
10	n-$C_8H_{17}I$	KBr	80	3.0	40[d]	Br
11	n-$C_6H_{13}CH(OSO_2Me)Me$	KBr	100	2.5	67[e]	n-$C_6H_{13}CHBrMe$

[a] 0.05 Molar equivalent.
[b] Saturated aqueous solution, 5 molar equivalents.
[c] The products were characterized by glc retention times.
[d] n-Octyl iodide, 60%; same values after 26 hr.
[e] 2-Octanol, 16%; octenes, 16%; 2-octyl mesylate, 1%.

were drawn from the data. The relative nucleophilicity of halide ions was $I^- > Br^- > Cl^- > F^-$ (Runs 1-7, Table VI); secondary substrates reacted more slowly than primary (Runs 3 and 11, Table VI); sodium and potassium salts reacted at about equal rates (Runs 1-6, Table VI); and iodide quantitatively displaced bromide in the presence of a large excess of KI (Runs 8 and 9, Table VI), whereas only a partial conversion was achieved for the reverse reaction using a large excess of KBr (Run 10, Table VI).

Cinquini *et al.* (1975) compared the catalytic activity of a number of macrocyclic polydentate ligands [(III), (V), (VI), (VIII)c, and (IX)c] in the reaction of octyl substrates with a variety of nucleophiles under liquid-liquid phase transfer conditions. For comparison, a quaternary phosphonium bromide (XIV)

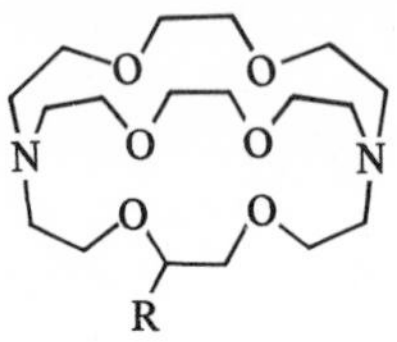

(IX)c R = H
(IX)c′ R = n-$C_{11}H_{23}$
(IX)c″ R = n-$C_{14}H_{29}$

n-$C_{16}H_{33}P^{+}(n\text{-}C_4H_9)_3$
Br^-

(XIV)

was also included. The results are summarized in Table VII. The data indicate that the order of effectiveness is (IX)c″ > (IX)c′ > (XIV) > (V) > (IX)c > (III)> (VI).

TABLE VII

Structural Dependence of Catalytic Activity of Macrocyclic Polydentate Ligands under Liquid-Liquid Phase Transfer Conditions

Substrate[a]	Reagent	Catalyst[b]	Temp. (°C)	Time (hr)	Yield (%)	Product
n-$C_8H_{17}Br$	KI[c]		80	24	4	n-C_8H_{17}-I
	KI[c]	(VI)	80	40	80	n-C_8H_{17}-I
	NaI[c]	(III)	80	21	80	n-C_8H_{17}-I
	KI[c]	(V)	80	3	100	n-C_8H_{17}-I
	KI[c]	(IX)c″	60	0.2	100	n-C_8H_{17}-I
	KI[c]	(IX)c′	60	0.5	92	n-C_8H_{17}-I
	KI[c]	(IX)c	60	14	90	n-C_8H_{17}-I
	KI[c]	(XIV)	60	1	93	n-C_8H_{17}-I
n-$C_8H_{17}Cl$	KI[c]	(IX)c″	80	5	77	n-C_8H_{17}-I
	KI[c]	(IX)c′	80	4	85	n-C_8H_{17}-I
	KI[c]	(XIV)	80	24	80	n-C_8H_{17}-I
n-$C_8H_{17}OSO_2Me$	KF[c]	(IX)c′[d]	120	4	85	n-$C_8H_{17}F$
	KF[c]	(XIV)[d]	120	2	94	n-$C_8H_{17}F$
n-$C_6H_{13}CH(Br)Me$	KI[c]	(IX)c″	80	3	86	n-$C_6H_{13}CH(I)Me$
	KI[c]	(XIV)	80	6	89	n-$C_6H_{13}CH(I)Me$

[a]The reactions were carried out with a saturated aqueous solution of the reagent.
[b]By glc analysis. Numbers refer to compounds in Fig. 1 and the text.
[c]5 Mol. equiv.
[d]0.1 Mol. equiv.

B. Carboxylate Ions

The poor nucleophilicity of acetate ion toward various substrates in condensed systems has been attributed to a combination of polarizability, basicity, and solvation factors. Liotta *et al.* (1974) reported that acetate solubilized as the potassium salt in acetonitrile or benzene containing (IV) became sufficiently nucleophilic to react smoothly and quantitatively, even at room temperature, with a wide variety of organic substrates. Displacement reactions at 1°, 2°, 3°, and benzylic positions along with competing elimination processes have been demonstrated with this reagent, which has been termed "naked" acetate. The data summarized in Table VIII deal specifically with the solvent acetonitrile. The same products were obtained in benzene but the reaction rates were slower. In the absence of crown, little or no reaction took place under identical conditions covering the same periods of time. For instance, in the case of benzyl bromide (Run 1, Table VIII), the most reactive substrate reported in Table VIII, less than 5% benzyl acetate was formed after several days with potassium

TABLE VIII

Reactions of Naked Acetate with Organic Substrates[a]

Run	Substrate	Products	Concentrations (*M*) (IV)	Concentrations (*M*) Substrate	Temp. (°C)	Time (hr)
1	Benzyl bromide	Benzyl acetate (100%)	0.16	3.4	25	2
2	n-$C_6H_{13}Br$	n-$C_6H_{13}OAc$ (100%)	0.10	1.4	25	150
3	n-$C_8H_{17}Br$	n-$C_8H_{17}OAc$ (96%)	0.10	1.4	83	3
4	1,2-Dibromoethane	Ethylene diacetate (90%)	0.09	1.3	83	3
5	1,2-Dibromoethane[b]	Ethylene diacetate (23%) 2-Bromoethyl acetate (77%)	0.06	1.9	83	3
6	2-Bromooctane	2-Acetoxyoctane (~90%) Octenes (~10%)	0.10	1.4	83	20
7	2-Chloro-2-methyl-cyclohexanone[c]	2-methylcyclohexenone (~10%) *cis*-2-Acetoxy-6-methyl-cyclohexanone (~54%) *trans*-2-Acetoxy-6-methyl-cyclohexanone (~10%) 2-Acetoxy-2-methyl-cyclohexanone (~25%)	0.15	2.9	83	1

[a] All reactions were run using at least a twofold excess of potassium acetate.

[b] Equimolar quantities of 1,2-dibromoethane and potassium acetate were used in this experiment.

[c] The acetate products were assigned structures based on evidence from a known mixture prepared by the method of House and Richey (1969).

acetate in the absence of crown, whereas the conversion was complete within 2 hr in the presence of crown.

The acetate reagent was prepared by dissolving (IV) in dry acetonitrile or dry benzene and then simply adding dry potassium acetate. After the heterogeneous system was stirred for 30 min, the organic substrate was added and the resulting mixture stirred until reaction was complete. In all cases, the crown was present in catalytic concentrations, indicating that the reactions shown in Table VIII were examples of phase transfer catalysis between solid and liquid phases.

The reaction products were primarily acetates. Small amounts of alkene products were sometimes observed. Highly activated bromides, such as benzyl bromide (Run 1, Table VIII), reacted rapidly with naked acetate at room temperature to produce benzyl acetate. Primary alkyl halides (Runs 2-5, Table VIII) required several days at ambient temperatures to reach completion. The rate, however, was substantially increased at reflux temperatures (~83°C). Interestingly, virtually no alkene product could be detected by glc or nmr techniques in these reactions. In order to determine the approximate relative rates of reaction for primary bromides, tosylates, and chlorides, 1-substituted hexanes were reacted with naked acetate in acetonitrile at 25°C. The rates of reaction were in the order Br > OTs > Cl (~4:2:1), suggesting that alkyl bromides are the best substrates for synthetic purposes. 2-Bromooctane (Run 6, Table VIII) produced the corresponding acetate in approximately 20 hr at reflux temperatures with formation of about 10-15% alkene products. Reaction of naked acetate with 1,2-dibromoethane (Runs 4 and 5, Table VIII) was conducted to yield either 1,2-diacetoxyethane or predominantly 2-bromoethyl acetate, depending on which reagent was present in excess. Reaction of 2-chloro-2-methylcyclohexanone (Run 7, Table VIII) with naked acetate in acetonitrile at reflux yielded a mixture of acetates with only 10% 2-methyl-

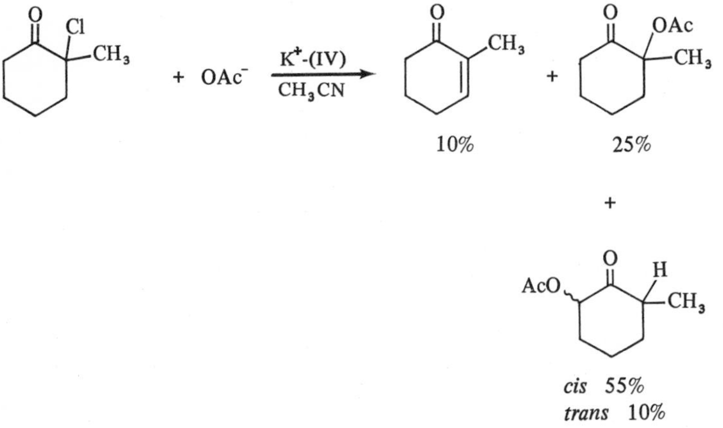

(10)

cyclohexenone [Reaction (10)]. These results appeared to be in direct contrast to those obtained in the reaction of naked fluoride with 1°, 2°, and 3° halides in that the fluoride reagent produced much larger quantities of alkene products. The authors suggested that naked fluoride is a stronger base than naked acetate.

The concentration of naked acetate in solution at 25°C has been determined from ^{1}H nmr analysis as a function of crown concentration. The protons of (IV) and the acetate appeared as singlets and were easily integrated. The results in benzene and acetonitrile-d_3 are shown in Table IX. It is clear that high concentrations of acetate were achieved in these solvents and that at least 80% of the crown was complexed with potassium acetate.

TABLE IX

Solubility of Potassium Acetate in Acetonitrile and Benzene Solutions Containing 18-Crown-6 (IV)

	(IV) (*M*)	Potassium acetate (*M*)
Benzene	0.55	0.4
	1.0	0.8
Acetonitrile-d_3	0.14	0.1

Durst (1974) has used the observations concerning naked acetate to synthesize a wide variety of *p*-bromophenacyl esters by the reaction of the potassium salt of a carboxylic acid with *p*-bromophenacyl bromide in acetonitrile and benzene using (V) as the solubilizing agent [Reaction (11)]. The results are summarized in Table X. Isolated yields ranging from 90 to 100% were reported.

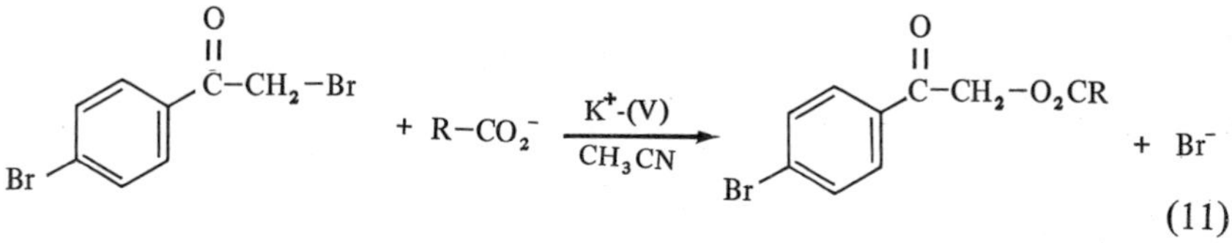

(11)

The reactions proceeded more rapidly in acetonitrile than in benzene, and crown was always present in catalytic quantities (0.05 mole %). It is interesting to note that potential steric problems (Runs 9 and 10, Table X) appeared to be completely absent. Sodium salts gave excellent yields of product but the reaction times were longer.

Durst *et al.* (1975) and Grushka *et al.* (1975) have subsequently reported the formation of phenacyl esters of fatty acids in essentially quantitative yields using

TABLE X

Reaction of Naked Carboxylates with *p*-Bromophenacyl Bromide

Run	Acid	*p*-Bromophenacyl derivative yield[a] (%)
1	Formic	93
2	Acetic	98
3	Propionic	98
4	Butyric	97
5	Heptanoic	99
6	Pivalic	95
7	Benzoic	93
8	2-Methylbenzoic	90
9	2-Iodobenzoic	92
10	Mesitoic (2,4,6-trimethylbenzoic)	98
11	4-*t*-Butylbenzoic	92

[a]Isolated yields.

crown ethers as catalysts. These esters absorbed ultraviolet (uv) radiation strongly at 254 nm, thus allowing the detection of as small a quantity as 1 ng of C_2 acid and 50 ng of C_{20} acid.

Knochel *et al.* (1975) have investigated the effect of a wide variety of macrocyclic polydentate ligands on the activation of acetate anion in its reaction with benzyl chloride in acetonitrile [Reaction (12)]. The approximate

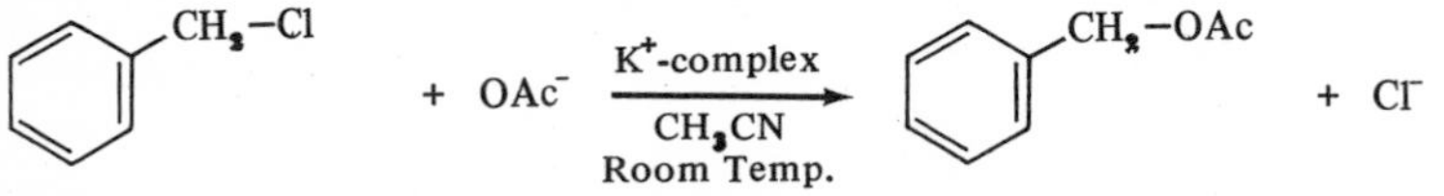

(12)

half-lives of the reactions as a function of macrocyclic polydentate ligand are summarized in Table XI. The characteristics of the ligand which influenced the rate were suggested to be (a) the stability of the metal-ligand complex, (b) the lipophilicity of the ligand, (c) the rigidity of the ligand, and (d) the reactivity of the ligand toward the substrate (the aza crowns).

Mack *et al.* (1975) have reported the facile synthesis of anhydrides and carboxylic carbonic anhydrides from the reaction of the potassium or sodium salts of carboxylic acids with activated halides (ethyl chloroformate, cyanuric chloride, and benzyl chloroformate) in acetonitrile in the presence of (IV) [Reaction (13)]. The results are summarized in Table XII. In general,

TABLE XI

Effect of Macrocyclic Polydentate Ligand on Rate of Reaction of Potassium Acetate with Benzyl Chloride in Acetonitrile

Ligand[a]	Approx. half-life (hr)	Ligand[a]	Approx. half-life (hr)
None	685[b]	(VIII)d	75
(IV)	3.5	(VIII)e	100
(VI)	9.5	(IX)a	8
(V)	1.5	(IX)b	0.8
(VIII)b	700[b]	(IX)c	5.5
(VIII)c	65		

[a]Numbers refer to compounds in Fig. 1.
[b]Extrapolated

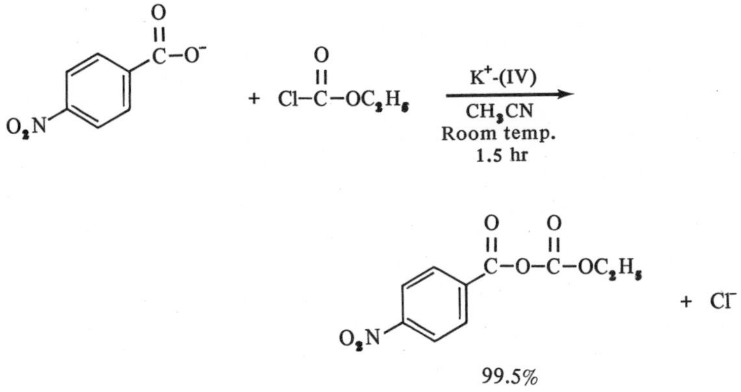

(13)

it was found that potassium salts were more efficient than the corresponding sodium salts.

Dehm and Padwa (1975) have reacted potassium phenylacetate with a series of 2-bromosubstituted carbonyl compounds in the presence of (IV) to form aldehydo or keto esters which were subsequently cyclized to five-membered unsaturated lactones on further heating [Reaction (14)]. The results are summarized in Table XIII. Two procedures were used. The first (Procedure A) involved heating a mixture of potassium phenylacetate, α-bromocarbonyl, and (IV) in acetonitrile for 1 hr, evaporating the solvent, adding DMSO and NaH, and heating at 70°C for 1-3 hr. A more facile alternate procedure (B) involved refluxing potassium phenylacetate, α-bromocarbonyl, and (IV) in acetonitrile

TABLE XII

Synthesis of Anhydrides and Carboxylic Carbonic Anhydrides

Acid, potassium salt	Activating halide	Product	Isolated yield[a] (%)
Cinnamic acid	Ethyl chloroformate	Cinnamic anhydride	95.5[b]
p-Nitrobenzoic acid	Ethyl chloroformate	*p*-Nitrobenzoic anhydride	99.5[b]
Benzoic acid	Ethyl chloroformate	Benzoic anhydride	90.6[b]
Acetic acid	Ethyl chloroformate	Acetic anhydride	85.0[b]
Propionic acid	Ethyl chloroformate	Propionic anhydride	86.1[b]
Benzoic acid	Cyanuric chloride	Benzoic anhydride	99.0[b]
Benzoic acid	Benzyl chloroformate	Benzoic-ethylcarbonic mixed anhydride	98.0[c]
p-Nitrobenzoic acid	Ethyl chloroformate	*p*-Nitrobenzoic-ethylcarbonic mixed anhydride	98.0[c]

[a] Molar ratio of 18-crown-6 (IV) to starting salts ranged from 1 : 20 to 1 : 50.
[b] Molar ratio of salt to activating halide is 2 : 1.
[c] Molar ratio of salt to activating halide is 1 : 2.

for 1-3 days to go directly to the furanone. The authors reported that no attempt was made to optimize the yields.

Hunter *et al.* (1974) have reported that at −45°C in dry THF sodium 3-(fluoren-9-ylidene)-2-phenylacrylate decarboxylated at an extremely slow rate—essentially less than 10% reaction in 200 min. Upon addition of (VI) at this temperature, reaction took place rapidly [Reaction (15)]. This observation

TABLE XIII

Synthesis of Trisubstituted 2(5H)-Furanones

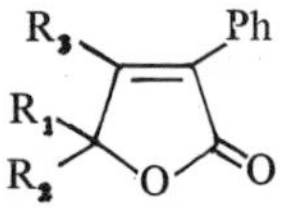

Product	Procedure	Yield (%)	Product	Procedure	Yield (%)
$R_1 = R_2 = Ph; R_3 = H$	B	90	$R_1 = Ph; R_3 = H$	B	62
$R_1 = Ph; R_3 = H$	B	81	$R_2 = p\text{-}BrC_6H_4$		
$R_2 = p\text{-}CH_3OC_6H_4$			$R_1 = Ph; R_3 = H$	B	20
$R_1 = Ph; R_2 = H$	A	80	$R_2 = p\text{-}CNC_6H_4$		
$R_3 = p\text{-}CH_3OC_6H_4$			$R_1 = R_3 = Ph$	A, B	80
$R_1 = H; R_3 = Ph$	A	75	$R_2 = H$		
$R_2 = p\text{-}CH_3OC_6H_4$			$R_1 = R_3 = CH_3$	B	90
			$R_2 = H$		

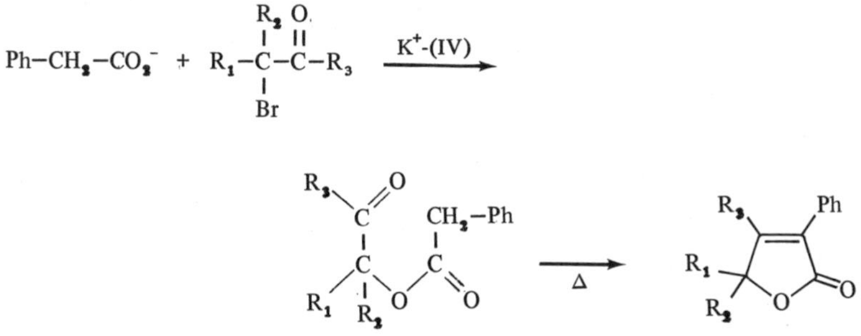

(14)

indicated that the nature of the ion pairing was an important consideration in this decarboxylation reaction. Thus, the crown ether effectively coordinated with the sodium ion, producing a highly reactive naked carboxylate from the much less reactive ion-paired species.

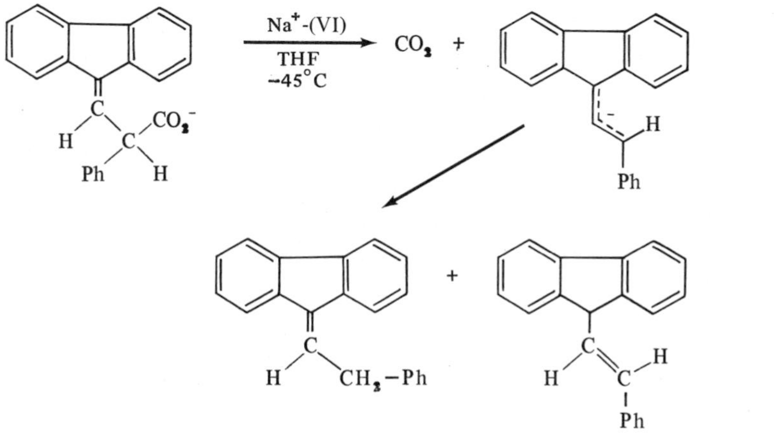

(15)

C. Cyanide Ion

Naked cyanide has been demonstrated to be a synthetically useful reagent in a wide variety of substitution, elimination, and addition processes. Some results reported by Cook *et al.* (1974) are summarized in Table XIV. The reactions were carried out by simply pouring a substrate-(IV) solution directly over excess dry KCN and stirring the two-phase system vigorously at ambient or reflux temperatures until reaction was complete. Little or no reaction was found to take place in the absence of crown under the same conditions covering the same periods of time (Runs 20 and 21, Table XIV). In all cases, the crown was present in catalytic quantities, indicating that it behaves as a solid-liquid phase transfer catalyst. In general, the reactions were more rapid in acetonitrile than in benzene.

TABLE XIV

Reactions of Naked Cyanide with Organic Substrates in the Presence of 18-Crown-6 (IV)

Substrate	Run	Solvent	Products[a] (Yield %)	Conc.[b] (*M*) (IV)	Conc.[b] (*M*) Substrate	Temp. (°C)	$t_{1/2}$[c] (hr)	t_{end} (hr)
1,3-Dibromopropane	1	CH_3CN	Glutaronitrile (97.3)[d]	0.151	1.80	83	7	25
	2	CH_3CN	Glutaronitrile (94.9)[d]	0.147	1.80	Ambient	10.5	48
	3	C_6H_6	Glutaronitrile (100)[e]	0.147	1.80	90	12	40
	4	C_6H_6	Glutaronitrile (100)[e]	0.147	1.80	Ambient	10	49
1-Bromo-3-chloropropane	5	CH_3CN	Glutaronitrile (100)[e]	0.152	1.80	83	1.1	30
1,3-Dichloropropane	6	CH_3CN	Glutaronitrile (96.8)[d]	0.154	1.80	83	0.23	1.5
1,4-Dibromobutane	7	CH_3CN	Adiponitrile (100)[e]	0.141	1.80	83	6.5	14.5
	8	CH_3CN	Adiponitrile (100)[e]	0.141	1.80	Ambient	11	57
1,4-Dichlorobutane	9	CH_3CN	Adiponitrile (94.7)[d]	0.144	1.80	83	0.2	0.7
	10	CH_3CN	Adiponitrile (86)[d]	0.144	1.80	Ambient	8	75
1-Bromohexane	11	CH_3CN	1-Cyanohexane (100)[e]	0.138	1.80	83	10	40
1-Chlorohexane	12	CH_3CN	1-Cyanohexane (90.6)[d]	0.139	1.80	83	0.5	2.2
2-Bromobutane	13	CH_3CN	2-Cyanobutane (69.7)[e]	0.141	2.00	83	16	32
	14	C_6H_6	2-Cyanobutane (43.4)[e]	0.140	2.01	90	58	66[f]
2-Bromooctane	15	CH_3CN	2-Cyanooctane (56)[d] (62)[e] Octenes (17)[e,g]	0.119	2.00	83	14	78
2-Chlorooctane	16	CH_3CN	2-Cyanooctane (77.5)[e,f] Octenes (3.1)[e,g,h]	0.122	2.00	83	59	244[h]
Cyclohexyl bromide	17	CH_3CN	Cyclohexene (46)[e,f]	0.132	1.80	83		53[f]
Cyclohexyl chloride	18	CH_3CN	Cyclohexene (32)[e,f]	0.146	1.80	83		122.5[f]
Benzyl bromide	19	CH_3CN	Benzyl cyanide (100)[e]	0.146	1.80	Ambient	13	25
Benzyl chloride	20	CH_3CN	Benzyl cyanide (94.2)[d]	0.147	1.80	Ambient	0.08	0.4
	21	CH_3CN	Benzyl cyanide (20)[e,f]	0.0	1.80	Ambient		75[f]
o-Dichlorobenzene	22	CH_3CN	No reaction[i]	0.145	1.80	83		109

Methacrylonitrile (Acetone cyanohydrin)[f]	23	CH_3CN	1,2-Dicyanopropane (91.9)[d]	0.132	1.51 (1.78)	83	0.3	0.6
	24	CH_3CN	1,2-Dicyanopropane (46)[e,f]	0.132	1.52	Ambient		189[f]
	25	CH_3CN	1,2-Dicyanopropane (77)[d,k]	0.0	1.51 (1.74)	83	6.1	15

[a]For the isolated products, nmr, ir, and mass spectral data confirmed the pure compound's identity. Also, where possible, these spectra and glc data of the products were compared to those of the commercial compounds (runs 1-12).

[b]The reactions were run in a total of 25 ml of solution, with a twofold excess of solid KCN per functional group for the substitution reactions, and a catalytic amount (approximately equivalent to 18-crown-6) of salt for the hydrocyanation.

[c]$t_{1/2}$ is defined as the time required for 50% of the starting material to react.

[d]Isolated yields. For these examples, glc and nmr analysis showed a quantitative conversion to the products.

[e]Calculated from glc and nmr data.

[f]In these runs, the reactions were stopped before completion.

[g]From glc analysis, 1-octene and both *cis-* and *trans*-2-octene were formed.

[h]There was difficulty in driving the reaction to completion. At 244 hr the composition consisted of 7.4% starting material, 89.1% substitution product, and 3.5% olefin by glc analysis. The yields given in Table I are thus based on reacted starting material.

[i]The absence of reaction of naked cyanide with *o*-dichlorobenzene was in contrast to the results recently reported for the analogous KOH-CH_3OH-crown system in which a 40-50% yield of *o*-chloroanisole was obtained.

[j]1.18 molar equivalents of acetone cyanohydrin. . ., etc. (see text).

[k]Run 25 was monitored frequently during the course of the reaction by glc, and thus a small quantity of product was lost in the transfers. The reaction was quantitative by glc and nmr analyses.

The reaction of naked cyanide with benzyl chloride proceeded quickly and quantitatively to product under mild conditions (Run 20, Table XIV). With primary substituted alkyl halides, the conversions to nitrile compounds were quantitative, with no elimination products detected (Runs 1-12, Table XIV). Displacement at secondary carbon produced primarily substitution products with only a small percentage of elimination products. (Runs 13-16, Table XIV). The results compare favorably to those obtained with naked acetate (Liotta *et al.*, 1974), but were in direct contrast to those of naked fluoride (Liotta and Harris, 1974), where large quantities of alkenes were obtained. It appeared, therefore, that naked fluoride was a stronger agent for promoting elimination processes than either naked cyanide or naked acetate. Cyclohexyl halides gave exclusively elimination product with naked cyanide (Runs 17 and 18, Table XIV). It is interesting to note that no reaction was observed with *o*-dichlorobenzene (Run 22, Table XIV). This result contrasts with the reported reaction of a KOH-CH_3OH-crown system with *o*-dichlorobenzene to produce a 40-50% yield of *o*-chloroanisole (Sam and Simmons, 1974).

Interestingly, primary chlorides reacted at much faster rates than the corresponding bromides under the reaction conditions. For example, benzyl chloride reacted about 100 times faster than benzyl bromide at room temperature (Runs 19 and 20, Table XIV) and 1-chlorohexane reacted about 20 times faster than 1-bromohexane at reflux temperatures (Runs 11 and 12, Table XIV). This observation is contrary to the normally accepted leaving-group order. It is also opposite to what has been observed with naked acetate (Liotta *et al.*, 1974) and naked fluoride (Liotta and Harris, 1974). With secondary halides, bromides reacted more rapidly than chlorides. For example, 2-bromooctane reacted approximately four times more rapidly than 2-chlorooctane at reflux temperatures (Runs 15 and 16, Table XIV); however, less alkene and a higher overall yield of substitution product was obtained with the chloride. The causes for these observations probably had their origin in the heterogeneous nature of the reaction system, the presence of catalytic quantities of crown ether, and the difference in solubility between KBr and KCl in the acetonitrile-crown medium.

Zubrick *et al.* (1975) have reported that under the conditions just described, naked cyanide reacted with benzyl chloride, *p*-nitrobenzyl chloride, *p*-chlorobenzyl chloride, and 3,4-dimethoxybenzyl chloride to give the corresponding nitrile in 85-90% yield. In addition, the facile synthesis of trimethylsilyl cyanide from displacement on trimethylsilyl chloride by naked cyanide was reported to proceed in 45% yield [Reaction (16)].

$$CH_3-\underset{CH_3}{\overset{CH_3}{\underset{|}{\overset{|}{Si}}}}-Cl + CN^- \xrightarrow{K^+\text{-(IV)}} CH_3-\underset{CH_3}{\overset{CH_3}{\underset{|}{\overset{|}{Si}}}}-CN + Cl^- \qquad (16)$$

Substitutions of halides by cyanide ion have been reported to occur (1) in ethanol-water mixtures under strenuous conditions (Mowry, 1948; Marvel and McColm, 1941; Allen, 1941; Adams and Thal, 1941); (2) in dipolar aprotic solvents such as DMSO (Smiley and Arnold, 1960; Friedman and Schechter, 1960); and (3) by the use of tetraalkylammonium or -phosphonium salts to transfer the cyanide ion across a water-organic interface (liquid-liquid phase transfer catalysis) (Starks, 1971, 1973; Starks and Owen, 1973; Sugimoto *et al.*, 1962). Comparatively, the naked cyanide system appeared to be superior to the ethanol-water method in reaction time, temperature, simplicity of workup, and yields (Mowry, 1948; Marvel and McColm, 1941; Allen, 1941; Adams and Thal, 1941; House, 1957). It also compared favorably to the dipolar aprotic solvents and phase transfer catalyst systems in both reaction time and yield for conversion of primary halides to nitriles (Smiley and Arnold, 1960; Freidman and Schechter, 1960; Starks, 1971; Starks and Owen, 1971; Sugimoto *et al.*, 1962). The advantages of the naked cyanide procedure over the latter methods were suggested to be lower reaction temperature and simplicity of workup. For secondary substrates, comparable yields to those presented in Table XIV were obtained in much shorter times in the dipolar aprotic solvent, while the phase transfer catalyst system gave a higher yield of substitution product (no reaction time was reported). The reaction temperatures, however, were higher in both the DMSO and the phase transfer catalyst systems.

Naked cyanide has been found to be an excellent nucleophile in displacement reactions at sterically hindered positions on substrates subject to Favorskii rearrangements (Liotta and Dabdoub, unpublished results). In the reaction of 2-chloro-2-methylcyclohexanone, the effect of temperature on the course of reaction was dramatically illustrated by the following reactions [Reaction (17)].

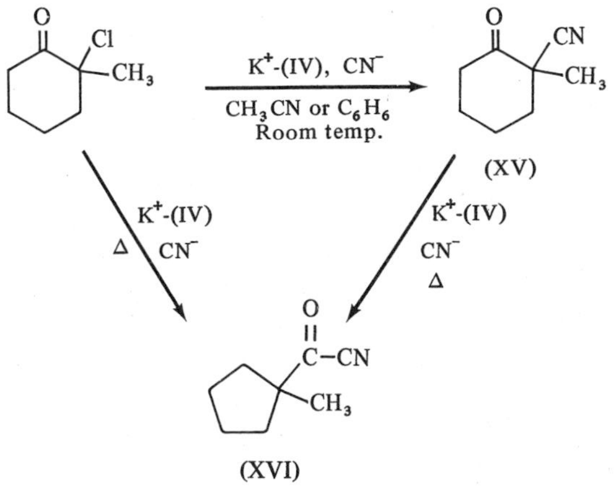

(17)

TABLE XV

Reactions of Cyanide Anion with 2-Chloro-2-Methylcyclohexanone and 2-Cyano-2-Methylcyclohexanone with and without 18-Crown-6 (IV)

Substrate	Run	Solvent[a]	Salt	Products	Yield[b]	Concentration[c] (*M*) (IV)	Concentration[c] (*M*) Substrate	Temp. (°C)	$t_{1/2}$[d]	t_{end}[e]
Chloroketone	1[f]	CH_3CN	KCN[g]	Cyanoketone[h] (XV)	[89.4]	0.18	2.60	Ambient	17 min	35 min.
Chloroketone	2[f]	C_6H_6	KCN	Cyanoketone (XV)	[88.0]	0.18	2.73	Ambient	25 min	153 min
Chloroketone	3[f]	CH_3CN	KCN	Cyanoketone (XV)	[90.2]	0.00	2.32	Ambient	30 min	135 min
Chloroketone	4	CH_3CN	KCN	Favorskii Rearrangement Product[i] (XVI)	[60.2]	0.18	2.33	83	16 hr	80 hr[j]
Cyanoketone	5	CH_3CN	KCN	Favorskii Rearrangement Product (XVI)	[66.0]	0.16	1.46	83	1.0 hr	10 hr[k]
Cyanoketone (XV)	6	CH_3CN	KCN	Cyanoketone (XV)	–	0.00	2.30	83	–	–
Chloroketone	7	CH_3CN	KCN	Cyanoketone (XV) and Favorskii Rearrangement Product (XVI)	[88.6]	0.18	2.60	Ambient	17 min	35 min
Cyanoketone	8	C_6H_6	KCN	Favorskii Rearrangement Product (XVI)	[47.3]	0.17	2.70	83	3 hr	31 hr
Chloroketone	9	CH_3CN	KCN	Cyanoketone (XV)	–[g]	0.21	2.65	83[l]	5 hr[m]	40 hr[m]
*Cyanoketone[d]	10	CH_3CN	KCN	Favorskii Rearrangement Product (XVI)	[46.4]	0.21	2.65	83	90 hr	–[n]
*Cyanoketone	11	CH_3CN	KCN	Favorskii Rearrangement Product (XVI)	[62.3]	0.19	2.43	83	7 hr	30 hr
*Cyanoketone	12	CH_3CN	–	Cyanoketone (XV)	–	0.18	2.37	83	–	–

[a]Runs 1-8 were carried out in the absence of acetone cyanohydrin, while Runs 9-12 were carried out in the presence of 0.15, 0.15, 0.13, and 0.22 *M* acetone cyanohydrin, respectively.

[b]Isolated yields. For all reactions with yields, glpc and nmr analysis showed a quantitative conversion to the products.

(cont.)

[c]The substrate solutions were prepared by weighing out the appropriate amount of reagent into a 25-ml volumetric flask and diluting to the mark with a stock solution (known concentration) of 18-crown-6 in CH_3CN or C_6H_6. The prepared solutions were then placed directly over solid, dry KCN (two- to threefold excess) and the system was stirred vigorously at ambient or reflux temperature.

[d]$t_{1/2}$ is the time required for 50% of the starting material to react.

[e]Estimated.

[f]Runs 1-3 were worked up at room temperature to avoid rearrangement.

[g]KCN was dried in a vacuum desiccator at 105°C.

[h]Physical properties of cyanoketone were as follows: B.P. 102-103°C at 23 mm Hg; ir (neat NaCl plates) 1725 cm^{-1} (C=O) and 2240 cm^{-1} (C≡N); mass spec. m/e 137, M^+ 136, and abundant fragments at 122, 109, 97, 94, 84, 78, 69, 68, 67, 55, 54, 43, 41, and 39. ^1H nmr (neat, internal TMS) at δ1.35 (sharp singlet, aliphatic CH, 3H's) at δ2.0-0.8 (multiplet 8H's). Integration gives a ratio of 3:2:2:4. Elemental analysis:

Calculated: C 70.04; H 8.08; N 10.21; O 11.66
Found: C 69.79; H 8.13; N 10.19; O 11.61

[i]Physical properties of the Favoriskii rearrangement product were as follows: B.P. 108-110°C at 23 mm Hg; ir (neat NaCl plates) 1720 cm^{-1} (C=O) abd 2220 cm^{-1} (conjugated C≡N); mass spec. m/e 137 and abundant fragments at 122, 94, 82, 68, 67, 55, 43, 42, 41, 39, 28. ^1H nmr (neat; internal TMS) at δ1.2 (sharp singlet, aliphatic CH, 3H's) at δ2.55-1.4 (multiplet 8H's). Elemental analysis:

Calculated: C 70.04; H 8.08; N 10.21; O 11.66
Found: C 70.02; H 8.15; N 10.22; O 11.61

[j]Reaction was about 80% (glpc) complete after 40 hr reflux.

[k]Reaction was about 95% (glpc) complete after 5 hr reflux.

[l]This reaction in the presence of acetone cyanohydrin does not proceed at room temperature.

[m]The reaction rate slows down greatly after $t_{1/2}$ is reached.

[n]The reaction was refluxed in the absence of KCN for 50 hr. and no product was observed.

[o]$t_{1/2}$ was found to be 90 hr and after ~200 hr the reaction was only about 60% complete by glpc analysis.

TABLE XVI

Hydrocyanation of Δ^4-Cholesten-3-one, 2, with Naked Cyanide and Acetone Cyanohydrin in the Presence and Absence of 18-Crown-6 (IV)

Run	Solvent	Yield[a]	Ratio[b] 5α/5β	KCN[c]	Concentration (*M*) (IV)	Concentration (*M*) Substrate	Reaction temp. (°C)	Reaction[d] time (hr)	t_{end}[e]	Results[f]
1	C_6H_6	–	–	Stoich.	0.0	0.15	Reflux	32	–	No reaction
2	C_6H_6	[83.6] 5α (7.0) 5β (70.0)	1/10[g]	2× stoich.[h]	0.19	0.14	Ambient	15	15	Reaction complete
3	C_6H_6	[82.8] 5α (13.2) 5β (62.1)	1/4.7	2× stoich.	0.17	0.16	52	10	10	Reaction complete
4	C_6H_6	[84.3] 5α (18.0) 5β (56.8)	1/3.16	Stoich.	0.19	0.16	Reflux	3	3	Reaction complete
5	C_6H_6	[85.1][i,j] 5α (15.0) 5β (60.1)	1/4.0	Catalytic	0.10	0.16	Reflux	5	5	Reaction complete
6	CH_3CN	–	–	Stoich.	0.0	0.16	Reflux	80	–	No reaction[k]
7	CH_3CN	[84.4] 5α (15.4) 5β (60.6)	1/3.9[e]	Stoich.	0.19	0.14	Reflux	5	5	Reaction complete
8	CH_3CN	[86.1] 5α (14.7) 5β (61.4)	1/4.2	Stoich	0.21	0.15	Reflux	10	5 1	Reaction was complete in 5 hr
9	CH_3CN	[43.8] 5α (6.1) 5β (32.1)	1/5.3	2× stoich.	0.20	0.13	52	20	–	Reaction ~60% complete[m]

10	CH_3CN	[41.2] 5α (6.9) 5β (29.7)	1/4.3	Catalytic	0.19	0.14	Reflux	7	–	Reaction ~50% complete[n]
11	C_6H_6	–	–	–[o]	0.20	0.15	Reflux	35	–	No reaction
12	C_6H_6	–	–	Stoich.[p]	0.19	0.16	Reflux	31	–	No reaction

[a]Numbers in square brackets refer to percent isolated yield as collected from column chromatography.

[b]Numbers in parantheses refer to percent isolated yield of each isomer separately after recrystallization.

[c]Catalytic KCN is ~10% KCN while stoichiometric KCN is 1.0 equivalent KCN to 1.0 equivalent of substrate to 1.2 equivalent acetone-cyanohydrin except in cases when 2 stoich. KCN was used, in which cases 1.2 (2 stoich.) acetone cyanohydrin was used.

[d]For those runs with reaction times >5 hr, an aliquot sample was analyzed by nmr techniques every 2 hr for the first 10 hr, and every 5-10 hr thereafter.

[e]Estimated.

[f]In all cases, progress of reaction was followed by nmr techniques.

[g]This is the best ratio of trans/cis products ever reported.

[h]When using 1X stoichiometric KCN the reaction was ~80% complete after stirring at ambient temperature for 30 hr.

[i]Infrareds, nmr's, ORD's, M.P.'s, and elemental analyses for the 5-cyano ketones isolated from all runs compare well with those reported in the literature.

[j]Run 5 was carried out under catalytic conditions and the reaction was complete in 5 hr.

[k]In the absence of crown ether no reaction occurs even after refluxing for prolonged times (Runs 1 and 6).

[l]Comparison of Runs 2 and 7 shows that the stereochemistry of the cyano group in the product is greatly influenced by the solvent. By simply changing the solvent from acetonitrile (D=37) to benzene (D=2) the ratio of cis to trans varies from 3.5/1 to 10/1.

[m]In Run 9, ~35% of the starting material was recovered.

[n]In Run 10, ~40% of the starting material was recovered.

[o]This run was carried out in the presence of crown ether and acetone cyanohydrin but in the absence of KCN. The reaction was refluxed for 35 hr but no reaction occurred.

[p]This run was carried out in the presence of crown ether and KCN but in the absence of acetone cyanohydrin. The reaction was refluxed for 31 hr but no reaction occurred.

The details of the experimental procedure and results are summarized in Table XV. At room temperature a quantitative conversion to simple displacement product (XV) was obtained, thus providing a potentially useful procedure for introducing a carbon functional group in a rather sterically hindered position. The reaction proceeded at a faster rate in acetonitrile than in benzene. Excellent yields of product were obtained in the absence of crown in acetonitrile but reaction times were quite long. At reflux temperatures the Favorskii product (XVI) was obtained in high yield; XV was converted to (XVI) by simply heating with the cyanide reagent. Reaction of naked cyanide with *cis*-2-chloro-4-methylcyclohexanone also proceeded smoothly to produce the simple displacement product [Reaction (18)] (Liotta and Dabdoub, unpublished results).

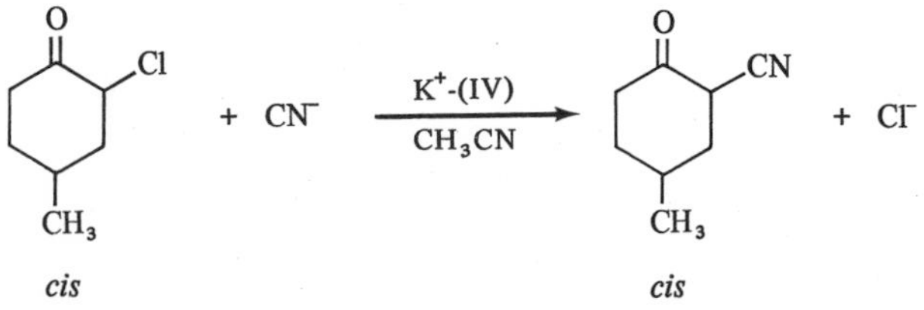

(18)

The versatility of the naked cyanide reagent was demonstrated in the quantitative hydrocyanation reaction of methacrylonitrile (Run 23, Table XIV). No reaction was observed in the absence of either (IV) or acetone cyanohydrin. These results are consistent with the following mechanistic sequence:

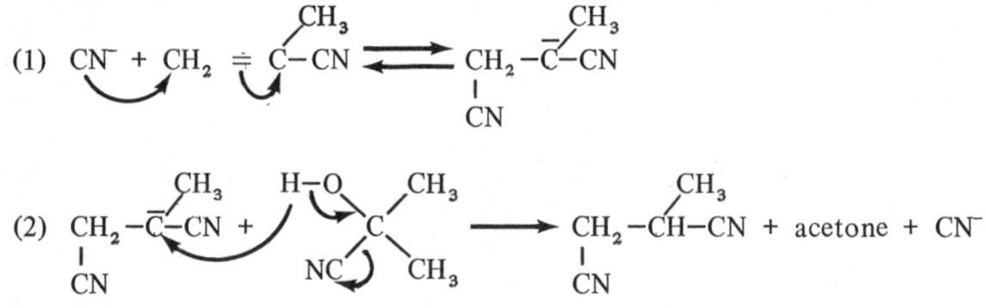

(3) Repeat (1) and (2)

In the absence of acetone cyanohydrin, the naked cyanide added reversibly to the carbon-carbon double bond and no product was observed. However, in the presence of acetone cyanohydrin, the reaction proceeded smoothly and quantitatively to products. This procedure has been applied to the hydrocyanation of cholestenone [Reaction (19)] (Liotta and Dabdoub, unpublished results). The results are summarized in Table XVI. Excellent yields of hydrocyanation products were obtained in benzene and acetonitrile and no cyanide hydrolysis products or dimers were detected. The lower the temperature, the greater was the preference for formation of the β isomer. At ambient temperatures, an α/β ratio of 1/10 was obtained in benzene. These stereochemical results

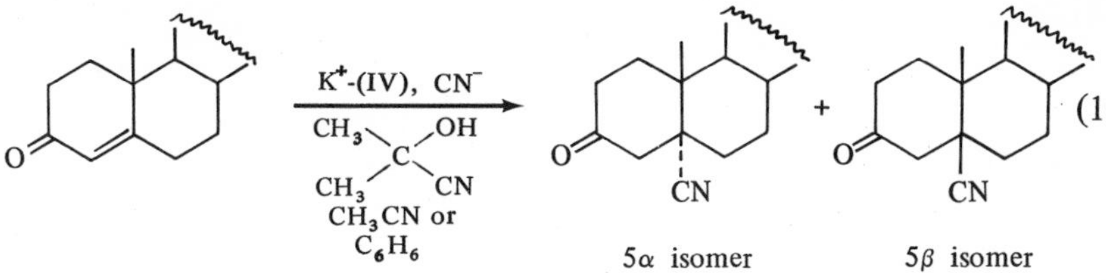

(19)

appeared to be equal to or superior to the results obtained with other hydrocyanating reagents (Nagata *et al.*, 1961, 1963a, b, 1967, 1972a-d; Rodig and Johnston, 1969; Meyer and Schnautz, 1962; Meyer and Wolfe, 1964; Bowers, 1961; Henbest and Jackson, 1967; Nagata, 1961; Marshall and Johnson, 1962; Johnson and Keane, 1963; Wenkert and Strike, 1964; Fishman and Torigoe, 1965; Barton *et al.*, 1968; Wieser *et al.*, 1969).

Cinquini *et al.* (1975) have reported the use of cryptate (IX)c, phosphonium salt (XIV), and crown (V) as liquid-liquid phase transfer catalysts in the reaction of potassium cyanide with 1-chlorooctane. The results are summarized in Table XVII. All three catalysts appear to be equally effective.

TABLE XVII

Reaction of 1-Chlorooctane with Potassium Cyanide under Liquid-Liquid Phase Transfer Catalytic Conditions

Substrate[a]	Reagent	Catalyst[b]	Temp. (°C)	Time (hr)	Yield (%)	Product
1-C_8H_{17}-Cl	KCN	(IX)c″	80	5	93	1-C_8H_{17}-CN
1-C_8H_{17}-Cl	KCN	(XIV)	80	5	94	1-C_8H_{17}-CN
1-C_8H_{17}-OSO_2CH_3	KCN	(V)	100	0.3	90	1-C_8H_{17}-CN

[a]The reactions were carried out with a saturated aqueous solution of the reagent.
[b]Numbers refer to compounds in Fig. 1 and in text.

D. Nitrite Ion

Zubrick *et al.* (1975) have reported the reaction of naked nitrite with primary alkyl halides to form nitro compounds as the major product, the major byproducts being nitrite esters. The results are summarized in Table XVIII. The reactions were run at temperatures ranging from 25 to 40°C with the molar ratio of halide:nitrite:crown of 20:22:1. The yields and product distributions were similar to, although slightly lower than, nitrite displacement in DMSO. Unfortunately, the authors failed to relate to the reader the nature of the nitrite salt used in the reaction, the reaction times, the reaction solvent, or the workup procedure.

TABLE XVIII

Reactions of Naked Nitrite with Organic Substrates in the Presence of 18-Crown-6 (IV)

Substrate	Product	Yield (%)
1-Bromooctane	1-Nitrooctane	65-70
1-Iodooctane	1-Nitrooctane	50-55
Cyclohexyl bromide	Nitrocyclohexane	0-3
2-Bromoethylbenzene	2-Nitroethylbenzene	32
3-Bromo-1-phenylpropane	3-Nitro-1-phenylpropane	51
Benzyl chloride	α-Nitrotoluene	34

E. Nitrate Ion

Knöchel *et al.* (1974) reported the use of (VI) in the reaction of an acetylated halosugar with a series of alcohols catalyzed by silver nitrate [Reaction (20)].

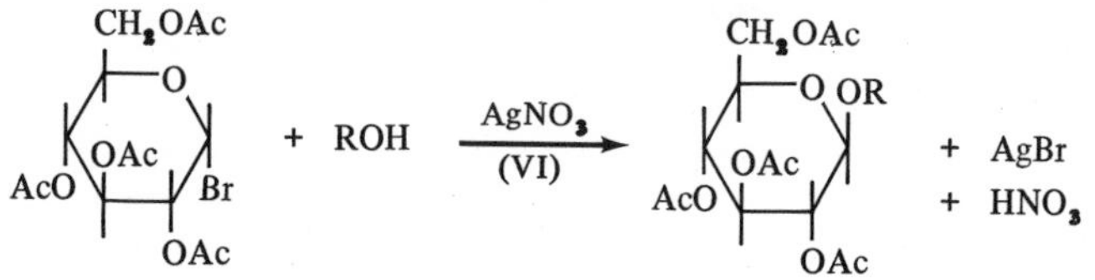

(20)

This Koenigs-Knorr reaction proceeded in good yield with inversion of configuration. The results are summarized in Table XIX.

TABLE XIX

Reaction of Acetylated Halosugar with a Series of Alcohols Catalyzed by Silver Nitrate and Dibenzo-18-Crown-6 (VI)

Alcohol	Temp (°C)	Time (min)	Yield (%)
Methanol	0	1-2	81
Isopropanol	0	1-2	65
t-Butanol	Ambient	3-5	57
Cyclohexanol	Ambient	3-5	43

In an extension of this work, Knöchel and Rudolph (1974) studied the comparative effect of macrocyclic polydentate ligands (VI), (VIII)c, and (IX)c on the ratio of β-glucoside to β-nitrate ester formed in the Koenigs-Knorr reaction

[Reaction (21)]. The results are summarized in Table XX. Only β-glucoside

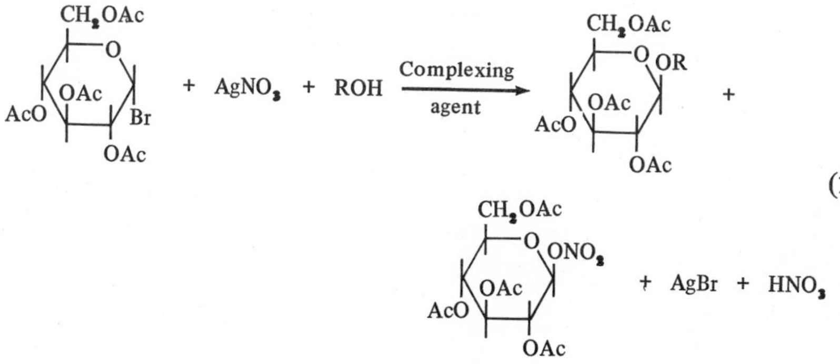

(21)

TABLE XX

Ratios of β-Glucoside to β-Nitrate Ester in the Reaction of an Acetylated Halosugar with a Series of Alcohols in the Presence of Silver Nitrate and Macrocyclic Polydentate Ligands

Run	Alcohol	Macrocyclic ligand[a]		
		(VI)	(VIII)c	(IX)c
1	Methanol	100/0	100/0	97.9/2.1
2	Isopropanol	100/0	100/0	88.8/11.2
3	*t*-Butanol	100/0	95.1/4.9	63.6/36.6
4	Cyclohexanol	100/0	97.7/2.3	73.1/26.9
5	Diglyme	–	0/100	0/100

[a]Numbers refer to compounds in Fig. 1.

was obtained when (VI) was used. Small quantities of nitrate ester were observed with (VIII)c, whereas substantial quantities of nitrate ester were obtained with (IX)c. In general, the more sterically hindered alcohols tended to give greater proportions of nitrate ester. This may be attributed to a steric inhibition of nucleophilicity. In diglyme, a nonnucleophilic solvent, nitrate ester was formed exclusively (Run 5, Table XX).

F. Oxygen Nucleophiles and Bases

Pedersen (1967b) reported that sterically hindered esters of 2,4,6-trimethylbenzoic acid were saponified with the potassium hydroxide-(V) complex in aromatic hydrocarbons [Reaction (22)]. The results are summarized in Table XXI. The same reaction carried out in hydrolytic solvents gave

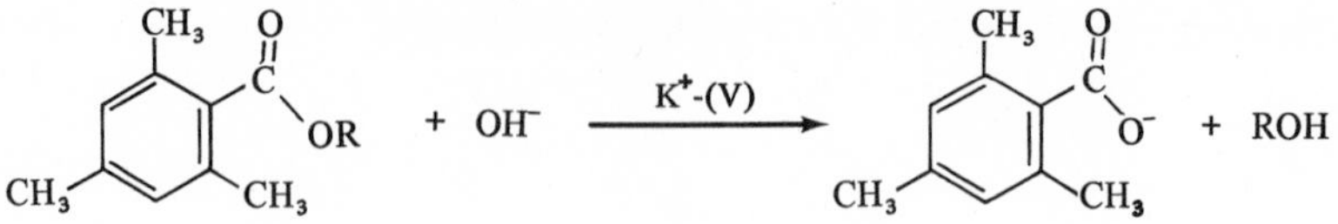

TABLE XXI

Saponification of Esters of 2,4,6-Trimethylbenzoic Acid with Potassium Hydroxide Complex of 18-Crown-6 (V)

		Concentration[a] (M)				
Solvent	Ester	Ester	Base	Temp. (°C)	Time (hr)	Hydrolyzed[b] (%)
Toluene	Methyl	0.122	0.122	73.8 ± 0.2	31	58.4
Benzene	Methyl	0.061	0.154	80-80.1	5	39
Toluene	Methyl	0.120	0.120	99.9 ± 0.1	2	53.0
Toluene	Methyl	0.061	0.154	105-111	5	93
Benzene	*t*-Butyl	0.060	0.154	76.6 ± 80.2	5	22
Toluene	*t*-Butyl	0.060	0.154	104-111	5	94
Benzene	Neopentyl	0.060	0.154	80.3-80.7	5	40

[a]Initial concentration.
[b]No saponification measurable by this method occurred when the methyl ester was refluxed at 75.5-77°C with excess of potassium hydroxide in 1-propanol for 5 hr.

essentially no product. That this reaction occurred by an acyl-oxygen cleavage mechanism is clear from the fact that the 1-menthol isolated from the hydrolysis of its 2,4,6-trimethylbenzoate ester with KOH-(V) was completely optically pure. While the KOH-(V) complex in toluene produced a 58% yield of hydrolysis products from methyl 2,4,6-trimethylbenzoate in 31 hr at 73.8°C, Dietrich and Lehn (1973) have shown that by replacing the crown by cryptate (IX)c a 70% yield was achieved in 12 hr at 25°C.

Sam and Simmons (1974) have reported that, using the procedure described by Pedersen for the formation of KOH-(V), only 11% of the anions in toluene were hydroxide. The major species was methoxide ion. Reaction of this solution with *o*-dichlorobenzene at 90°C for 16 hr produced a 40-50% yield of *o*-chloroanisole [Reaction (23)]. No phenols, diphenyl ethers, or *m*-chloroanisole

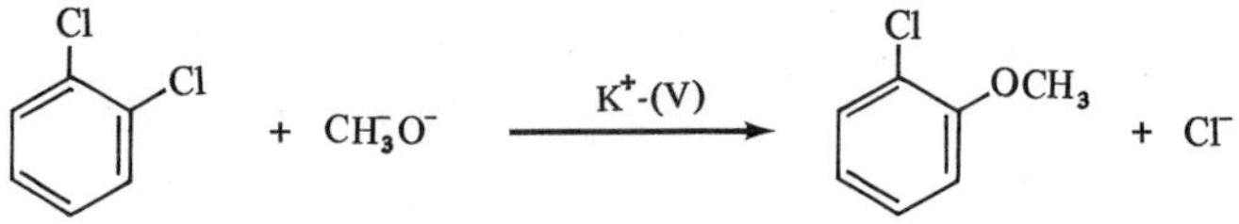

(23)

were detected. Reaction with *m*-dichlorobenzene produced only *m*-chloroanisole in low conversions [Reaction (24)]. No benzyne products were observed. The

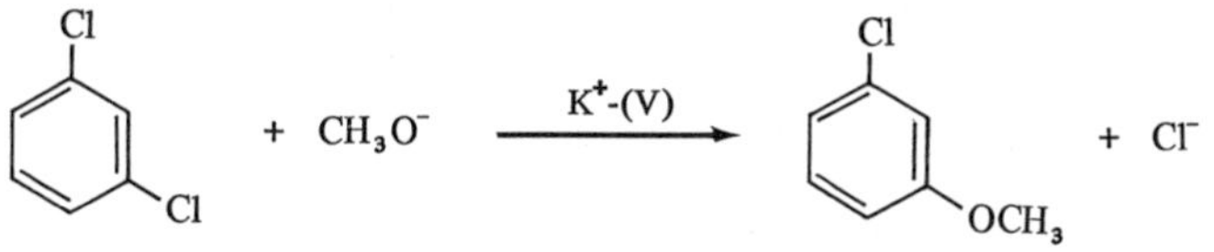

(24)

authors concluded that the reactions proceeded via an addition-elimination aromatic nucleophilic substitution mechanism [Reaction (25)]. In the absence of (V) no reaction took place.

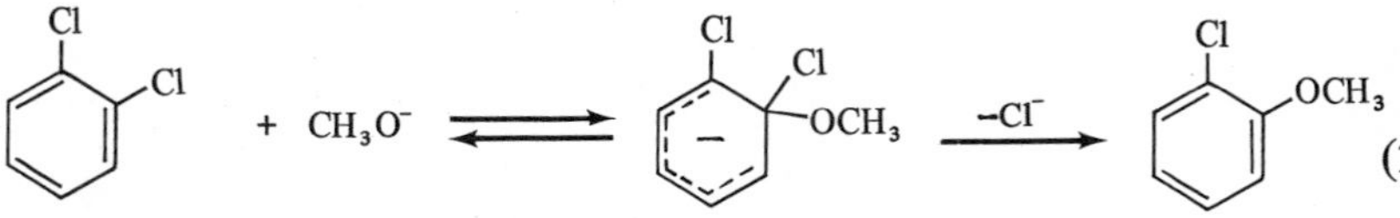

(25)

Thomassen *et al.* (1971) have investigated the rates of alkylation of potassium phenoxide with 1-bromobutane in dioxane at 25°C in the presence of a variety of linear and cyclic polyether additives [Reaction (26)]. The results are

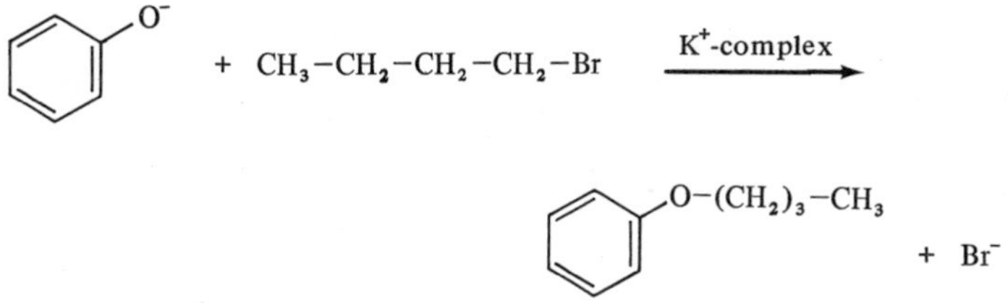

(26)

summarized in Table XXII. Addition of tetraglyme (tetra-EGDME) produced only a modest increase in rate (11) compared to pure dioxane (Runs 1-5, Table XXII) and dimethylsulfoxide had only a negligible effect (Run 32, Table XXII). In contrast to this, all of the crown ethers reported in this study had a considerably greater effect on the rate. In a comparison of the dibenzo-3*n*-crown-*n* series (where *n* is the number of oxygen atoms in the ring) at the same crown/1-bromobutane ratio, the rate of reaction varied with *n* in the following way: $4 < 5 < 6 >> 8$ (Runs 6, 8, 13, and 17, Table XXII). This indicated that the 18-membered crown has the best cavity size for accommodating potassium ion, a conclusion supported by a host of other studies. In a comparison of the various 18-membered crowns reported in this study, it was observed that the effectiveness of the crown varies as follows: dibenzo < monobenzo < dicyclohexo (Runs 13, 20, and 21, Table XXII). This is the same

TABLE XXII

Effect of Polyether Additives in the Reaction Between Potassium Phenoxide and Butyl Bromide in Dioxane[a]

$$C_6H_3O^-K^+ + n\text{-BuBr} \rightarrow C_6H_5OBu + KBr$$

Run	Additive	$\frac{[\text{Additive}]}{[\text{phenoxide}]_0}$ (%)	$r_i[\text{BuBr}]_0$ ($\times 10^5$ sec^{-1})
1	None	–	0.01^5
2	Tetra-EGDME	5.4	0.015
3		20	0.030
4		62	0.071
5		100	0.11
6	Dibenzo-12-crown-4	5.4	0.023
7	Dibenzo-15-crown-5	1	0.011
8		5.4	0.023
9		20	0.062
10		50	0.15
11		76	0.22
12	Dibenzo-18-crown-6 (VI)	1	0.11
13		5.4	0.37
14		20	1.9
15		30	3.1
16	Dibenzo-24-crown-8	1	0.032
17		5.4	0.075
18		15	0.20
19	Benzo-15-crown-5 (III)	5.4	0.043
20	Benzo-18-crown-6	5.4	1.00
21	Dicyclohexo-18-crown-6 (V)	5.4	2.80
22		30	15.0
23		49	33.0
24		100	87.0
25		128	104
26		153	112
27		206	117
28	Bis(*t*-butylbenzo)-18-crown-6	5.4	0.33
29		100	13.0
30		153	15.0
31		201	15.0
32	Dimethyl Sulfoxide	100	0.014

[a]Temperature: 25°C $[\text{BuBr}]_0$ = 0.05 or 0.10 mole/liter; $[C_6H_5OK]_0$ = 0.020 mole/liter; r_i is the initial consumption of phenoxide in moles per liter per second.

direction as that predicted for the increase of base strength of the oxygens of the crown. These results, of course, are in agreement with those reported by Pedersen (1967a, b) and by Pedersen and Frensdorff (1972) several years earlier. The rate of Reaction (26) in pure dioxane solvent was compared with that in pure tetraglyme. The results are shown in Table XXIII. It is interesting that the same rate enhancement was achieved in dioxane in the presence of 0.05 *M* (V) (Run 3, Table XXIII). Of equal interest was that in the presence of an equivalent amount of tetra-*n*-butylammonium bromide, the rate was increased to an even greater degree (Run 4, Table XXIII).

TABLE XXIII

Effect of Solvents and Additives in the Alkylation of K-Phenoxide with *n*-Butyl Bromide[a]

Run	Solvent	Additive[b]	$r_i[BuBr]_0$ ($\times 10^5$ sec^{-1})
1	Dioxane	–	0.01
2	Tetra-EGDME	–	150
3	Dioxane	(V)	150
4	Dioxane	Bu_4NBr[c]	400

[a] $[Phenoxide]_0$ = 0.025 mole/liter; temp. 25°C; r_i is the initial consumption of phenoxide in moles per liter per second.

[b] Concentration of additive is 0.05 *M*.

[c] Additions of a slight excess of Bu_4NBr to a solution of K-phenoxide in dioxane leads to formation of Bu_4N-phenoxide and precipitation of KBr.

It has been reported that in the reaction of *t*-BuOK with 2-nitrochlorobenzene and 4-nitrochlorobenzene in *t*-BuOH the *ortho* isomer reacts approximately 100 times faster than the *para* isomer. The *ortho-para* rate ratio is reported to be even greater for the corresponding fluoro compounds (300). Cima *et al.* (1973) have advanced the following explanation to account for these facts. Since *t*-BuOH is a solvent of low polarity, ion pairing becomes an important mechanism for stabilizing developing charges during the course of reaction. If it is assumed that these aromatic nucleophilic substitutions proceed via an anionic sigma complex (Meisenheimer complex), then the intermediate in *ortho* substitution is stabilized by an ion-pair interaction with the proximate *o*-nitro substituent (XVII). This

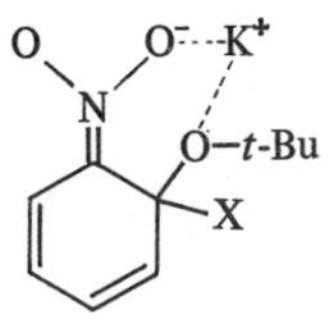

(XVII)

type of stabilization is not available during the *para* substitution process. That this explanation is quite reasonable is substantiated by experiments carried out in the presence and in the absence of (V). The results are summarized in Table XXIV. Upon addition of a molar equivalent of crown, the rate of reaction of

TABLE XXIV

Second-Order Rate Coefficient[a] for Reactions of 1-Fluoro-2- or 1-Fluoro-4-Nitrobenzene with Potassium Methoxide or *t*-Butoxide in the Presence or Absence of Dicyclohexo-18-Crown-6 (V) or with Benzyltrimethylammonium *t*-Butoxide in the Corresponding Lyate Solvents[b]

Run	[Substrate]/*M*	[Nucleophile]/M	(V)*M*	10^4k 1 mol^{-1} sec^{-1}	k_o/k_p
1	*o*-F, 0.35	$KOBu^t$, 0.35		54	3.6×10^2
2	*p*-F, 0.44	$KOBu^t$, 0.44		0.15	
3	*o*-F, 0.22	$KOBu^t$, 0.22	0.22	1.7×10^2	0.8
4	*p*-F, 0.21	$KOBu^t$, 0.21	0.21	2.2×10^2	
5	*o*-F, 0.33	$PhCH_2NMe_3OBu^t$, 0.32		1.7×10^2	7.4
6	*p*-F, 0.34	$PhCH_2NMe_3OBu^t$, 0.33		23	
7	*o*-F, 0.24	KOMe, 0.24		5.0	1.0
8	*p*-F, 0.24	KOMe, 0.24		4.7	
9	*o*-F, 0.24	KOMe, 0.24	0.25	4.2	0.6
10	*p*-F, 0.24	KOMe, 0.24	0.25	7.0	

[a]*k* = rate/[substrate] [nucleophile].
[b]Temperature 31°C unless otherwise stated.

o-nitrofluorobenzene increased by a small factor (≈3) (Runs 1 and 3, Table XXIV). In contrast to this, the corresponding reaction of the *para* isomer increased by a factor of 2000 (Runs 2 and 4, Table XXIV). The small rate increase in the *ortho* case was easily rationalized. The potassium ion was effectively solvated by either the *o*-nitro substituent (XVII) or the crown, both processes enhancing the reactivity of the *t*-butoxide. With the *para* substitution case, on the other hand, the ion-paired potassium *t*-butoxide was the reactive species in the absence of crown, whereas the more naked (and more reactive) *t*-butoxide anion was the reactive species in the presence of crown. The *ortho-para* rate ratio was greater than 300 in the absence of crown (Runs 1 and 2, Table XXIV) and only 0.8 in the presence of crown (Runs 3 and 4, Table XXIV), an observation consistent with the bridged ion-pair mechanism discussed earlier. When a quaternary ammonium ion was used in place of the metal, the *ortho para* rate ratio was 7.4 (Runs 5 and 6, Table XXIV). This value was considerably smaller than that observed with potassium ion. It can be seen from Table XXIV that crown had very little effect on the *ortho-para* rate ratios of potassium methoxide in methanol (Runs 7-10, Table XXIV). This was understandable in that methanol is a solvent of somewhat greater polarity than *t*-butanol and, as a result, free ions would be the predominant reactive species.

The data in Table XXV are instructive with regard to the notion of inherent nucleophilicities of alkoxide ions in their lyate solvents. A comparison was

TABLE XXV

Relative Reactivities of Some *t*-Butoxide Salts and Potassium Methoxide, in their Lyate Solvents, with 1-Fluoro-2- or 1-Fluoro-4-Nitrobenzene

	Cation		
Substrate	K^+	Dibenzo-18-crown-6 (V)	$PhCH_2NMe_3^+$
o-$NO_2C_6H_4F$	5	30	30
p-$NO_2C_6H_4F$	0.03	30	5

made between the rates of reaction of *t*-BuO^- with 2-nitrofluorobenzene and 4-nitrofluorobenzene with potassium, potassium-(V), and quaternary ammonium as the counter cations. Because each of the alcohol solvents varied dramatically in polarity, the extent of ion pairing and/or ion aggregation varied, thus masking the true nucleophilicity of the alkoxide ion. Potassium methoxide was 30 times more reactive than KO-*t*-Bu but in the presence of crown the reverse was true. When a quaternary ammonium ion was used in place of the metal, the difference was somewhat less (5), implying that even with the quaternary ammonium ion, aggregation might have been a contributing factor.

The rates of reaction of a number of oxygen nucleophiles with *p*-nitrophenyl phosphinate and with *p*-nitrophenyl benzoate in toluene containing (V) have been reported by Curci and DiFuria (1975). The anions studied included *t*-Bu-O-OK, *t*-BuOK, *n*-Bu-O-OK, and *p*-CH_3-C_6H_4-OK. Rates of nucleophilic displacement on *p*-nitrophenyl phosphinate by H-O-O^- and *t*-Bu-O-O^- were also determined in water. In all cases the nucleophilic rate was greater in toluene than in water. Comparison of the nucleophilic reactivity of the α anion *t*-Bu-O-O^- with oxygen nucleophiles of comparable base strength toward the foregoing substrates in toluene and water suggested that solvation was not a major factor in determining the α effect of the *t*-Bu-O-O^- anion.

Maskornick (1972) has investigated the isomerization of 2-methylbicyclo-[2.2.1]hepta-2,5-diene to 5-methylenebicyclo[2.2.1]hept-2-ene using excess KO-*t*-Bu in dimethylsulfoxide in the presence and in the absence of an equivalent concentration of (IV) [Reaction (27)]. In the absence of crown, the rate in KO-*t*-Bu changed from zeroth order at high concentrations to first order at more dilute concentrations. In the presence of an equivalent amount of crown,

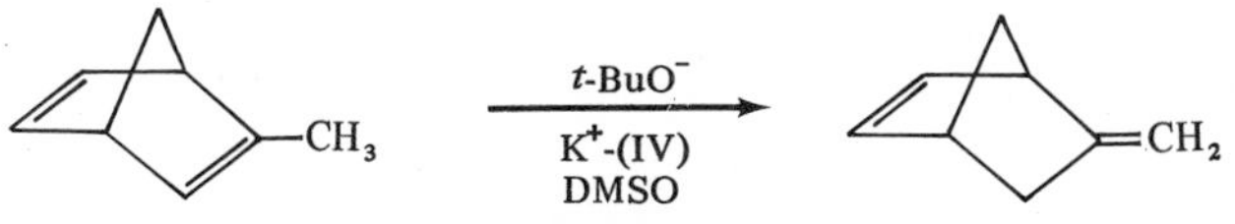

(27)

the rate in base remained first order at all concentrations studied. Addition of an equimolar amount of tetraglyme in place of the crown produced results similar to those obtained in pure dimethylsulfoxide. Maskornick interpreted these results in terms of the crown's ability to break up the ion aggregates of KO-*t*-Bu to form monomeric base species. These results vividly demonstrated that even in a highly solvating medium for cations, crown effectively improved the activity of this potassium salt.

Dietrich and Lehn (1973) have generated a number of highly basic species in dipolar and nonpolar aprotic solvents using the cryptate (IX)c. Reactions (28)-(30) illustrate the use of these systems in generating delocalized carbanions.

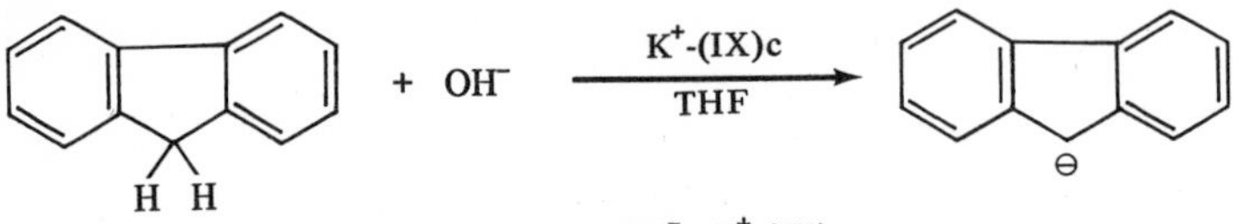

(28)

$$Ph_3C{-}H \xrightarrow[\text{THF}]{OH^-,\ K^+\text{-(IX)c}} \quad \xrightarrow[\text{THF}]{NH_2^-,\ K^+\text{-(IX)c}} \quad \xrightarrow[C_6H_6 \text{ or } C_6H_5\text{-}CH_3]{t\text{-amylate}^-,\ Na^+\text{-(IX)c}} \quad \xrightarrow[\text{Hexane}]{n\text{-Bu}^-,\ Li^+\text{-(IX)c}} Ph_3C^{\ominus} \qquad (29)$$

$$Ph_2CH_2 \xrightarrow{\text{Same as (29)}} Ph_2CH^{\ominus} \qquad (30)$$

The anions were easily trapped by reaction with benzyl chloride. The authors also reported the preparation of an 0.8 *M* dimethylsulfoxide solution of potassium hydroxide using (IX)c and the use of this solution in saponifying methyl mesitoate.

The use of oxyanions and related ionic bases in promoting stereospecific and regiospecific elimination reactions on organic substrates represents an important and useful synthetic approach for the introduction of unsaturation into organic molecules. For several decades a large number of workers have been concerned with problems related to these elimination processes. They have concluded that the chemical course of reaction is dictated, to a large extent, by the state of aggregation of the basic ionic species, and that this was dependent on concentration, solvent, and temperature. With the advent of crown ethers and their use in probing the mechanisms of elimination reactions, new and interesting insights have emerged to aid in our understanding of this fundamental process.

The effect of ionic association upon positional and geometrical orientation in base-promoted β-elimination processes from 2-butyl halides and *p*-toluenesulfonate systems in solvents of low polarity has been extensively explored by Bartsch (1975). The results of the reactions of 2-bromobutane with a number of base-solvent systems to produce 1-butene and *cis*- and *trans*-2-butene [Reactions (31)] are summarized in Table XXVI (Bartsch *et al.*, 1973). In the

TABLE XXVI

Alkene Products from Reactions of 2-Bromobutane with Potassium Alkoxides in Various Solvents at 50.0°C

Run	Base-Solvent	[Base] (*M*)	[2-BuBr] (*M*)	% of 1-butene of total butenes	*trans*-2-Butene: *cis*-2-Butene
1	MeOK-MeOH	0.25	0.10	15.4	3.34
2	MeOK-MeOH	0.50	0.10	15.4	3.34
3	MeOK-MeOH	1.00	0.10	15.4	3.41
4	EtOK-EtOH	0.25	0.10	17.9	3.23
5	EtOK-EtOH	0.50	0.10	17.9	3.21
6	EtOK-EtOH	1.00	0.10	18.1	3.22
7	*t*-BuOK-*t*-BuOH	0.10	0.10	37.7	1.86
8	*t*-BuOK-*t*-BuOH	0.25	0.10	41.6	1.78
9	*t*-BuOK-*t*-BuOH	0.50	0.10	44.1	1.66
10	*t*-BuOK-*t*-BuOH	1.00	0.10	50.6	1.47
11	*t*-BuOK-*t*-BuOH	1.00	0.25	49.8	1.47
12	*t*-BuOK-*t*-BuOH	1.00	0.51	49.9	1.48
13	*t*-BuOK-DMSO	0.10	0.09	30.5	3.04
14	*t*-BuOK-DMSO	0.25	0.09	30.3	3.12
15	*t*-BuOK-DMSO	0.50	0.09	30.6	3.16
16	*t*-BuOK-DMSO	1.03	0.09	30.5	2.99

reaction of *t*-BuOK in *t*-BuOH, the positional and geometrical distribution of products was dependent on base concentration (Runs 7-12, Table XXVI), while

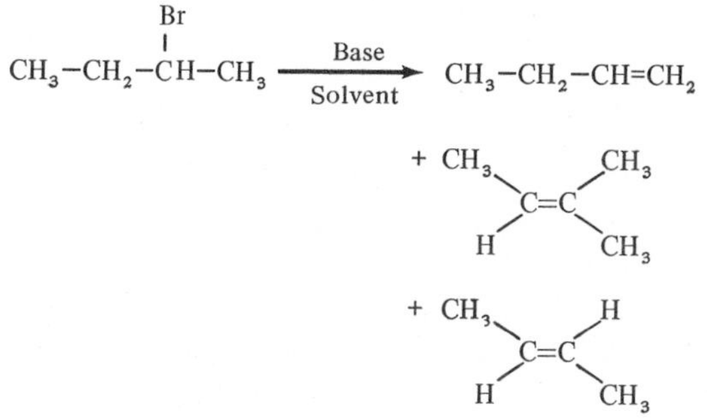

(31)

the corresponding reactions promoted by MeOK-MeOH, EtOK, EtOH, or *t*-BuOK-DMSO were essentially insensitive to base concentration (Runs 1-6 and 13-16, Table XXVI). The results were rationalized in terms of base association. In solvents of low dielectric constant, such as *t*-BuOH (D=11.2 at 30°), the ionic species existed as a combination of free ions, ion pairs, and ion aggregates [Reaction (32)], each of which was heavily solvated by *t*-BuOH. As

$$RO^- + M^+ \longrightarrow RO^-M^+ \longrightarrow (RO^-M^+)_n \tag{32}$$

the concentration of base increased, the state of aggregation increased and the effective size of the basic species promoting the β elimination increased. This, in effect, caused the change in the distribution of positional and geometrical products summarized in Table XXVI. That this was a most reasonable rationalization was supported by the data in Table XXVII, in which various additives of different donicity number were added to the *t*-BuOK-*t*-BuOH base-solvent system (Bartsch *et al.*, 1973). The more effectively the additive coordinated

TABLE XXVII

Alkene Products from Reaction of 2-Bromobutane with 0.50 *M tert*-Butoxide in *t*-BuOH at 50.0°C

Run	Base	Additive	Donicity no. of additive	% of 1-butene of total butenes	*trans*-2-Butene: *cis*-2-butene
1	*t*-BuOK	None		44.1	1.66
2	*t*-BuOK	0.28 *M* Dicyclohexo-18-crown-6 (V)		32.5	2.92
3	*t*-BuOK	0.50 *M* Hexamethyl-phosphortriamide	38.8	36.3	2.62
4	*t*-BuOK	0.50 *M* Dimethyl-formamide	30.9	37.7	2.26
5	*t*-BuOK	0.50 *M* Dimethyl-sulfoxide	29.8	40.5	2.42
6	*t*-BuOK	0.50 *M* Tetramethylene sulfone	14.8	43.4	1.92
7	*t*-BuOK	None		41.6	1.78
8	*t*-BuON(*n*-Pr)$_4$	None		31.3	2.99

with the cation, the more the distribution of products was shifted toward the distribution observed in solvents of higher polarity, where aggregates became less important. From Table XXVII it appears that dicyclohexo-18-crown-6 effectively coordinates with the potassium ion to break up aggregates, shifting Reaction (32) to the left (Run 2, Table XXVII). It is interesting to note that the distribution of products was almost identical for the quarternary ammonium salt of *t*-BuO and for *t*-BuOK with crown (Runs 2 and 8, Table XXVII). The effect of ionic aggregation in solvents of varying dielectric constant and in the presence of crown ether is most drastically illustrated by the examples in Table XXVIII (Bartsch *et al.*, 1973). For solvents of high dielectric constant, such as

TABLE XXVIII

Alkene Products from Reaction of 2-Bromobutane with 0.25 *M* ROK in ROH at 50.0°C in the Presence and Absence of 0.25 *M* Dicyclohexo-18-Crown-6 (V)

	(V) Absent		(V) Present	
ROH of ROK-ROH	% of 1-butene of total butenes	*trans*-2-butene: *cis*-2-butene	% of 1-butene of total butenes	*trans*-2-butene: *cis*-2-butene
Methanol	15.4	3.34	16.3	3.43
Ethanol	17.9	3.23	17.6	3.31
1-Propanol	19.1	3.07		
1-Hexanol	20.3	2.92		
1-Octanol	22.7	2.75		
2-Propanol	23.7	2.72	23.8	2.95
2-Butanol	25.9	2.69		
3-Pentanol	30.7	2.50		
2-Methyl-2-propanol	41.6	1.78	31.2	3.20
2-Methyl-2-butanol	48.4	1.46		
3-Methyl-3-pentanol	57.4	1.02		
3-Ethyl-3-pentanol	63.1	1.04	37.3	2.82

methanol, there appears to be little difference in product distribution with or without crown. However, as the solvent system becomes less polar, rather startling differences begin to emerge, indicating that crown is acting effectively in breaking up ion aggregates [shifting Reaction (32) to the left].

The reaction of 2-butyl *p*-toluenesulfonate with a variety of base-solvent systems is summarized in Table XXIX. In solvents of high dielectric constant (EtOH, DMSO), *trans*- to *cis*-2-butene ratios were high, whereas in *t*-BuOH this ratio was quite small. The presence of crown, however, produces a ratio similar to that observed in the solvents of high dielectric solvent. A steric argument has been presented by Bartsch to rationalize the *trans* to *cis* ratio trends.

That *t*-BuOK in *t*-BuOH exists as aggregates was suggested by a number of studies. Partial molar volume measurements indicated that *t*-BuOK is monomeric only at concentrations below 10^{-3} *M* (Liotta *et al.*, unpublished results). The conductivity of *t*-BuOK in *t*-BuOH and benzene increased markedly upon addition of crown (Saunders *et al.*, 1968).

TABLE XXIX

Alkene Products from Reactions of 2-Butyl *p*-Toluenesulfonate with Various Base-Solvent Systems at 50.0°

Base	Solvent	% of 1-butene of total butenes	*trans*-2-butene: *cis*-2-butene
EtOK	EtOH	35	1.95
t-BuOK	*t*-BuOH	63.5	0.40
t-BuOK[a]	*t*-BuOH	53.6	1.88
t-BuOK	DMSO	57.3	2.21

[a]0.29 *M* Dicyclohexo-18-crown-6 (V).

The reactions of the 2-butyl systems discussed earlier represent concerted anti-β-elimination processes. However, substrates are known which undergo competing syn-β-elimination processes. Sicher (1972) has proposed the following transition state for the syn-elimination process. This process involves a concerted six-centered transition state in which the departing anionic leaving group is coordinated with the cationic portion of the base [Structure (XVIIa)]. It would be anticipated that such a transition state should become important in solvents of low polarity where the interactions of the solvent molecules with the departing anion would be rather weak, thus requiring the assistance of the cation in stabilizing the incipient negative ion. The anti process [Structure

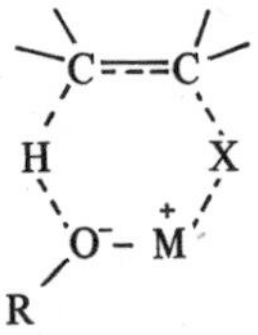

(XVII)a

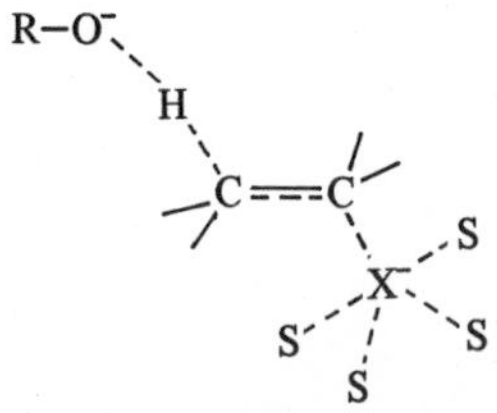

(XVII)b

(XVIIb)], on the other hand, should tend to dominate in solvents of high polarity where the leaving group would be strongly solvated, thus requiring little help by the cation. Bartsch and Wiegers (1972) and Bartsch *et al.* (1974) have investigated competition between syn and anti elimination from *trans*-2-phenylcyclopentyl tosylate with *t*-BuOK in *t*-BuOH as illustrated in Reaction (33). The 1-phenylcyclopentene was produced by a syn elimination, while the 3-phenylcyclopentene was produced via an anti pathway. The results are summarized in Table XXX. Using 0.1 *M* *t*-BuOK, 89% syn elimination was

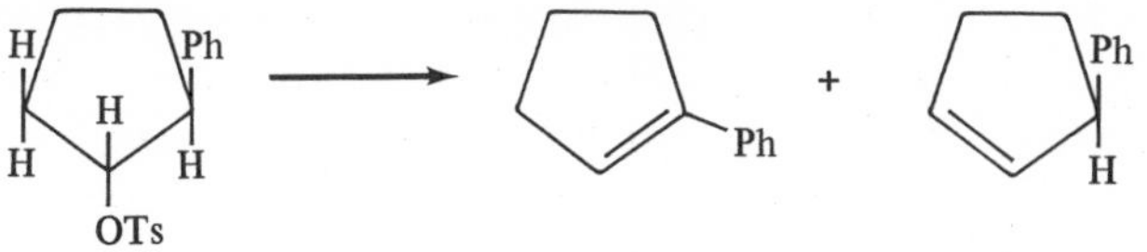

(33)

TABLE XXX

Effect of Dicyclohexo-18-Crown-6 (V) and Tetramethyl-12-Crown-4 on the Relative Proportions of Isomeric Phenylcyclopentenes Formed in Reactions of 0.025 *M* Tosylate with 0.10 *M* *t*-BuOK-*t*-BuOH at 50.0°

Crown ether	[Crown ether] (*M*)	Total phenylcyclopentenes (%)	
		1-Phenylcyclopentene	3-Phenylcyclopentene
		89.2 ± 0.5	10.8 ± 0.5
(V)	0.031	46.5 ± 0.3	53.5 ± 0.3
(V)	0.049	33.0 ± 0.2	67.0 ± 0.2
(V)	0.10	30.1 ± 0.7	69.9 ± 0.7
(V)	0.17	29.5 ± 0.3	70.5 ± 0.3
(V)	0.22	30.8 ± 0.4	69.2 ± 0.4
Tetramethyl-12-crown-4	0.10	90.7 ± 0.4	9.3 ± 0.4

realized; in the presence of an equivalent quantity of crown, however, the syn product was reduced to 30%. Addition of an equal number of moles of tetramethyl-12-crown-4—a crown whose cavity size is particularly well suited for complexation with Li^+—yielded a percentage of syn product nearly identical with that formed in the absence of crown.

Reaction of *exo*-2-norbornyl-exo-3-d tosylate with sodium salt of 2-cyclohexylcyclohexanol in triglyme [Reaction (34)] produced norbornene containing

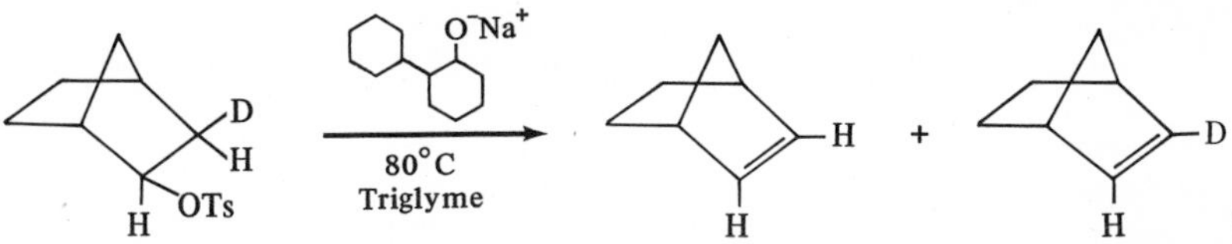

(34)

no deuterium. This represented an exclusive syn-elimination process. In the presence of (IV), 27% of the product contained deuterium (Bartsch and Kayser, 1974). Assuming an isotope effect of 6, the authors calculated a syn-anti rate ratio of 15:1. This contrasted sharply with a ratio of 100:1 for associated base. These experiments vividly demonstrated the effect of base association on syn- versus anti-elimination processes.

Base association effects have been studied using conformationally mobile systems. Reaction of 4,4,7,7-tetramethylcyclodecyl bromide and tosylate with *t*-BuOK in benzene, *t*-BuOH, and DMF produced alkenes which arose from both syn- and anti-elimination processes (Svoboda *et al.,* 1972). The cis products arose from anti elimination while the trans products arose from syn processes [Reaction (35a)]. The results are summarized in Table XXXI. In each of the

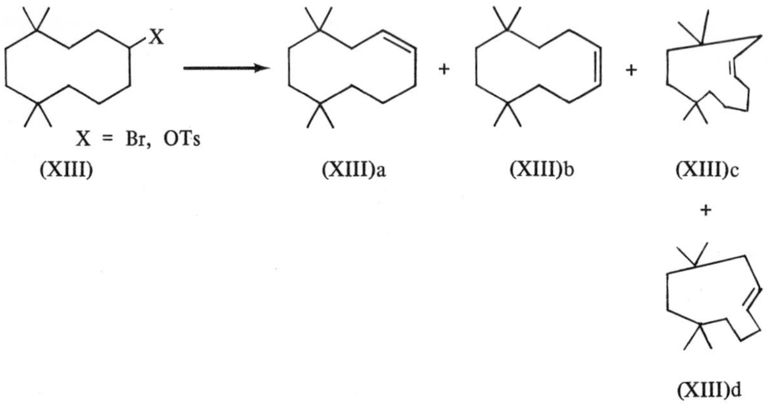

(35a)

TABLE XXXI

Effect of Dicyclohexo-18-Crown-6 (V) on Alkene Composition in Reaction of 4,4,7,7,-Tetramethylcyclodecyl Bromide and Tosylate with Potassium *t*-Butoxide in Different Solvents

X	Solvent	(XIII)d[a] (% trans)	(XIII)b[a] (% cis)	trans/cis	(XIII)c[a] (% trans)	(XIII)a[a] (% cis)	trans/cis
Br	Benzene	83.0	1.5	55.0	14.0	1.4	9.3
	Benzene + (V)	9.4	75.8	0.12	0.5	14.3	0.03
Br	*t*-BuOH	76.0	8.4	9.0	12.2	3.4	3.6
	t-BuOH + (V)	8.9	75.3	0.12	1.1	14.7	0.07
Br	DMF	7.0	72.0	0.10	1.0	20.0	0.05
	DMF + (V)	3.7	79.4	0.05	0.1	16.8	0.006
OTs	Benzene	88.8	1.4	63.4	8.8	1.0	8.8
	Benzene + (V)	12.0	58.4	0.20	0.5	29.1	0.02
OTs	*t*-BuOH	68.8	7.0	9.8	12.6	11.6	1.1
	t-BuOH + (V)	22.2	54.6	0.40	2.8	20.4	0.14
OTs	DMF	40.2	43.0	0.94	2.8	14.0	0.20
	DMF + (V)	5.8	68.7	0.08	0.3	25.2	0.01

[a]Numbers refer to structure in Reaction (35a).

nonpolar solvents studied (benzene and *t*-BuOH) (XIII) underwent predominantly syn elimination, whereas in the presence of crown the predominant elimination pathway was anti. This is dramatically represented by the trans/cis columns in Table XXXI. In the more polar solvent DMF the anti process dominated even in the absence of crown. Upon addition of crown, however, the anti pathway became even more important.

Reaction of cyclodecyl chloride with a number of base-solvent systems has been reported [Reaction (35b)] (Traynham *et al.*, 1967). The results are summarized in Table XXXII. Using the *t*-BuOK-DMSO system the anti pathway

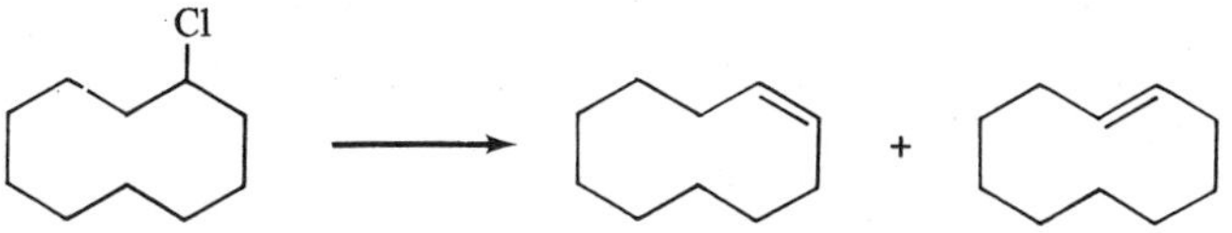

(35b)

TABLE XXXII

Olefinic Products from Reaction of Chlorocyclodecane with Various Base-Solvent Systems

Base-solvent	Conditions	% of total cyclodecenes: *trans*-Cyclodecene	% of total cyclodecenes: *cis*-Cyclodecene
t-BuOK-DMSO	5 min, room temp.	18 ± 1	82 ± 1
t-BuOK-DMSO	15 min, room temp.	6 ± 1	94 ± 1
t-BuOK-DMSO	2 hr, room temp.	4 ± 1	96 ± 1
$LiN(Cy)_2$-ether-hexane	24 hr, reflux	87 ± 2	13 ± 2
$LiN(Cy)_2$-ether-hexane-tetramethyl-12-crown-4 (0.3 *M*)	3 hr, reflux	69 ± 1	31 ± 1
$LiN(Cy)_2$-ether-hexane-tetramethyl-12-crown-4 (0.3 *M*)	24 hr, reflux	68 ± 3	32 ± 3
t-BuOK-*t*-BuOH	24 hr, 50°C	56 ± 2	44 ± 2
t-BuOK-*t*-BuOH-dicyclohexo-18-crown-6	2 hr, 50°C	62 ± 1	38 ± 1

dominated. It is interesting to note that the cis/trans product ratio changed with time, indicating a subsequent isomerization process. Reaction of cyclic chloride with lithium cyclohexyl amide in ethyl ether-hexane produced 13% anti-elimination product but this increased to 32% in the presence of a tetramethyl-12-crown-4 (Bartsch and Shelly, 1973). The presence of (V) with the KO-*t*-Bu-*t*-BuOH system produced a small inverse effect; the percent of product formed from the anti-elimination process actually decreased slightly. In general, however, it appeared that base association facilitates syn-elimination processes.

Zavada *et al.* (1972) have reported the reaction of *t*-BuOK with 5-decyl tosylate in benzene, *t*-BuOH, and DMF in the presence and absence of (V) [Reaction (36)]. The results are summarized in Table XXXIII. In the absence of

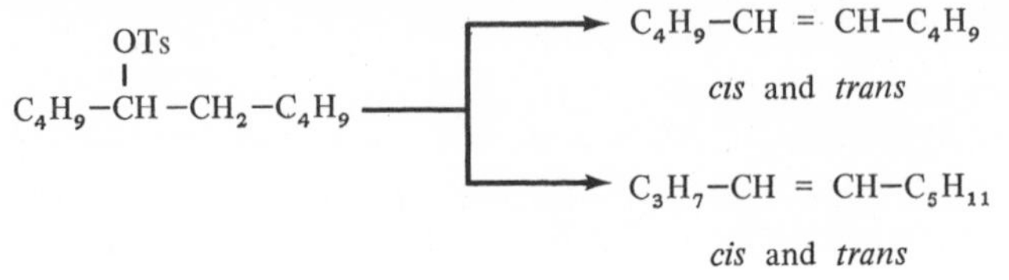

(36)

TABLE XXXIII

trans- to *cis*-Alkene Ratios[a] in Reaction of 5-Decyl Tosylate with Uncomplexed and Complexed Potassium-*t*-Butoxide in Different Solvents

Base	Benzene	*t*-Butanol	Dimethyl-formamide
t-BuOK	0.85	0.41	3.16
t-BuOK + (V)	2.12	2.54	3.16

[a] trans-4-Decene and 5-decene/cis-4- and 5-decene.

crown, the trans isomers were favored, whereas in the presence of an equimolar quantity of crown the cis isomers dominated. The trans/cis ratio in DMF appeared to be independent of added crown. It was concluded that ion-pair aggregates favored formation of trans products, whereas free ions produced primarily cis products. The contribution of the anti- and syn-mechanistic pathways were also investigated using this system. The results for the formation of *trans*- and *cis*-5-decene are summarized in Table XXXIV. It may be seen that the anti-

TABLE XXXIV

Effect of (V) on Contributions of syn and anti Pathways to *cis*- and *trans*-5-Decene Formation in Reaction of 5-Decyl Tosylate with Potassium-*t*-Butoxide

	Anti Pathway			Syn Pathway		
Conditions	% trans	% cis	trans/cis	% trans	% cis	trans/cis
Benzene	33.6	50.4	0.67	12.4	3.6	3.4
Benzene + (V)	63.9	29.2	2.19	4.1	2.8	1.5
t-BuOH	24.8	68.2	0.36	4.2	2.8	1.5
t-BuOH + (V)	67.1	26.7	2.51	4.7	1.5	3.1
DMF	73.2	22.6	3.24	2.8	1.4	2.0

elimination was the major process in all the solvent systems reported and that the cis product was favored with aggregated base, whereas trans product was favored in the presence of crown or in the dipolar aprotic solvent dimethylformamide. When a comparison was made between trans product formed from anti and syn processes, it was clear that the percentage of syn elimination decreased upon addition of crown to either benzene or *t*-butyl alcohol. In DMF, the percentage of syn process was very small and independent of added crown, in agreement with the aggregation arguements presented earlier. These results are summarized in Table XXXV (Bartsch, 1975). In contrast to this, a similar

TABLE XXXV

Stereochemistry of *trans*-5-Decene Formation in Eliminations from 5-Decyl-6-*d* Tosylate

Base-solvent system	Syn elimination (%)
t-BuOK-benzene	27
t-BuOK-benzene-(V)	6
t-BuOK-*t*-BuOH	15
t-BuOK-*t*-BuOH-(V)	7
t-BuOK-DMF	4
t-BuOK-DMF-(V)	4

comparison of anti and syn processes for the formation of the cis isomer (Table XXXVI) showed that syn elimination actually increased slightly with addition of crown.

TABLE XXXVI

Stereochemistry of *cis*-5-Decene Formation in Eliminations from 5-Decyl-6-*d* Tosylate

Base-solvent system	Syn elimination (%)
t-BuOK-benzene	6.67
t-BuOK-benzene-(V)	8.75
t-BuOK-*t*-BuOH	3.94
t-BuOK-*t*-BuOH-(V)	5.32
DMF	5.84

The stereochemical course of reaction of *cis*- and *trans*-1-methoxy-d_3-2-*d*-acenaphthene (XVIII and XIX) with *t*-BuO^- in *t*-BuOH has been shown to be dependent on the nature of the counter cation. The results are summarized in Table XXXVII

(Hunter and Shearing, 1971). The amount of syn elimination decreased in the

TABLE XXXVII

Effect of Cation on Relative Exchange and Elimination Rates of Deuterium-Labeled 1-Methoxyacenaphthene

Base-solvent system	Exchange k_{cis}/k_{trans}	Elimination k_{syn}/k_{anti}
t-BuOLi-*t*-BuOH	≥16	≥13
t-BuOK-*t*-BuOH	2.2	≥7
t-BuOCs-*t*-BuOH	0.9	≥5
t-BuON$(CH_3)_4$-*t*-BuOH	0.5	0.3
t-BuOK-*t*-BuOH-(V)	0.6	0.3
MeOK-MeOH	1.0	0.6
MeOK-MeOH-(V)	1.0	–

order $Li^+ > K^+ > Cs^+ > N(CH_3)_4^+ = K^+$-crown. Base association was also expected to decrease in this order.

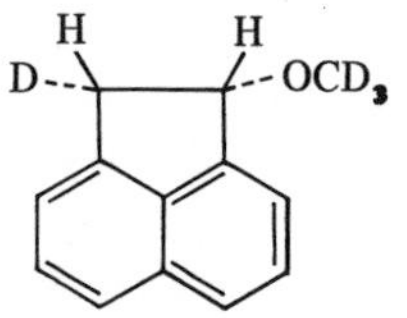

(XVIII)

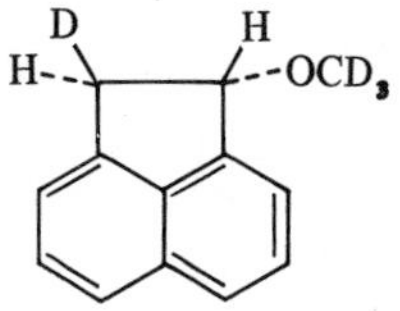

(XIX)

Table XXXVII also contains rates of cis and trans exchange on 1-methoxy-d_3-1,2,2-trideuterioacenaphene (XX). It is evident that these rate ratios

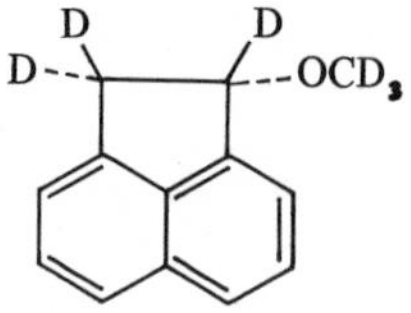

(XX)

paralleled the corresponding syn- to anti-elimination ratios. The authors have interpreted these results in terms of an elimination mechanism proceeding through a carbanion intermediate (ElcB). This is illustrated as follows:

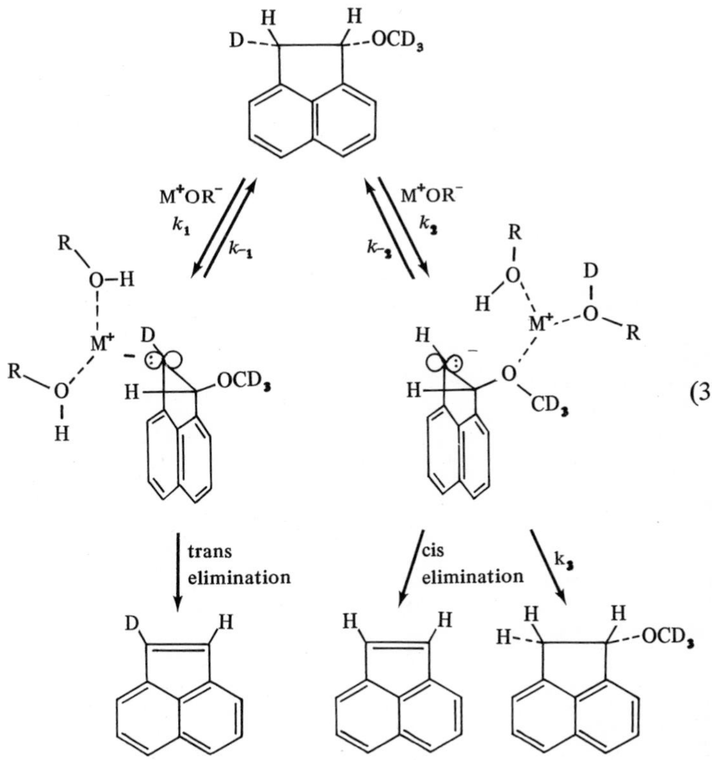
M⁺OR⁻
k₁
k₋₁
k₂
k₋₂
trans elimination
cis elimination
k₃
OCD₃

(37)

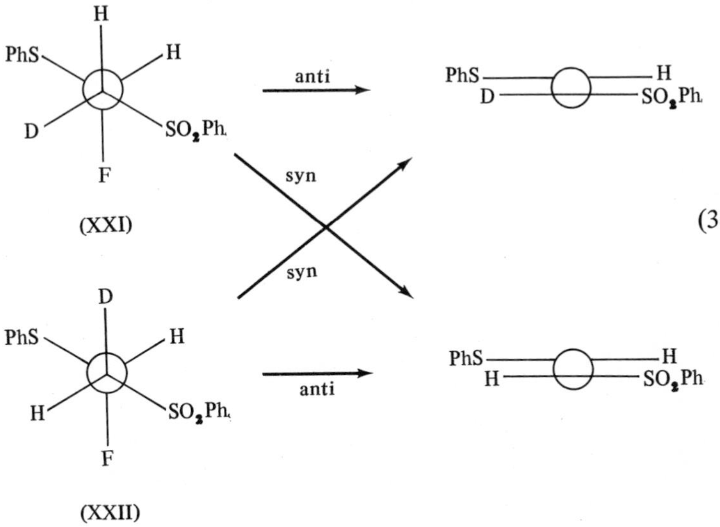
PhS
SO₂Ph
anti
syn
(XXI)
(XXII)

(38)

Fiandanese *et al.* (1973) have reported the effect of crown on the base-promoted syn- and anti-elimination pathways on the diastereomeric 1-deutero-2-fluoro-2-phenylthioethyl phenyl sulfones (XXI) ("threo") and (XXII) ("erythro"). The various mechanistic modes of elimination are illustrated in Reaction (38). The deuterium was retained in the anti-elimination process involving the threo isomer, while the deuterium was lost for the same process involving the diasteromeric erythro isomer. The exact opposite was true for the syn pathways. The data are summarized in Table XXXVIII. In the absence of crown the

TABLE XXXVIII

Stereochemical Course of Reactions of Diastereomeric 1-Deutero-2-Fluoro-2-Phenylthioethyl Phenyl Sulfones (XXI) and (XXII) with Potassium Phenoxide and Sodium *t*-Butoxide at 25°C

Substrate	Base	Solvent	Products anti	Products syn
(XXI)	PhO^-K^+	Dioxane	13	87
(XXI)	PhO^-K^+-(V)	Dioxane	58	42
(XXII)	PhO^-K^+	Dioxane	4	96
(XXII)	PhO^-K^+-(V)	Dioxane	50	50
(XXI)	*t*-BuO−Na^+	C_6H_6-*t*-BuOH (80-20)	77	23
(XXI)	*t*-BuO−Na^+-(V)	C_6H_6-*t*-BuOH (80-20)	79	21
(XXII)	*t*-BuO−Na^+-(V)	C_6H_6-*t*-BuOH (80-20)	33	67
(XXII)	*t*-BuO−Na^+-(V)	C_6H_6-*t*-BuOH (80-20)	38	62

[a]Numbers refer to structures in text.

syn-elimination pathway dominated, while in the presence of crown the *anti* pathway became more important. This suggested that the crown caused a separation of the potassium phenoxide ion pair. These results supported the six-centered transition state (XXIII) for the syn-elimination pathway. In

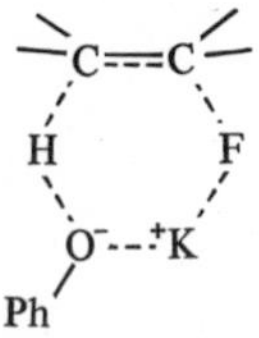

(XXIII)

contrast to this, the corresponding reaction with NaO-*t*-Bu showed little or no effect of crown on the stereochemical course of reaction. The authors concluded that "the lack of any significant crown ether effect suggests that ion-pairing and consequently base-leaving group interaction cannot be considered as the sole factor responsible for *syn*-contribution." Although this may be true, it was not entirely clear that the crown ether, under the conditions of the experiments, successfully broke up Na^+O-*t*-Bu ion pairing to any significant extent. Further work must be carried out to explore this point.

The reaction of 2-butyltrimethylammonium tosylate with *t*-BuOK in *t*-BuOH produced a distribution of products which appeared to be independent of the base concentration and the presence of crown. This sharp contrast with the 2-butyl bromide and tosylate systems was rationalized in the terms of the following equilibrium (39) (Borchardt and Saunders, 1974).

$$RO^-M^+ + R'N(CH_3)_3{}^+X^- \rightleftharpoons RO^- + R'N(CH_3)_3^+ + M^+X^- \qquad (39)$$

G. Superoxide

In spite of the availability of superoxide radical anion from commercial sources in the form of inexpensive potassium superoxide, KO_2, or sodium superoxide, NaO_2, and from electrochemical reduction of O_2 [Reaction (40)],

$$O_2 + e^- \rightarrow O_2^{\dot{-}} \quad -0.75v\ (SCE) \qquad (40)$$

the use of this species as a reagent for synthetic transformations has been limited (Merritt and Sawyer, 1970). This is mainly due to solubility problems associated with the K^+ and Na^+ salts and to the complex nature of the electrochemical technique, although the preparation of dialkyl peroxides has been reported using the latter approach [Reaction (41)] (Dietz *et al.*, 1970). Recently, in the study of the reaction of superoxide anion with a copper (II) complex in

$$2R{-}Br + 2O_2^{\dot{-}} \longrightarrow R{-}O{-}O{-}R + 2Br^- + O_2 \qquad (41)$$

aprotic solvents, Valentine and Curtis (1975) reported the successful solubilization of KO_2 in DMSO containing (V). In the absence of crown, KO_2 was only sparingly soluble in DMSO. Johnson and Nidy (1975) have used (IV), (V), and (VI) to dissolve KO_2 in benzene, tetrahydrofuran, and dimethylformamide and have reacted the resulting solutions of "naked" superoxide with primary and secondary alkyl bromides to form dialkyl peroxides. The results are summarized in Table XXXIX. The yields of dialkyl peroxides ranged from 40 to 80% and the major side products were reported to be alcohols and alkenes. In the reported procedure, equivalent quantities of alkyl bromide, KO_2, and crown were employed, although the authors did report similar results using 0.1 equivalent of crown at longer reaction times. The mechanism for the formation of dialkyl peroxides may be visualized as

TABLE XXXIX

Reactions of Alkyl Bromides and Sulfonate Esters with KO_2 in the Presence of 18-Crown-6 and Dicyclohexo-18-Crown-6

	Products (%)		
Substrate (R-X)	Peroxide (ROOR)	Alcohol (ROH)	Olefins
n-C_5H_{11}-Br	53		
n-C_6H_{13}-Br	54		
n-C_7H_{15}-Br	56		
n-$C_{16}H_{33}$-Br	44	21	
n-$C_{18}H_{37}$-Br	77	21	
n-$C_{18}H_{37}$-Br	61	18	
c-C_6H_{11}-Br			67
c-C_5H_9-Br	42		24
$C_6H_{13}CH(CH_3)$-Br	55		37
n-$C_{18}H_{37}$-OTos	50	42	
n-$C_{18}H_{37}$-OMs	46	40	
$C_6H_{13}CH(CH_3)$-OTs	52	13	16
$C_6H_{13}CH(CH_3)$-OMs	44	19	14

$$R{-}Br + O_2^{\dot{-}} \longrightarrow R{-}O{-}O\cdot + Br^- \qquad (42)$$

$$R{-}O{-}O\cdot + O_2^{\dot{-}} \longrightarrow R{-}O{-}O^- + O_2 \qquad (43)$$

$$R{-}O{-}O^- + RBr \longrightarrow R{-}O{-}OR + Br^- \qquad (44)$$

That steps (42) and (44) proceeded via back-side SN_2 pathways was demonstrated by Johnson and Nidy (1975) using the reaction sequence outlined in Reaction (45). Since it had been well established that the

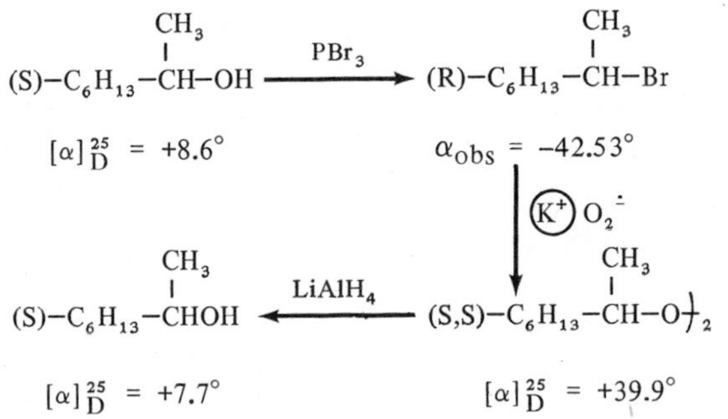

(45)

first step proceeded with inversion of configuration, the sum total of the three steps was calculated to proceed with a 94% net retention of configuration, a result consistent with the SN_2 stereochemistry.

Corey *et al.* (1975) have reported the nucleophilic reactions of $O_2^{\dot{-}}$ with a wide variety of organic substrates using KO_2 in DMSO, DMF, DME, and diethyl ether (and combinations of these solvents) containing (IV). Reactions (46)-(48) demonstrated the nucleophilic character of $O_2^{\dot{-}}$ and provided a simple means of converting bromides to alcohols. In each of these cases, 4 equivalents of KO_2 and 2 equivalents of crown were employed.

$$C_9H_{19}{-}CH_2{-}Br + (K^+) + O_2^{\dot{-}} \xrightarrow[25^\circ,\ 2\ \text{hr}\ \ 80\%]{\text{DMSO}} C_9H_{19}{-}CH_2{-}OH \qquad (46)$$

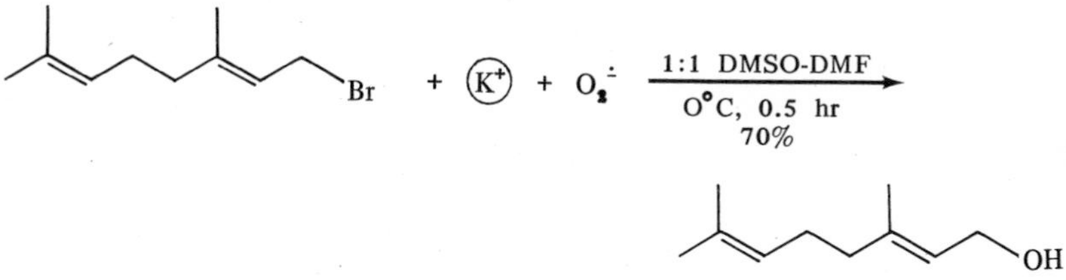

(47)

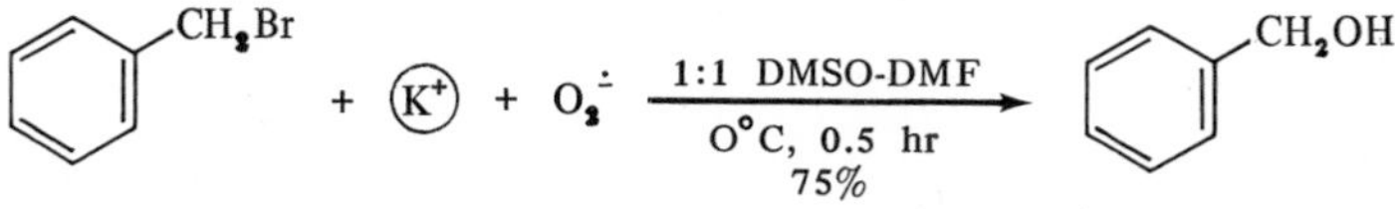

(48)

Naked superoxide was shown to be effective in providing a means of inverting the configuration of an oxygen functionality at a carbon center. Thus, the unnatural 15-R-prostanoid structure was converted to the natural 15-S-prostanoid structure, as illustrated in Reaction (49). As illustrated in Reactions

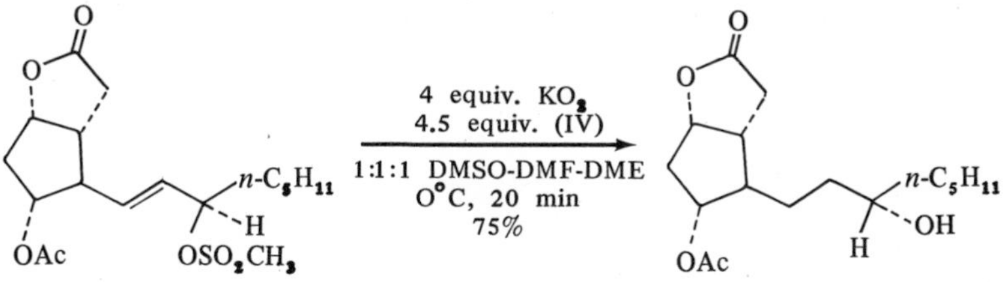

(49)

(50) and (51), *cis*- and *trans*-4-*t*-butylcyclohexanol were interconverted in high yield and the *p*-toluenesulfonate of cholesterol was converted to 3-*epi*-

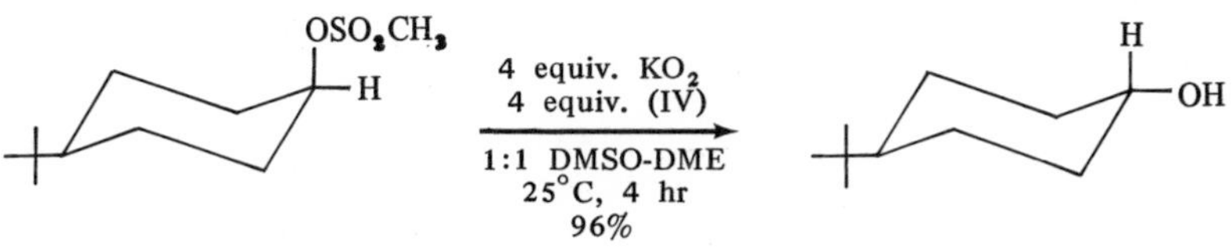

(50)

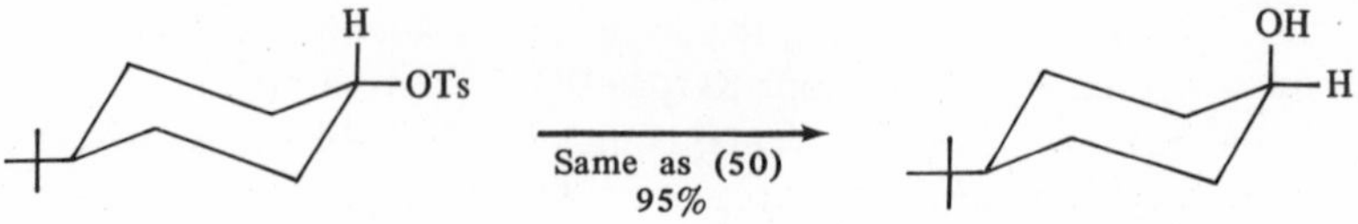

(51)

cholesterol in moderate yield, as illustrated in Reaction (52). In the latter reaction a small quantity of cholestadiene was observed. Some interesting and

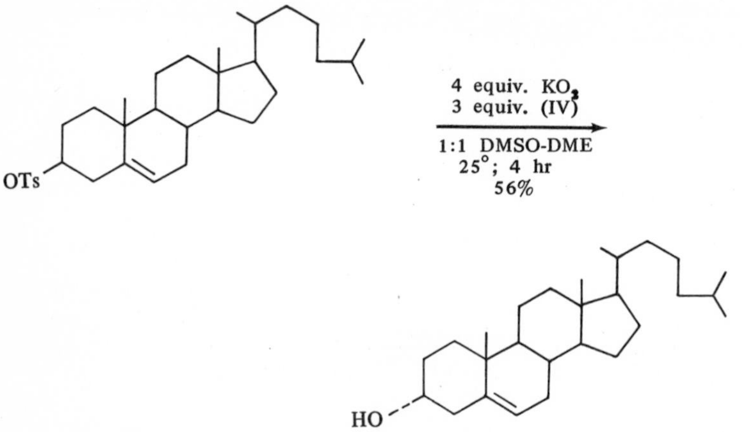

(52)

potentially useful reactions of naked superoxide are illustrated in Reactions (53)-(55). The relative amounts of *t*-BuOH and isobutylene were not given.

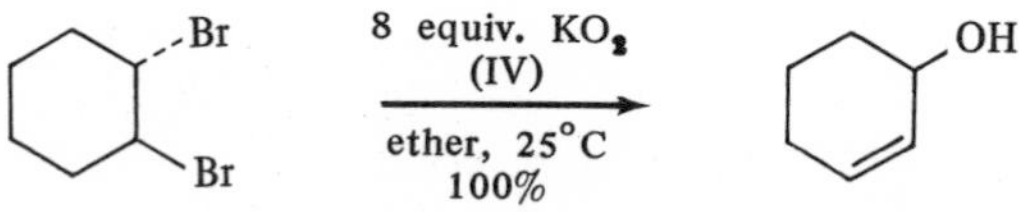

(53)

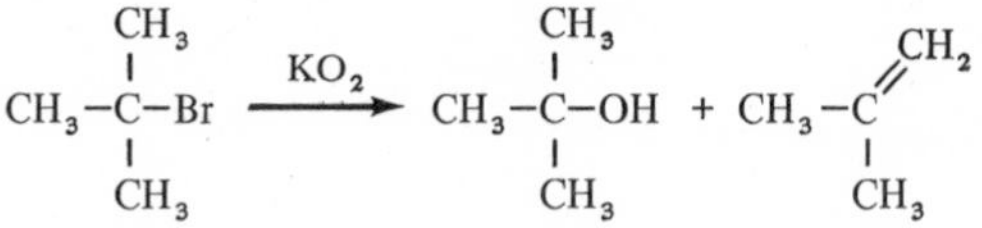

(54)

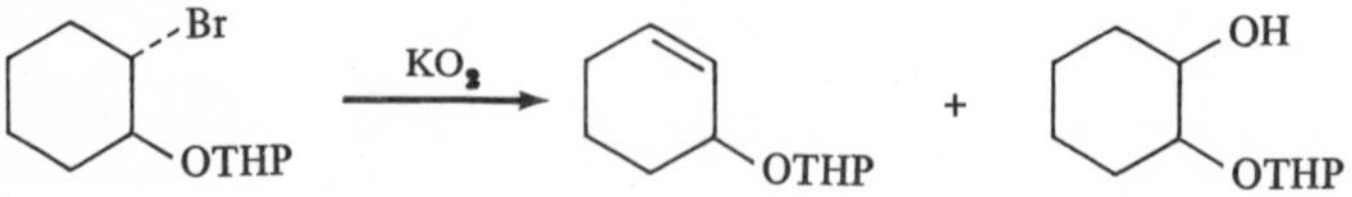

(55)

67% 29%

A novel reaction of superoxide and one related to the work reported by Johnson and Nidy (1975) is shown in Reaction (56).

San Filippo *et al.* (1975) have also demonstrated that organic halides and tosylates may be converted to the corresponding alcohols in the presence of (IV)

TABLE XL

Reaction of Potassium Superoxide with Various Organic Halides and Tosylates[a]

Substrate	Products[b] (%)	Relative reactivity
1-$C_8H_{17}I$	1-Octanol (46)	4.5
	1-Octene (3)	
	Octanal (11)	
2-$C_8H_{17}I$	2-Octanol (48)	3.3
	Octenes (48)	
	2-Octanone (<1)	
1-$C_8H_{17}Br$	1-Octanol (63)	1.0
	1-Octene (<1)	
	Octanal (12)	
2-$C_8H_{17}Br$	2-Octanol (51)	0.98
	Octenes (34)	
	2-Octanone (<1)	
$CH_3(CH_2)_2C{-}(CH_3)_2Br$	2-Methyl-2-pentanol (20)	0.90
	2-Methylpentenes (30)	
1-$C_8H_{17}Cl$	1-Octanol (34)	0.089
	1-Octene (~1)	
	Octanal (5)	
2-$C_8H_{17}Cl$	2-Octanol (36)	0.020
	Octenes (12)	
	2-Octanone (<1)	
1-$C_8H_{17}OTs$	1-Octanol (75)	1.0
	1-Octene (<1)	
	Octanal (1)	
2-$C_8H_{17}OTs$	2-Octanol (75)	
	Octenes (23)	—
	2-Octanone (<1)	
$C_6H_5CH_2Cl$	Benzyl alcohol (41)	2.9
	Benzyladehyde (6)	

[a]Most reactions were carried out by adding 3.33 mmole of halide or tosylate to a vigorously stirred mixture of KO_2 (10.0 mmole) and 18-crown-6 (IV) (1.0 mmole) in dry DMSO (20 ml) at ambient temperature. Reaction time varied from 75 min for the alkyl bromides, iodides, tosylates, and benzyl chloride to 3 hr for the alkyl chlorides. Yields did not improve with increased reaction times.

[b]Yields are based on alkyl halide or tosylate and were determined by glc analysis using the internal standard procedure.

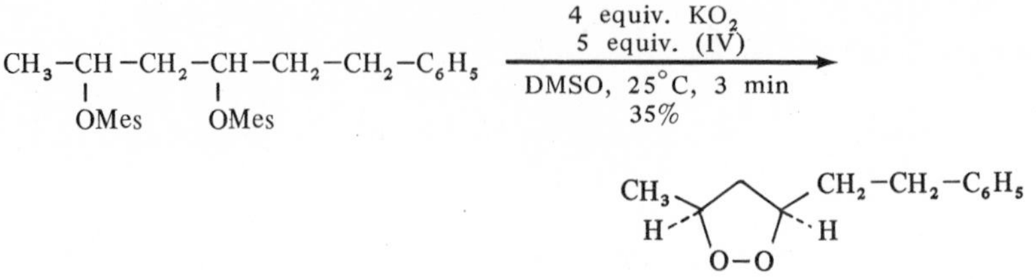

(56)

in dimethylsulfoxide at ambient temperatures. The results are summarized in Table XL. Reactions were also carried out in hexamethylphosphoramide (HMPA) and benzene but the reaction times were much longer. In all cases, the yields of alcohol products were moderate, primary tosylates giving the best yields (75%). The reactions were usually accompanied by substantial elimination and oxidation products. Phenyl halides failed to react. The authors commented that optimum yields were obtained when the halide:superoxide ratio was greater than three. In general, substrate reactivity varied as follows: benzyl $> 1° > 2° > 3° >$ aryl and I $>$ Br $>$ OTs $>$ Cl. This was in agreement with the general order of reactivity reported for the reaction of naked acetate with organic substrates. The authors clearly showed that the formation of alcohol product took place with inversion of configuration at a chiral center. Thus, the reaction of the tosylate of (+)-(S)-2-octanol with naked superoxide produced (−)-(R)-2-octanol, which corresponds to a stereoselectivity of 97% [Reaction (57)]. A mechanism consistent with the results reported above may be illustrated by Scheme I.

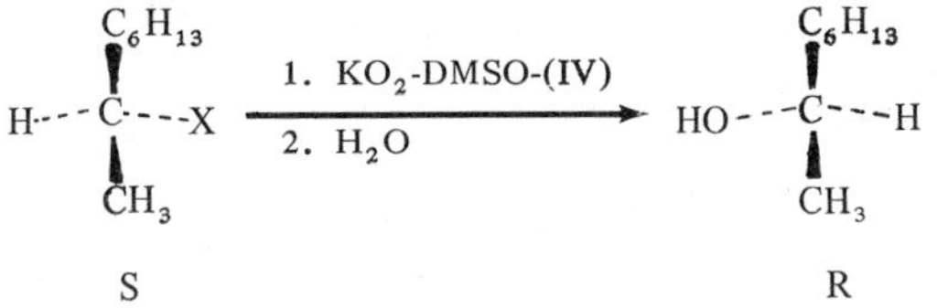

(57)

$$\begin{array}{ll}
1. & R{-}X + O_2^{\dot{-}} \longrightarrow R{-}O{-}O\cdot + X^- \\
2. & R{-}O{-}O\cdot + O_2^{\dot{-}} \longrightarrow R{-}O{-}O^- + O_2 \\
3. & R{-}O{-}O^- + RX \longrightarrow R{-}O{-}O{-}R + X^- \\
4. & R{-}O{-}O{-}R + O_2^{\dot{-}} \longrightarrow R{-}O{-}O{-}O\cdot + RO^- \\
5. & R{-}O{-}O{-}O\cdot \longrightarrow R{-}O\cdot + O_2 \\
6. & R{-}O\cdot + O_2^{\dot{-}} \longrightarrow R{-}O^- + O_2
\end{array}$$

Scheme I

San Filippo *et al.* (1976) have reported the reaction of carboxylic esters with approximately a threefold excess of potassium superoxide solublized in benzene with (IV) to produce, after acidic aqueous workup, the corresponding alcohol and carboxylic acid in moderate to excellent yields [Reaction (58)]. The results are summarized in Table XLI. Esters of primary, secondary, and tertiary

$$\mathrm{R{-}\overset{O}{\overset{\|}{C}}{-}O{-}R'} \xrightarrow[\text{2. } H_3O^+]{\text{1. } KO_2\text{-(VI)-}C_6H_6} \mathrm{R{-}\overset{O}{\overset{\|}{C}}{-}OH + R'{-}OH} \qquad (58)$$

TABLE XLI

Reaction of Potassium Superoxide with Various Esters[a]

Run	Substrate	Products[b] (%)	Reaction time (hr)
1	1-$C_7H_{15}CO_2CH_3$	1-Octanoic acid (98) (68)[c]	24
2	$(CH_3)_3CCO_2CH_3$	Pivalic acid (81)	72
3	1-$C_7H_{15}CO_2CH_2C_6H_5$	1-Octanoic acid (100) Benzoic acid (55)	24
4	$C_6H_5CO_2CH_2CH_3$	Benzoic acid (88)	24
5	1-$C_7H_{15}CO_2CH(CH_3)_2$	1-Octanoic acid (98)	72
6	$CH_3CO_2CH(CH_3)C_6H_{13}$	2-Octanol (89)	72
7	1-$C_9H_{19}CO_2C_6H_5$	1-Decanoic acid (84) Phenol (70)	8
8	1-$C_7H_{15}CO_2C(CH_3)_3$	1-Octanoic acid (96)	140
9	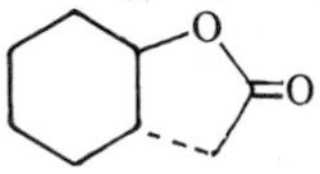	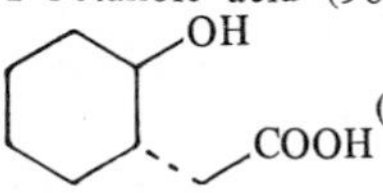 (85)	28
10	1-$C_7H_{15}C(O)SC_4H_9$	1-Octanoic acid (93) 1-Butyl disulfide (33)	8
11	$C_6H_5CH_2CH(NHCOCH_3)$- $CO_2CH_2CH_3$	*N*-Acetylphenylalanine (56)	24
12	$(C_6H_5O)_3P(O)$	Phenol (30)	24
13	$(1\text{-}C_8H_{17}O)_3P(O)$	1-Octanol (<1)	72

[a]Most reactions were carried out at ambient temperature by adding a solution of ester (3.33 mmole) and 18-crown-6 (IV) (1.0 mmole) in dry benzene (20 ml) to powdered potassium superoxide. Vigorous stirring was maintained during addition and throughout the course of the reaction.

[b]Yields were determined by glc.

[c]Carried out in dry Me_2SO.

alcohols were investigated as well as phenols (Run 7, Table XLI) and thiols (Run 10, Table XLI). Phosphate esters of phenol were also cleaved by superoxide (Run 12, Table XLI) but the corresponding esters of alkyl alcohols showed no appreciable reactivity (Run 13, Table XLI). The use of dimethylsulfoxide as solvent produced shorter reaction times but the yields were much reduced (Run 1, Table XLI). Reaction of superoxide with the acetate ester of (−)-(R)-2-octanol gave only (−)-(R)-2-octanol, indicating that there was net retention of configuration at the chiral carbon and that the reaction proceeded by means of an acyl oxygen cleavage. This mechanism was also in agreement with the influence that the structure of the alcohol portion of the ester has on the relative rates of reaction (Ph > 1° > 2° > 3°). The authors also reported that simple amides and nitriles did not react with superoxide under the conditions of ester cleavage.

San Filippo *et al.* (1976) report that potassium superoxide dissolved in benzene with (IV) is an effective reagent for promoting the oxidative cleavage of α-keto, α-hydroxy, and α-halo ketones, esters, and carboxylic acids to carboxylic acids [Reactions (59) and (60)]. The results are summarized in Table XLII.

$$\mathrm{R{-}\overset{O}{\overset{\|}{C}}{-}\overset{O}{\overset{\|}{C}}{-}R'} \xrightarrow[\text{2. } \mathrm{H_3O^+}]{\text{1. } \mathrm{KO_2\text{-}(IV),\ C_6H_6,\ 25^\circ C}} \mathrm{R{-}CO_2H + R'CO_2H} \qquad (59)$$

$$\mathrm{R{-}\overset{X}{\overset{|}{C}H}{-}\overset{O}{\overset{\|}{C}}{-}R'} \xrightarrow[\text{2. } \mathrm{H_3O^+}]{\text{1. } \mathrm{KO_2\text{-}(IV),\ C_6H_6,\ 25^\circ}} \mathrm{R{-}CO_2H + R'CO_2H} \qquad (60)$$

R′ = OH, OR, alkyl, aryl; X = OH, Cl, Br

TABLE XLII

Reactions of Potassium Superoxide with Various α-Keto, α-Hydroxy, and α-Halo Ketones, Esters, and Carboxylic Acids

Substrate	KO_2/substrate[a] (m*M*)	Product	Yield (%)
Benzil	3/1	Benzoic acid	87
Camphoroquinone	4/1	Camphoric acid	87
1,2-Cyclohexadione	4/1	Adipic acid	53
2-Ketoglutaric acid	4/1	Succinic acid	42
2-Ketophenylacetic acid	4/1	Benzoic acid	93
Ethyl 2-ketophenylacetate	4/1	Benzoic acid	93
Benzoin	3/1	Benzoic acid	98
2-Hydroxycyclohexanone	4/1	Adipic acid	69
Mandelic acid	4/1	Benzoic acid	94
2-Hydroxystearic acid	12/1	Heptadecanoic acid	77
1-Hydroxycycloheptanecarboxylic acid	4/1	1-Hydroxycycloheptanecarboxylic acid	–
1-Cyclohexymandelic acid	4/1	1-Cyclohexylmandelic acid	–
Ethyl mandelate	4/1	Benzoic acid	93
2-Chlorocyclohexanone	4/1	Adipic acid	60
2-Chlorocyclooctanone	4/1	Octanedioic acid	62
3-Bromocamphor	4/1	Camphoric acid	54
Phenacyl chloride	4/1	Benzoic acid	72
2-Bromo-2-phenyl-acetic acid	4/1	Benzoic acid	90
2-Bromooctanoic acid	4/1	Heptanoic acid	58
Methyl 2-bromo-2-cyclohexane acetate	4/1	Cyclohexanecarboxylic acid	54

[a]Most reactions were carried out for 24 hr using a 1:10 ratio of 18-crown-6 (IV) to KO_2. Reaction times were noticeably shorter at higher ratios.

Corey *et al.* (1975) have also reported that dissolved superoxide was an effective reagent for converting esters and lactones to carboxylic acids and ketones to α-diketones.

H. Hydride Reductions

Matsuda and Koida (1973) have reported studies related to the use of sodium borohydride in toluene in the reduction of a variety of ketone substrates in the presence of equivalent amounts of (VI), diglyme, or dimethyoxyethane [Reaction (61)]. In all cases the ketone was allowed to reflux with the sodium

$$>C=O + BH_4^- \xrightarrow[\text{2. Hydrolysis}]{\text{1. Na-(VI), } C_6H_5CH_3} >C(OH)H \qquad (61)$$

borohydride and the additive in toluene for 5 hr. The results are summarized in Table XLIII. The authors reported that the use of potassium borohydride

TABLE XLIII

Effect of Ethereal Additives on the Reduction of Ketones in Toluene

Ketone	Products (%)		
	(VI)	Diglyme	Dimethoxyethane
Acetophenone			
Alcohol	49	42	23
Ketone	0	38	40
Residue	28	12	13
Cyclohexanone			
Alcohol	50	28	19
Ketone	0	32	56
Residue	37	17	8
Methyl *n*-amyl ketone			
Alcohol	41	27	14
Ketone	3	9	34
Residue	29	22	18
Methyl *iso*-propyl ketone			
Alcohol	23	11	0
Ketone	63	74	78
Residue	10	6	5

required prolonged reaction times. From Table XLIII it appears that crown is slightly better than the other ethereal additives in terms of yield of reduced product. Unfortunately, competing base-catalyzed condensation reactions accompanied the reduction [Reaction (62)]. This was especially true for the reactions carried out using crown ether. Table XLIV summarizes the isomer

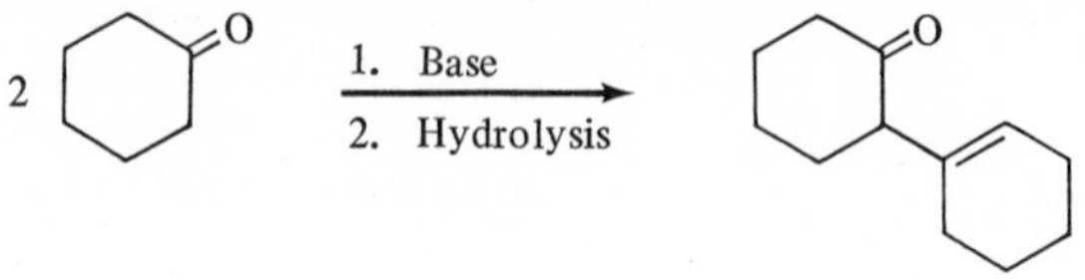 (62)

TABLE XLIV

Stereochemical Outcome in the Reduction of Cyclic Ketones

	Conditions			Products[a]		
Ketone	$NaBH_4$ (VI)	Time (hr)	Solvent[b]	Ketone	cis exo[c]	trans (endo)[c]
4-Methylcyclohexanone	1	5	T	0	26	74
4-*t*-Butylcyclohexanone	1	5	T	3	22	75
	0.5			0	21	79
3,3,5-Trimethylcyclo-	1[d]	5	T	77	10	13
hexanone	1			31	52	17
	0.1			23	62	13
	0.05			27	56	16
2-Norbornanone	1	5	T	84	5	11
	1	9	T	78	6	16
	1	3.5	X	41	20	39
	1	5	X	4	31	65

[a]Weight percent for distillate.
[b]T is toluene, X *o*-xylene.
[c]For 2-norbornanone.
[d]This reaction was carried out at 100°C; all the others were carried out under reflux.

distribution in the reduction of substituted cyclohexanones and 2-norbornanone in the presence of (VI). It is difficult to evaluate what the foregoing results actually mean, since it is not clear how much reduction takes place in the aprotic solvent and how much takes place during the hydrolytic workup. This question appears to have been completely overlooked by these workers.

Pierre and Handel (1974) have investigated the fundamental role of lithium and sodium cations in the reduction of cyclohexanone by aluminum hydride ion (AlH_4^-) and borohydride ion (BH_4^-) using cryptates (IX)b and (IX)c for lithium and sodium cations, respectively. In diglyme, cyclohexanone was quantitatively reduced to cyclohexanol in the presence of either 2 moles or 0.25 mole of lithium aluminum hydride ($LiAlH_4$). In the presence of an equivalent amount of (IX)b all of the $LiAlH_4$ went into solution but the ketone was recovered unchanged after hydrolysis. If, however, an excess of LiI or NaI (25 equivalents) was added in addition to (IX)b, the reaction proceeded quantitatively to completion before hydrolysis. The authors concluded that an alkali metal cation was necessary for reaction and that it needed only be present in catalytic quantities. A mechanism involving lithium cation in an electrophilic capacity was advanced [Reaction (63)]. In diglyme, cyclohexanone was slowly reduced to cyclohexanol

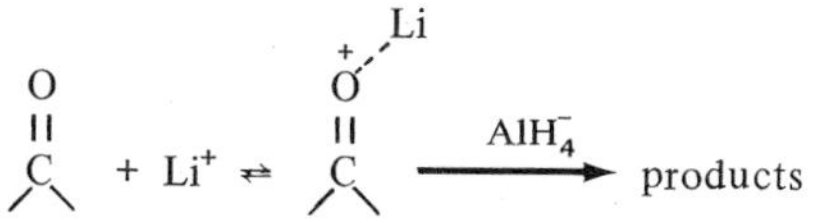

(63)

in the presence of 2 moles of sodium borohydride ($NaBH_4$). Upon addition of an equimolar quantity of (IX)c, all of the $NaBH_4$ went into solution but no reaction took place before hydrolysis. The reaction, however, did proceed during the hydrolytic workup. This was in direct contrast to similar experiments with $LiAlH_4$ where unreacted cyclohexanone was recovered from the reaction mixture. If an excess of $LiClO_4$ was added in addition to (IX)c, reaction was very rapid before hydrolysis. Using excess $NaClO_4$ (25 equivalents) in place of $LiClO_4$, reaction proceeds slowly before hydrolysis. When methanol was used as solvent the reaction proceeded quantitatively in the presence or absence of (IX)c. In at least one respect the behavior of $NaBH_4$ was different from that of $LiAlH_4$. In the absence of a coordinating electrophilic cation, reduction by BH_4^- always takes place in the hydrolyzing medium. To account for this, the following transition state (XXIV) was suggested.

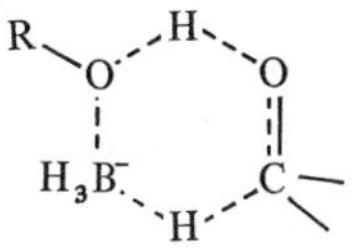

R = H, CH_3

(XXIV)

Handel and Pierre (1975) have studied the stereochemical course reduction of *N-t*-butyl-2-acetyl aziridine with $LiAlH_4$ in diglyme and $NaBH_4$ in diglyme and methanol [Reaction (64)] in the presence and absence of a number of additives. The results are summarized in Table XLV. In the AlH_4^- reductions, electrophilic catalysis by Li^+ was essential for reaction to proceed. Reduction was effected even in the presence of catalytic quantities of Li^+. In all cases reported, the SR configuration was favored. In the BH_4^- reductions, Li^+ was more effective than Na^+ in terms of rate and stereoselectivity. When the Na^+ was coordinated with an equimolar quantity of (IX)b, reaction took place during hydrolysis

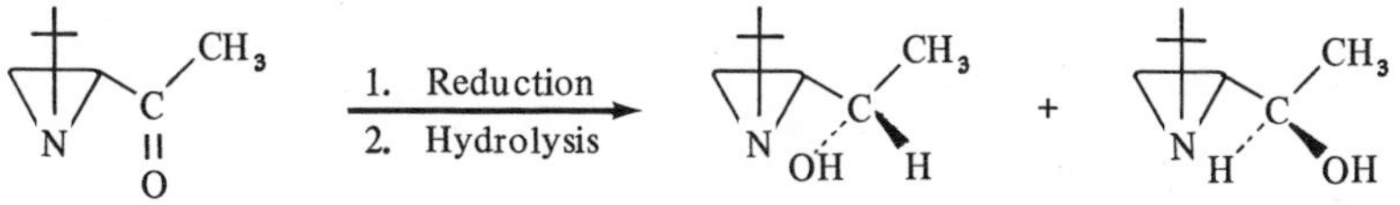

(64)

SR RR

TABLE XLV

Reduction of *N-t*-Butyl-2-acetyl Aziridine

Solvent	Reducing agent (equiv.)	Alcohol SR obtained (%)
Diglyme	$LiAlH_4$ (1)	98
Diglyme	$LiAlH_4$ (1/4)	90
Diglyme	$LiAlH_4$ (2) + (IX)a (2)	–
Diglyme	$LiAlH_4$ (4) + (IX)a (2) + NaI (25)	80
Diglyme	$LiAlH_4$ (2) + (IX)a (2) + LiI (25)	98
Diglyme	$NaBH_4$ (2)	70
Diglyme	$LiBH_4$ (2)	92
Diglyme	$NaBH_4$ (2) + (IX)b (2)	35
Diglyme	$NaBH_4$ (1) + (IX)b (2) + $NaClO_4$ (1)	35
Diglyme	$NaBH_4$ (2) + (IX)b (2) + $NaClO_4$ (25)	70
Diglyme	$NaBH_4$ (2) + (IX)b (2) + $LiClO_4$ (25)	90
MeOH	$NaBH_4$ (2)	40
MeOH	$NaBH_4$ (2) + (IX)b (2)	25
MeOH	$NaBH_4$ (2) + $NaClO_4$ (30)	65
MeOH	$NaBH_4$ (2) + $LiClO_4$ (30)	90

and the RR configuration dominated. If excess $NaClO_4$ or $LiClO_4$ was added in addition to the cryptate, the reaction proceeded with predominant formation of the SR configuration. If only 1 equivalent of $NaClO_4$ was employed, the reaction took place only during the hydrolysis step and the RR configuration dominated. In methanol, BH_4^- gave slightly greater quantities of the RR isomer in the presence or absence of (IX)b. However, when 30 equivalents of $NaClO_4$ or $LiClO_4$ was added, the SR configuration became dominant. In all cases, Li^+ was more effective than Na^+ in producing the SR isomer. The authors suggested that the SR isomer was formed via a transition state resembling (XXV), where the metal cation was coordinated to both the aziridine nitrogen and the carbonyl,

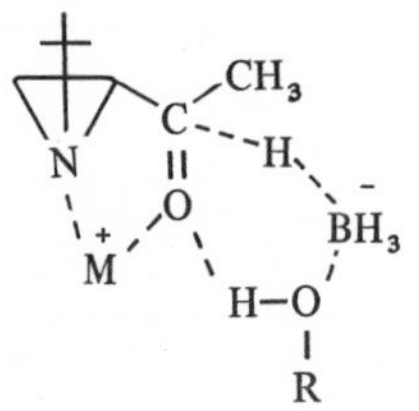

(XXV)

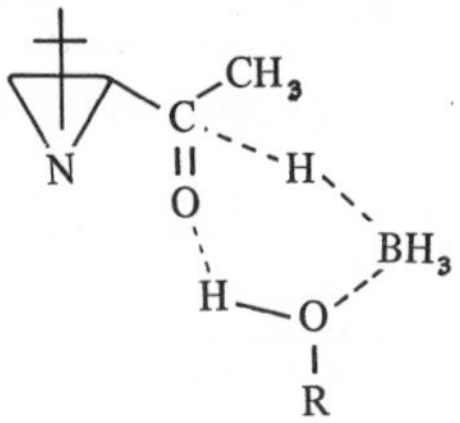

(XXVI)

thus orienting the reaction of the BH_4^--solvent molecule couple with the carbonyl group. In contrast to this, the RR configuration was formed via a transition state resembling (XXVI), where a solvent molecule and the BH_4^- are involved and the metal cation is unimportant.

Durst *et al.* (1974) have reported the facile reduction of alkoxysulfonium salts dissolved in methylene chloride with $NaBH_3CN$ dissolved in methanol or ethanol at 0°C in the presence of (IV) [Reaction (65)]. The results are summarized

$$R_2\overset{+}{S}{-}O{-}CH_3 + NaBH_3CN \xrightarrow[CH_2Cl_2,\ CH_3OH \text{ or } CH_2Cl_2,\ C_2H_5OH]{(IV)} R_2S + CH_3OH \qquad (65)$$

in Table XLVI. The yields of sulfides were quantitative based on recovery of starting material. The reaction was employed in the presence of aldehydes or ketones and gave only sulfoxide reduction.

TABLE XLVI

Reduction of Alkoxysulfonium Salts with Sodium Cyanoborohydride

Sulfoxide	Yield of sulfide[a] (%)
Dibutylsulfoxide	85
Diphenylsulfoxide	77
Dibenzylsulfoxide	91
Tetramethylenesulfoxide	87

[a]Isolated yield after workup.

Cinquini *et al.* (1975) have reported the reduction of 2-octanone with 1.5 molar equivalents of $NaBH_4$ in aqueous 2 *N* sodium hydroxide and benzene using cryptate (IX)c, crown (V), and phosphonium salt (XIV) as liquid-liquid phase transfer catalysts [Reaction (66a)]. Excellent yields of reduced product

$$n{-}C_6H_{13}{-}\overset{O}{\overset{\|}{C}}{-}CH_3 + BH_4^- \xrightarrow[C_6H_6\text{-}H_2O,\ NaOH]{Na^+\text{-cat.}} n{-}C_6H_{13}{-}\overset{OH}{\overset{|}{C}H}{-}CH_3 \qquad (66a)$$

were reported, the phosphonium salt being the most effective catalyst.

I. Oxidation Reactions

Sam and Simmons (1974) have reported the solubilization of potassium permanganate in benzene using (V). The solid complex was isolated but was found to be more stable when present in solution. At 25°C, the half-life of the complex in solution was 48 hr; the main product of decomposition was adipic acid. It was demonstrated that $KMnO_4$ solublized in benzene provided a convenient, mild, and efficient oxidant for a wide variety of organic compounds. Table XLVII summarizes the reported results. Since the potassium salts of the acid products were insoluble in benzene, they were apparently not subject to further oxidation. Thus, in the oxidation of α-pinene to *cis*-pinonic acid, a 90% yield was observed [Reaction (66b)]. This compared favorably to the 40-60% yield reported in the literature. The authors suggested that the mechanism for

TABLE XLVII

Oxidations by Potassium Permanganate in Benzene (25°C)

Starting material	Product	Isolated yields (%)
trans-Stilbene	Benzoic acid	100
Cyclohexene	Adipic acid	100
α-Pinene	Pinonic acid	90
3-Methylenecyclobutane-carboxylic acid	3-Oxocyclobutane-carboxylic acid	Incomplete
1-Heptanol	Heptanoic acid	70
Benzyl alcohol	Benzoic acid	100
Benzhydrol	Benzophenone	100
Toluene (solvent)	Benzoic acid	78
p-Xylene (solvent)	Toluic acid	100
Benzaldehyde	Benzoic acid	100

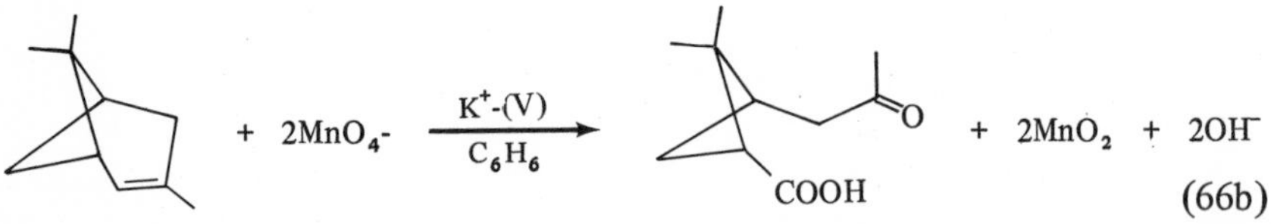

(66b)

permanganate oxidation involved an initial thermally allowed [2 + 4] cycloaddition followed an electron transfer and then by a thermally allowed [2 + 2 + 2] cheletropic change to form products (Scheme II).

Cardillo *et al.* (1976) have reported the reaction of potassium chromate with primary alkyl halides in hexamethylphosphoramide containing crowns (V) or (VI) at 100°C to produce aldehydes in good yields [Reaction (67)]. The

$$R{-}CH_2{-}X + K_2CrO_4 \xrightarrow[\text{HMPA, } 100^\circ C]{\text{(V) or (VI)}} R{-}CH_2{-}OCrO_3^-K^+ \rightarrow R{-}CHO \qquad (67)$$

results are summarized in Table XLVIII. Here the chromate ion was believed to

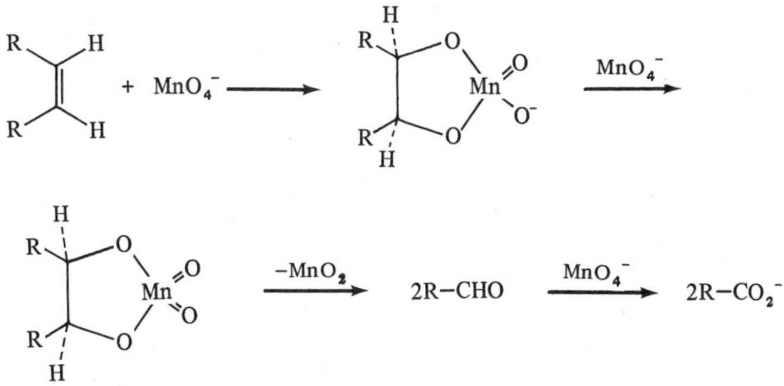

Scheme II

TABLE XLVIII

Chromate Oxidation of Alkyl Halides

Alkyl halide	Time (hr)	Aldehyde yield[a] (%)
γ,γ-Dimethylallyl bromide	2	78
Geranyl bromide	2	82
Farnesyl bromide	2	80
Benzyl chloride	6	80
Octyl bromide	6	20

[a]Determined by glc analysis.

act as both a nucleophile and an oxidant. In the reaction with geranyl bromide, a 20% yield of geraniol was isolated. Landini *et al.* (1974) have reported the oxidation of 1-octene to *n*-heptanoic acid in benzene solution by stirring with a saturated aqueous solution of $KMnO_4$ at room temperature in the presence of (VI). An 80% yield of product was realized in 45 min.

Dietrich and Lehn (1973) have reported the formation of the fluorenyl anion from the reaction of fluorene with potassium hydroxide solubilized in toluene with cryptate (IX)c. Reaction of the anion with molecular oxygen produced fluorenone in good yield [Reaction (68)].

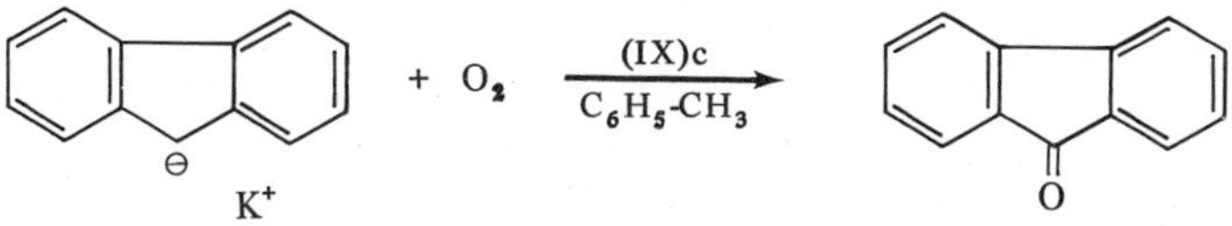

(68)

Boden (1975) has reported the homogeneous photosensitization of oxygen by solubilizing the anionic dyes Rose Bengal and Eosin Y in methylene chloride and carbon disulfide using crown (IV) and quaternary ammonium salt Alaquat

336. The presence of singlet oxygen was demonstrated by trapping with anthracene [Reaction (69)] and tetramethylethylene [Reaction (70)].

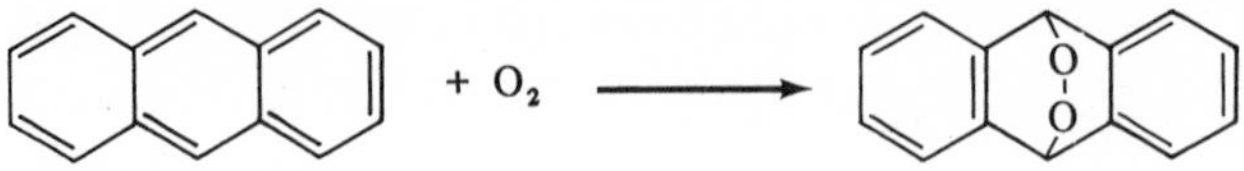

(69)

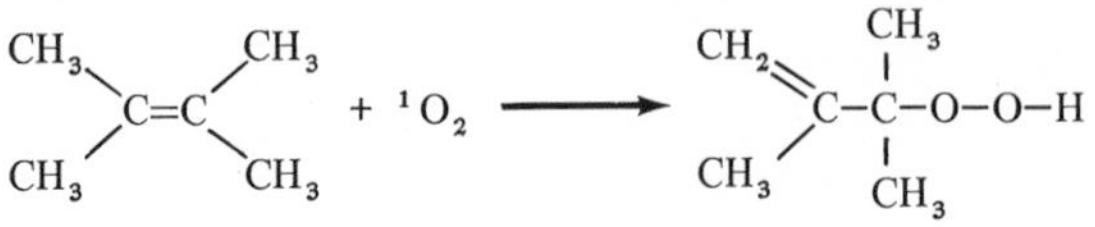

(70)

The results with anthracene and tetramethylethylene are summarized in Tables XLIX and L, respectively. In both cases, Rose Bengal was superior to Eosin Y.

TABLE XLIX

Photooxidation of Anthracene

Sensitizer	Solvent	Catalyst	Yield (%)
Rose Bengal	CS_2	*a*	71
	CH_2Cl_2	*a*	86
	CS_2	*b*	44
	CH_2Cl_2	*b*	85
Eosin Y	CS_2	*a*	10
	CH_2Cl_2	*a*	34
	CS_2	*b*	32
	CH_2Cl_2	*b*	34

[a] 18-Crown-6 (IV).
[b] Alaquat 336.

TABLE L

Photooxidation of Tetramethylethylene

Sensitizer	Solvent	Catalyst	Yield (%)
Rose Bengal	CS_2	*a*	36
	CH_2Cl_2	*a*	74
	CS_2	*b*	71
	CH_2Cl_2	*b*	28
Eosin Y	CS_2	*a*	—
	CH_2Cl_2	*a*	3
	CS_2	*b*	20
	CH_2Cl_2	*b*	17

[a] 18-Crown-6 (IV).
[b] Alaquat 336.

J. Bromine Addition Reactions

Shchori and Jaqur-Grodzinski (1972a-c) have studied the effect of ethereal additives on the rates of bromination of stilbene in the absence and in the presence of a slight excess of hydrogen bromide [Reaction (71)]. The results are summarized in Tables LI and LII. In the absence of HBr, the rates of bromination

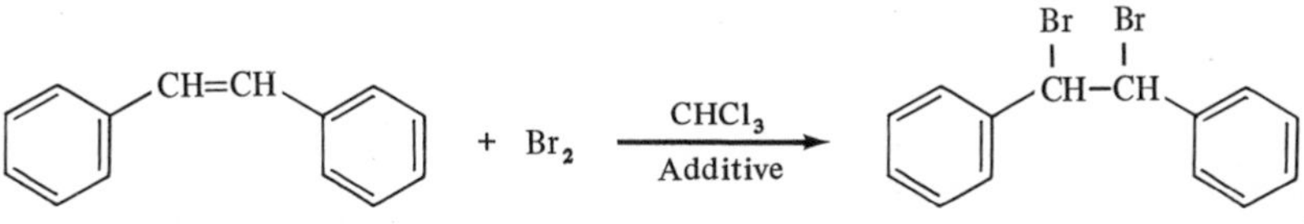

(71)

TABLE LI

Bromination of Stilbene in Chloroform at 29°C in the Presence of Ethers

$[Br_2]_0$ (m*M*)	[Ether] (*M*)	*k* (sec^{-1} M^{-2})
7.6	none	4.5
7.6	(Dioxane) = 0.060	4.5
6.7	(THF) = 0.120	4.4
4.3	(V) = 0.020	5.7

TABLE LII

Bromination of Stilbene in Chloroform at 29°C in the Presence of HBr[a]

$[HBr]_0$ (m*M*)	(V) (m*M*)	$\dfrac{10^3 \times \ln [Br_2]_0/[Br_2]_t}{[\text{Stilbene}] \cdot t}$ (sec^{-1} M^{-1})	
49	None	82	
37	None	65	k_1 = 1.7 sec^{-1} M^{-2}
18	None	30	
74	20	1.7	
36	20	1.5	
13	20	1.5	
6.5	20	1.5	k_1 = 1.7 × 10^3
36	40	1.8	sec^{-1} M^{-1}
36	20	1.5	
36	10	2.0	
36	5	1.7	
35	THF 120 m*M*	>500	

[a] $[Br_2]_0$ = 6.7-8.1 m*M*; $[\text{stilbene}]_0$ = 0.100 *M*.

appeared to be independent of the ethereal additive. The very slight increase in rate observed in the presence of (V) was probably due to crown scavanging traces of HBr which acted as an inhibitor. This, in all probability, was formed as a result of interaction between bromine and (V) contaminated with minute amounts of peroxide. In the presence of a slight excess of HBr, the presence of crown caused a dramatic decrease in rate, whereas addition of tetrahydrofuran caused a marked rate increase. This was rationalized in terms of the strong complex (XXVII) formed by (V), HBr, and Br_2, which was inactive toward

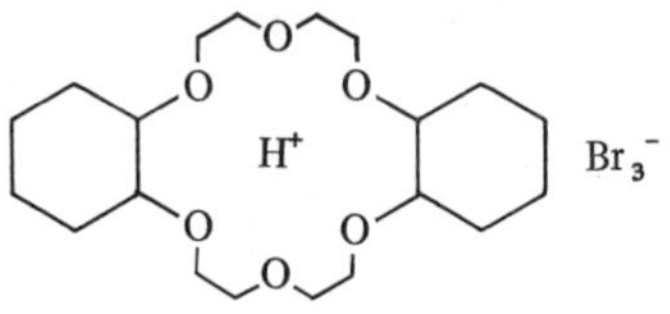

(XXVII)

stilbene as a brominating agent. In tetrahydrofuran, on the other hand, the tribromide was much less stable.

K. Sulfur Nucleophiles

Cinquini *et al.* (1975) have effectively used thiophenoxide ion as a nucleophile under liquid-liquid phase transfer catalytic conditions in reactions with 1-bromooctane in the presence of (IX)c″ and (XIV). Reaction took place in a few minutes at 20°C, and produced thioether product in quantitative yield.

Herriott and Picker (1975) have studied the catalytic effects of a wide variety of quaternary ammonium and phosphonium salts on the rates of reaction of thiophenoxide with 1-bromooctane under liquid-liquid phase transfer catalytic conditions. They presented excellent evidence that crown ether (V) serves as an efficient phase transfer catalyst in the two-phase benzene-water system. The rate constant in the presence of the crown was almost identical to that of the fastest quaternary ion catalyst.

Liotta *et al.* (1975) have reported the reaction of benzyl tosylate with potassium thiocyanate in acetonitrile containing (IV), forming a quantitative yield of benzyl thiocyanate.

L. Carbenes

Moss and Pilkiewicz (1974) have reported the generation of phenylbromo- and phenylchlorocarbenic species by a variety of methods and have examined the "freeness" of these species by comparing their selectivities toward a number of alkyl-substituted olefins by competition experiments [Reaction (72)]. The results are summarized in Table LIII. The selectivities of the carbenic species

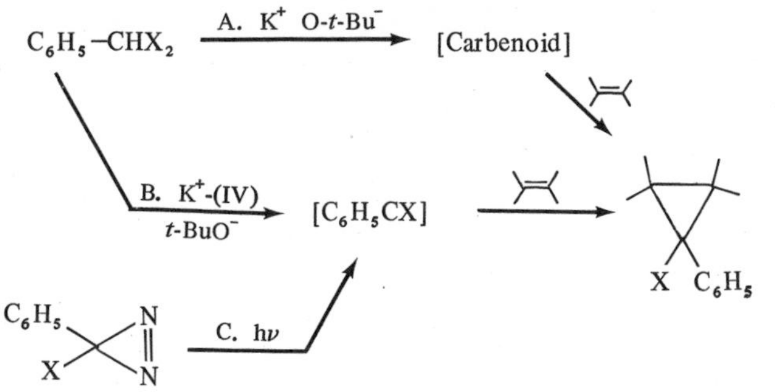

(72)

TABLE LIII

Relative Reactivities of Alkenes toward Phenylbromocarbene at 25°C

TME[a]/olefin	(k_{TME}/k_i)Base[b]	(k_{TME}/k_i)hν[c]	(k_{TME}/k_i)Crown[d]
TME-trimethylethylene	$1.28 \pm 0.09_3$	$1.74 \pm 0.04_2$	1.72
TME-isobutene	$1.65 \pm 0.021_7$	$4.44 \pm 0.18_2$	$4.11 \pm 0.14_2$
TME-*cis*-butene	$5.79 \pm 0.11_2$	$8.34 \pm 0.04_2$	8.24
TME-*trans*-butene	$11.3 \pm 1.6_4$	$17.5 \pm 0.8_2$	$17.1 \pm 0.4_2$
TME-isobutene	$2.6 \pm 0.1_3$	$5.0 \pm 0.3_6$	4.8

[a]TME is tetramethylethylene.
[b]α,α-dihalotoluene + KO-t-Bu.
[c]phenyl halo diazirine
[d]α,α-dihalotoluene ± KO-t-Bu + 18-crown-6 (IV)

generated from photolysis of the phenyl halo diazirines and those generated from KO-t-Bu in the presence of (IV) were found to be identical. The results were interpreted to mean that pathways B and C proceeded via free carbenic species, whereas pathway A proceeded via a carbenoid intermediate ($C_6H_5CX{\cdot}KX$). In a similar experiment CCl_2 was generated from $HCCl_3$ and KO-t-Bu in the presence and absence of (IV) and competitively reacted with methylenecyclohexane and cyclohexene. The relative olefin reactivities were 4.79 and 5.51, respectively, implying that free CCl_2 was generated in both cases. The authors emphasized that the results of these experiments suggested that free carbenic species may be generated by α-elimination processes on halide precursors in the presence of (IV) when the requisite diazoalkane or diazirine precursors are not readily available.

Moss *et al.* (1975) also reported an extension of the foregoing study to phenylfluorocarbene and chloromethylthiocarbene generated from α-bromo-α-fluorotoluene and α,α-dichlorodimethylsulfide, respectively, by the action of KO-t-Bu in the absence and the presence of (IV). The results are summarized in Table LIV. Competition experiments showed that in the absence of crown the

TABLE LIV

Relative Reactivities of Alkenes toward Carbenic Species at 25°

Alkene	C_6H_5CBr		C_6H_5CCl		C_6H_5CF		CH_3SCCl	
	Carbenoid[a]	Carbene[b]	Carbenoid[c]	Carbene[d]	Carbenoid[e]	Carbene[f]	No crown[g]	Crown[h]
$Me_2C{=}CMe_2$	1.6	4.4	2.6	5.1	2.7	5.8	7.4	7.6
$Me_2C{=}CHMe$	1.3	2.5	1.6	3.2	1.2	3.0	1.9	2.1
$Me_2C{=}CH_2$	1.00	1.00	1.00	1.00	1.00	1.00	1.00	1.00
c-MeCH=CHMe	0.29	0.53	0.31	0.37	0.12	0.28	0.29	0.21
t-MeCH=CHMe	0.15	0.25	0.11	0.20	0.10	0.20	0.21	0.18
m_{CXY}^{obsd}	0.59	0.70	0.75	0.83	0.87	0.89	0.88	0.93

[a] ($C_6H_5CHBr_2$ + KO-*t*-Bu).
[b] ($C_6H_5CBrN_2$ + $h\nu$).
[c] ($C_6H_5CHCl_2$ + KO-*t*-Bu).
[d] ($C_6H_5CClN_2$ + $h\nu$).
[e] (C_6H_5CHBrF + KO-*t*-Bu).
[f] Same as *e*, with 18-crown-6 (IV).
[g] (CH_3SCHCl_2 + KO-*t*-Bu).
[h] Same as *g*, with 18-crown-6 (IV).

phenylfluorocarbenic species was carbenoid in nature but in the presence of crown the free carbenic species was generated. In sharp contrast to this, chloromethylthiocarbene was a free species in the presence or absence of crown. Thus, the use of crown ether enabled the authors to develop a deeper understanding of those factors which control the selectivity in the reaction of carbenic species with olefins.

Moss *et al.* (1975) have tabulated values (Table LIV) of m_{CXY}^{obsd}, the least squares slope of $\log(k_i/k_{isobutylene})$ for CXY versus $\log(k_i/k_{isobutylene})$ for CCl_2 at 25°C, for "free" CH_3CCl, C_6H_5CBr, C_6H_5CCl, CCl_2, CFCl, and CF_2, and found that these are correlated by the two-parameter equations

$$m_{CXY} = -0.94 \sum_{X,Y} \sigma_R^+ + 0.69 \sum_{X,Y} \sigma_I - 0.27 \qquad (73)$$

It was suggested that m_{CXY} is a measure of carbenic selectivity relative to CCl_2 ($m = 1.0$). The equation quantitatively reflects the resonance and inductive components contributing to the internal stabilization of the carbenic species.

Sepp *et al.* (1974) reported the preparation of diazomethane in 48% yield from the reaction of hydrazine hydrate, chloroform, and potassium hydroxide in ether solvent in the presence of catalytic quantities of phase transfer catalyst (IV). Cinquini *et al.* (1975) reported the reaction of styrene with dichlorocarbene produced from aqueous $HCCl_3$ containing 2.5 molar equivalents of NaOH in the presence of phase transfer catalyst IXc″. 1,1-Dichloro-2-phenylcyclopropane was produced in 60% yield at 20°C. The reaction time was 24 hr. Kostikov and Molchanov (1975) converted cyclohexene and *trans*-stilbene to the respective *gem*-dihalocyclopropanes in 35-75% yield by treatment with 50% aqueous NaOH and $HCCl_3$ or HBr_3 in the presence of (VI) at 40°C.

M. Carbanions and Anion Rearrangements

Fraenkel and Pechhold (1970) have reported the solubilization of potassium phenylazoformate in tetrahydrofuran using dicyclohexyl-18-crown-6 (V). Upon heating to reflux temperatures, the salt decomposed to form benzene, potassium benzoate, carbon dioxide, and nitrogen [Reaction (74)]. When the reaction was

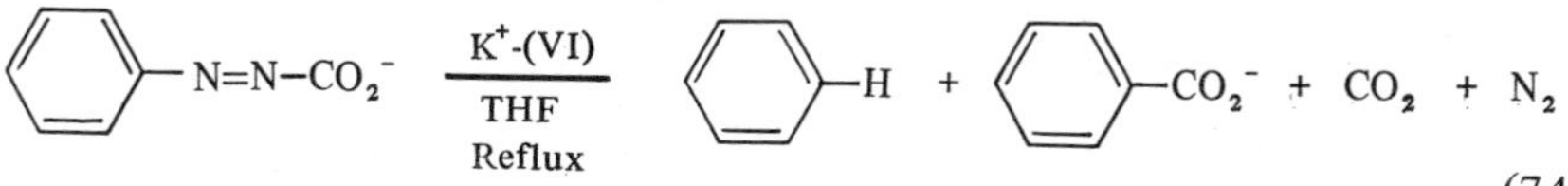

(74)

1 Closed system 50% 20%
2 He ebullition 60% 10%

carried out while helium swept through the reaction mixture, the benzene: potassium benzoate ratio increased. In the presence of amyl acetate small quantities of acetophenone (0.5%) were isolated. The authors concluded that the results are consistent with initial formation of phenylpotassium. Indeed, when phenylpotassium made by another route was dissolved in tetrahydrofuran using dicyclohexo-18-crown-6, proton abstraction took place with the formation of benzene.

Reaction of benzoyl-*t*-butyldiimide with an equivalent of a 1:1 ratio of potassium methoxide and dicyclohexo-18-crown-6 produced an immediate evolution of nitrogen together with a complex mixture of organic products after hydrolysis [Reaction (75)]. A similar distribution of products was obtained in

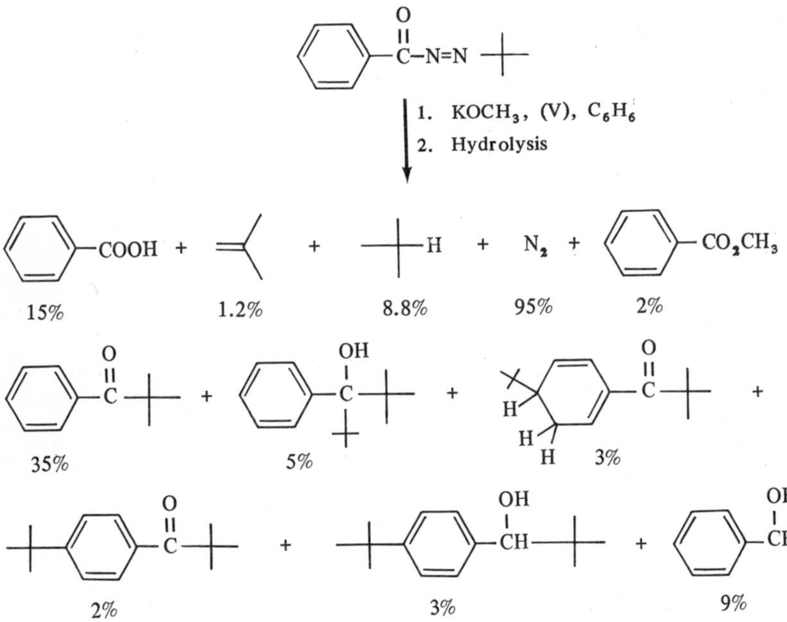

(75)

the reaction of *t*-butyllithium with methyl benzoate. Fraenkel and Pechhold (1970) suggested the following scheme (III) to rationalize the products formed before hydrolysis.

Staley and Erdman (1970) have shown that the 6-methyl-6-phenylcyclohexadienyl anion in liquid ammonia can undergo a variety of competitive inter- and intramolecular reactions at ambient temperatures and that these reactions are greatly influenced by the nature of the counter cation [Reaction (76)]. The results are summarized in Table LV. In the presence of an equivalent amount of dicyclohexo-18-crown-6, the reaction of potassium amide increased the yield of *o*-phenyltoluene to 22%, similar to that for the lithium cation. Staley and Erdman (1970) concluded that this product arose from a solvated lithium cation, since the addition of crown to the case involving potassium ion caused a dramatic increase in yield. The cyclopentadiene products were believed

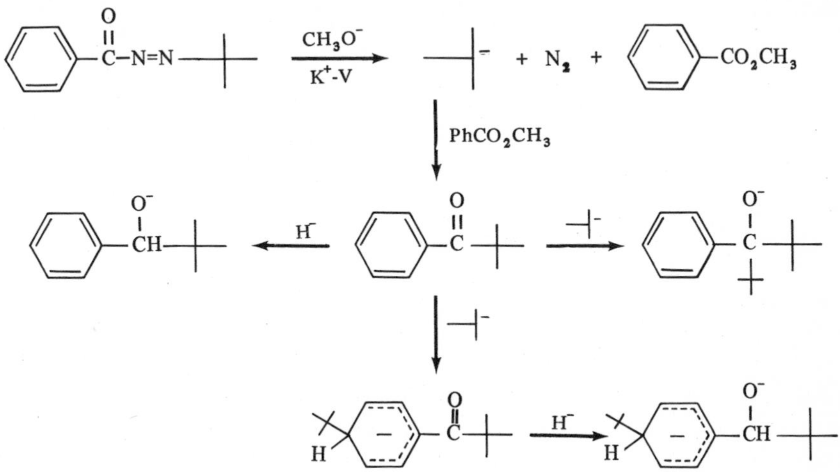

Scheme III

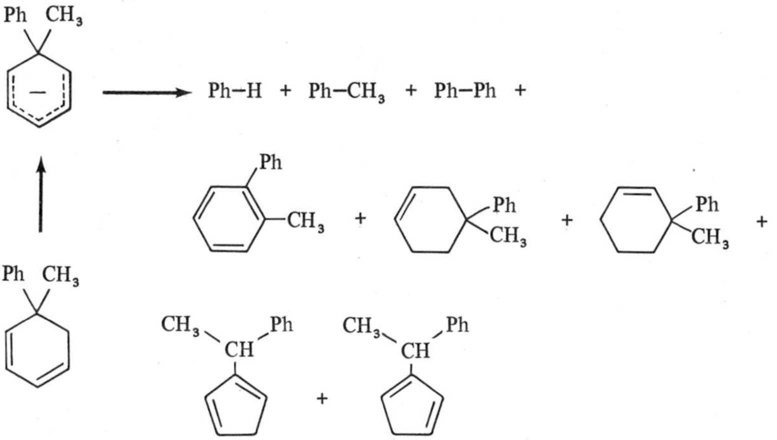

(76)

TABLE LV

Relative Yields of Major Products from the Reaction of 0.08 *M* 5-Methyl-5-Phenyl-1,3-Cyclohexadiene with 0.8 *M* Alkali Metal Amide[a]

Product	$LiNH_2$	$NaNH_2$	KNH_2	$CsNH_2$	KNH_2-(V)
Toluene	53	62	60	60	59
Biphenyl	18	26	20	7	9
Phenyl migration	24	6	6	3	22
Cyclopentadienes	5	6	14	30	10

[a]In liquid ammonia at 25°C for 49 hr.

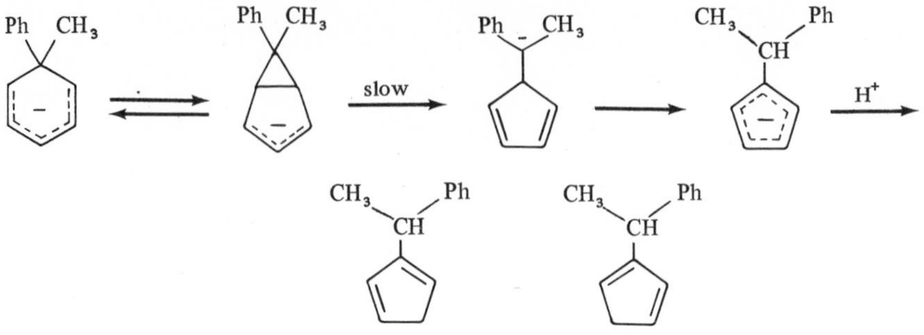

Scheme IV

to arise from the pathway shown in Scheme IV. The opening of the cyclopropyl ring was believed to occur through an intimate ion pair. The authors put forth a hard-soft acid-base explanation to account for the results.

Grovenstein and Williamson (1975) have observed that the reaction of 2,2,3-triphenylpropyllithium in tetrahydrofuran at 0°C rearranges with at least 98% 1,2-migration of benzyl. The corresponding cesium compound in tetrahydrofuran at 65°C gives 96% 1,2-migration of phenyl. The reaction pathways which have been established for these two processes are shown in Scheme V.

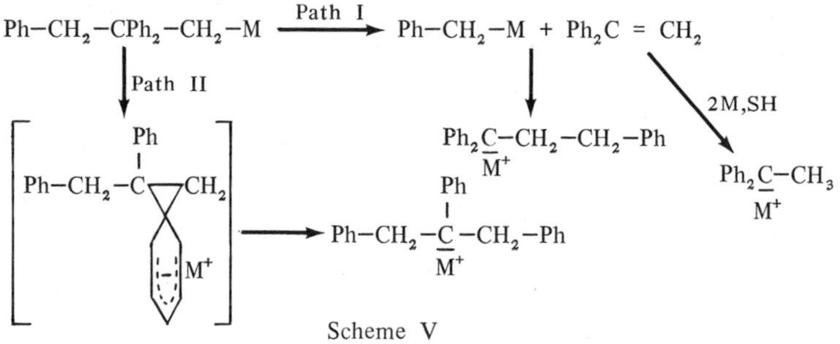

Scheme V

In order to understand further the conditions which favor each of these pathways, Grovenstein and Williamson (1975) have studied the rearrangement of these metal anionic species under a variety of conditions. The results are summarized in Table LVI. The function of the alkali metal *t*-butoxide can be visualized by the following metathesis reaction [Reaction (77)]. That this is reasonable is

$$\begin{aligned} &\mathrm{Ph{-}CH_2{-}CPh_2{-}CH_2{-}Li + M{-}O\text{-}}t\mathrm{\text{-}Bu} \rightleftharpoons \\ &\mathrm{Ph{-}CH_2{-}CPh_2{-}CH_2{-}M + Li{-}O\text{-}}t\mathrm{\text{-}Bu} \end{aligned} \tag{77}$$

suggested by the fact that reaction of the chloride precursor with cesium metal at −75°C gives approximately the same product distribution as the reaction of the lithium carbanion with CsO-*t*-Bu at the same temperature. It is

TABLE LVI

Rearrangements of 2,2,3-Triphenylpropyl Alkali Metal Compounds

Conditions	Temp. (°C)	Products[a] (rel. mol. %)			
		Starting material	Benzyl migration	Phenyl migration	Styrene reduction
$PhCH_2CPh_2CH_2Li$, 7 hr, THF	−75	100	0	0	0
$PhCH_2CPh_2CH_2Li$ + 2(IV), 3.3 hr. THF	−75	100	0	0	0
$PhCH_2CPh_2CH_2Li$, 30 min, THF	0	0	100	0	0
$PhCH_2CPh_2CH_2Li$, 3 hr, Et_2O	+35	0	0	100	0
$PhCH_2CPh_2CH_2Li$ + 2NaO-*t*-Bu, 30 min, THF	−75	33	58	0	9
$PhCH_2CPh_2CH_2Li$ + 2KO-*t*-Bu, 30 min, THF	−75	0	63	37	0
$PhCH_2CPh_2CH_2Li$ + 2CsO-*t*-Bu, 30 min, THF	−75	0	25	72	3
$PhCH_2CPh_2CH_2Li$ + 2KO-*t*-Bu + 2(IV), 30 min, THF	−75	0	100	0	0
$PhCH_2CPh_2CH_2Li$ + 2CsO-*t*-Bu + 2(IV), 30 min, THF	−75	20	77	0	3
$PhCH_2CPh_2CH_2Cl$, K, THF	+65	0	10	90	0
$PhCH_2CPh_2CH_2Cl$, Cs, THF	+65	0	2	96	2
$PhCH_2CPh_2CH_2Cl$, Cs, THF	−75	0	5	67	28
$PhCH_2CPh_2CH_2Cl$ + 2 (IV), excess K, THF	−75		2	0	98
$PhCH_2CPh_2CH_2Cl$ + 2 (IV), excess Cs, THF	−75	0	0	<8	>92

[a]Yields are based only on acidic products from carbonation.

clear from Table LVI that pathway I is favored by low temperatures, good cation-coordinating solvents, effective coordination of alkali metal cation with 18-crown-6, and lithium and sodium as the counter cations. Pathway II, on the other hand, is favored by low temperatures, poor cation-coordinating solvents, the absence of macrocyclic polydentate ligands such as (IV), and large (K^+ and Cs^+) alkali metal counter cations. Another way of stating this is that pathway I is favored by conditions which favor loose or separated ion pairs whereas pathway II is favored by conditions which favor tight or contact ion pairs.

Biellmann and Schmitt (1973) have shown that the sulfur carbanion (XXIX), formed from dibenzylthioether and *n*-butyl lithium in THF containing tetramethylethylenediamine, was stable at −78°C. Upon increasing the temperature, products arising from both Sommelet- and Stevens-type rearrangements were observed [Reaction (78)]. As the temperature of reaction was increased, the

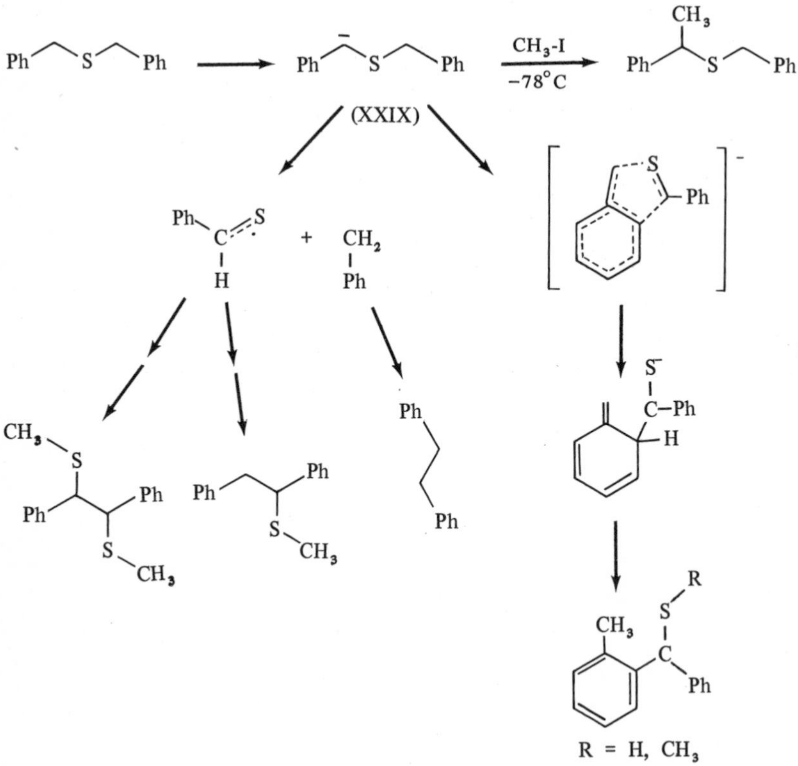

(78)

ratio of Stevens to Sommelet rearrangement increased. Upon addition of 1 equivalent of (IX)c, the Sommelet products dominated, whereas addition of 9-14 equivalents of LiBr resulted in more Stevens rearrangement products. The authors concluded that intimate ion pairs and solvated ions give mostly Stevens-type rearrangement, whereas solvent-separated ion pairs or free ions result in Sommelet-type rearrangement.

Almy *et al.* (1970) have reported studies related to the simultaneous base-catalyzed racemization, isotopic exchange, and isomerization of optically pure (−)-3-*t*-butyl-1-methylindene-1-h (XXVII) and its deuterated counterpart in the one position (XXVIII) to 1-*t*-butyl-3-methylindene under a variety of conditions [Reaction (79)]. With potassium methoxide as catalyst [in CH_3OD

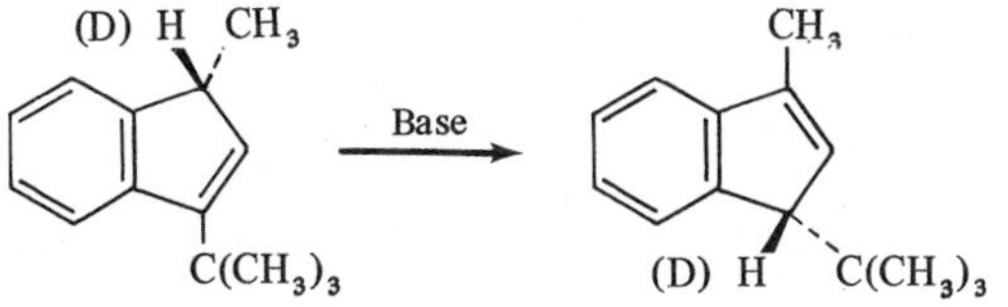

(79)

(XXVII) (H)
(XXVIII) (D)

with (XXVII) and CH_3OH with (XXVIII)], the initial rates of isotopic exchange and loss of optical activity were equal within experimental error. Addition of (VI) produced little change in the rates of the three reactions of (XXVII). With potassium phenoxide in 20% phenol-75% benzene (by volume) the rate of exchange was eight times greater than the rate of loss of optical activity of (XXVII). This rate ratio was decreased to two in the presence of (VI). In addition, the rate of isomerization was decreased by a factor of eight in the presence of crown. With *n*-propylamine-*N*-d_2 in tetrahydrofuran the rate of exchange was 30 times faster than the rate of loss of optical activity. In the presence of (VI), however, this rate ratio fell to 0.6. Crown was observed to have little effect on the rate of isomerization. These observations were rationalized in terms of mechanisms involving contact, solvent-separated, and dissociated ions.

Roitman and Cram (1971) have reported additional stereochemical studies using crown ether. (−)-4-Biphenylylphenylmethoxydeuteriomethane was reported to undergo isotopic exchange and racemization with potassium *t*-butoxide in *t*-butanol in the presence of (VI) at about equal rates [Reaction (80)]. In the

$$p\text{-}C_6H_5{-}C_6H_4{-}\overset{*}{\underset{\displaystyle OCH_3}{\overset{\displaystyle C_6H_5}{C}}}{-}D \xrightarrow[(CH_3)_3C{-}OH]{(CH_3)_3C{-}O^-K^+} p\text{-}C_6H_5{-}C_6H_4{-}\underset{\displaystyle OCH_3}{\overset{\displaystyle C_6H_5}{C}}{-}H \qquad (80)$$

absence of (VI), the exchange rate was 46 times the rate of racemization. The presence of crown, however, had an overall rate-increasing effect on both processes. The cleavage of (+)-4-phenyl-3,4-dimethyl-3-hexanol to 2-phenylbutane with potassium *t*-butoxide in *t*-butanol in the presence of crown proceeded with 15% net retention of configuration [Reaction (81)]. In the absence

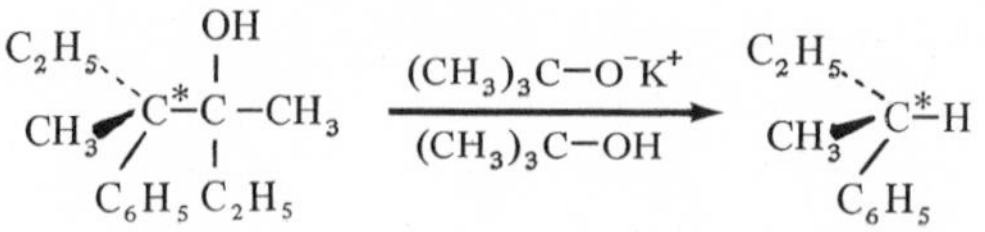

(81)

of crown, however, the reaction proceeded with approximately 90% net retention of configuration. The authors suggested that the organizing capacity of the potassium ion is modified by coordination with the crown. The rates of isotopic exchange and racemization of (+)-2-methyl-2,3-dihydro-2-deuteriobenzo[b]-thiophene-1,1-dioxide with potassium methoxide in methanol was found to be approximately the same in the presence and absence of crown [Reaction (82)].

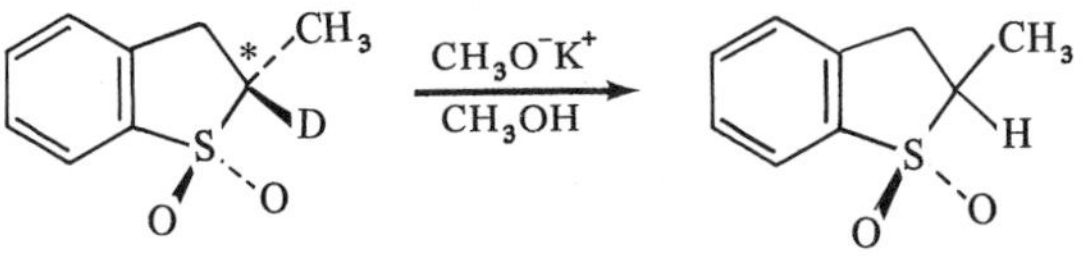

(82)

It was suggested that in media of high polarity, crown ligands tend to have little influence on the stereochemical course of reaction.

Luteri and Ford (1975) have reported the reaction of 2-phenylallylmagnesium phenoxide and *trans*-stilbene in tetrahydrofuran in the absence and in the presence of a variety of cation-coordinating agents to produce moderate yields of r-1, *t*-2, *c*-4-triphenylcyclopentane [Reaction (83)]. The order of effectiveness

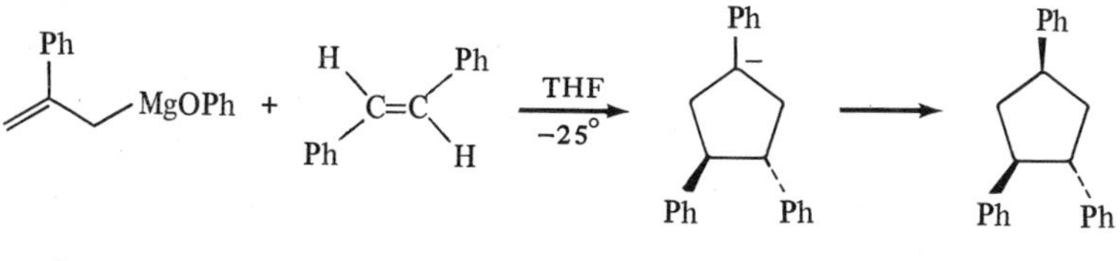

(83)

of the added ligand in catalyzing this reaction is [2.1.1] cryptate > HMPA > dimethyldiaza-15-crown-5 ≈ 15-crown-5 > dimethyldiaza-18-crown-6 > tetramethylcyclam > 12-crown-4. No reaction takes place in the absence of added ligand or in the presence of 18-crown-6 and tetramethylethylenediamine.

Evans and Golob (1975) have shown that the rate of the anionic oxy-Cope rearrangement of (XXX) depends on the nature of the counter cation [Reaction (84)]. When Li^+ or $MgBr^+$ were employed as the counter cations, no rearrangement took place in refluxing THF. When Na^+ was used, the half-life for the rearrangement was 1.2 hr, whereas the K^+ salt underwent rearrangement in several minutes. The yield of methoxy ketone was ≥98%. Upon addition of 1-3 equivalents of (IV) to the K^+ salt in THF at 0°C, rearrangement took place rapidly. Rapid rearrangement of the K^+ salt was also observed in HMPA at 10°C. The authors concluded that ion pair dissociation resulted in maximal rate acceleration and that the rate dependence upon solvent dielectric was negligible.

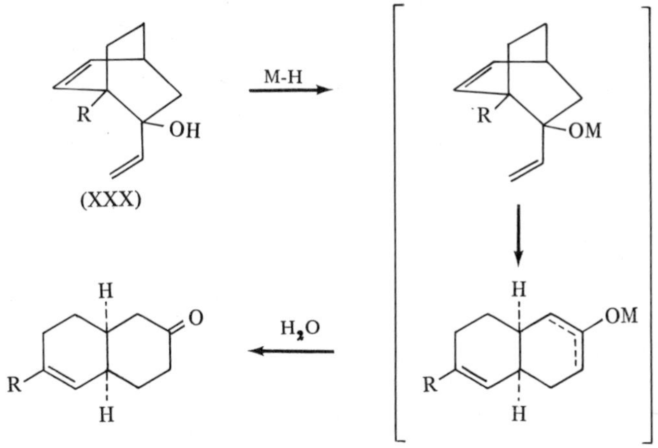

(84)

Martin and Bioch (1971) have solubilized the potassium salt of diazacyclopentadiene-2-carboxylic acid in *o*-xylene using dicyclohexyl-18-crown-6 and thermally generated the dehydrocyclopentadienyl anion, a benzyne analogue, in the presence of tetraphenylcyclopentadienone (tetracyclone) and 3,6-diphenyl-*sym*-tetrazine [Reaction (85)]. The same reactions were observed in the absence

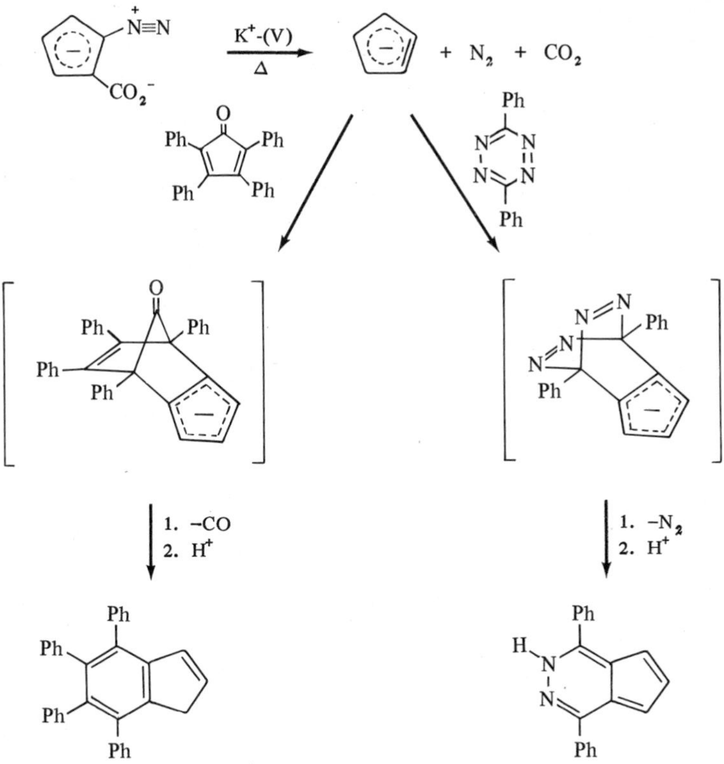

(85)

of solvent from heating a paste of the potassium-crown salt and the trapping agent.

Tabner and Walker (1973) have observed that the lithium, sodium, and potassium salts of 4H-cyclopenta[def] phenanthrene radical anion are considerably more stable in the presence of (VI), especially when sodium or potassium is the counter cation. In this study, the selectivity of the crown (VI) toward alkali metal ions was found to be $Na^+ > K^+ > Li^+$. It was also observed that the order of radical anion decay changed from two (in the absence of crown) to one (with a crown to radical anion ratio of 0.50). The authors suggested that the first-order decay may be due to steric hindrance to a bimolecular decay resulting from the large size of the crown ether.

Cinquini *et al.* (1975) have used the [222] cryptate, dicyclohexyl-18-crown-6 and $n\text{-}C_{16}H_{33}P^+Bu_3Br^-$ as catalysts in the alkylation of phenylacetone with 1.2 mole equivalents of *n*-butyl bromide in 50% aqueous NaOH solution [Reaction (86)]. Greater than 90% yield of product was observed. The catalytic effectiveness varied in the order (IX)c (0.75 hr), (V) (1.5 hr), and (XIV) (2.5 hr).

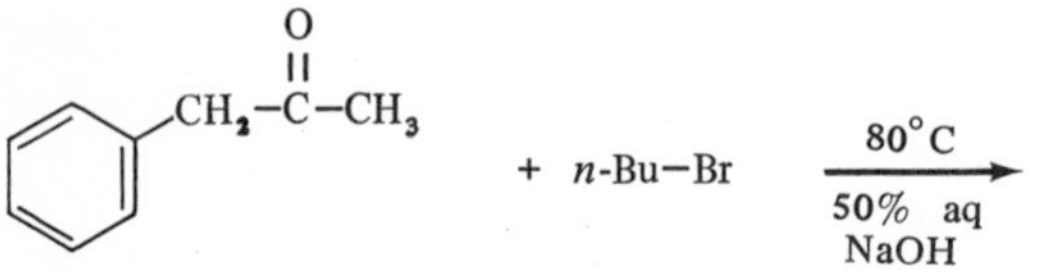

(86)

Makosza and Ludwikow (1974) have reported the use of dibenzo-18-crown-6 as a liquid-liquid phase transfer catalyst in the generation and reaction of carbanions and halocarbenes. Addition of 1 mole % of crown to the two-phase system consisting of an organic phase and a concentrated aqueous NaOH phase resulted in carbanionic sodium species which were soluble in the organic phase, where they underwent further transformation. Among the reactions studied were (a) the alkylation of C-H acids [Reaction (87)], (b) reactions involving trichloromethyl anions and dichloromethylene [Reaction (88)], (c) reaction of carbanions with nitro compounds [Reaction (89)], and (d) Darzens condensation [Reaction (90)].

Smith and Milligan (1968) have reported that the reaction of sodium 9-fluorenone oximate with methyl iodide in 33.5% acetonitrile and 66.5% *t*-butyl alcohol solvent at 25°C gave a concentration-dependent second-order rate

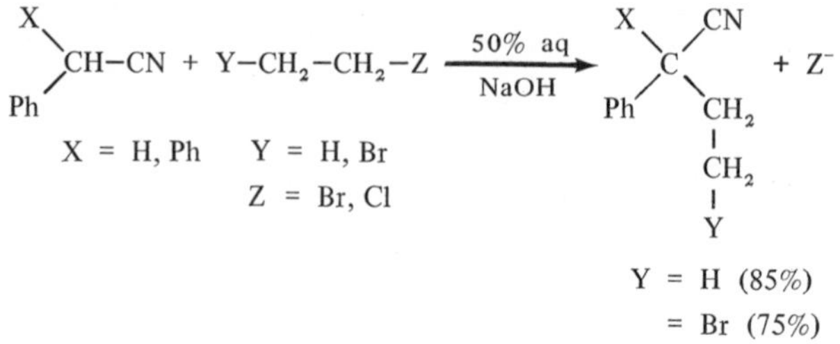

(87)

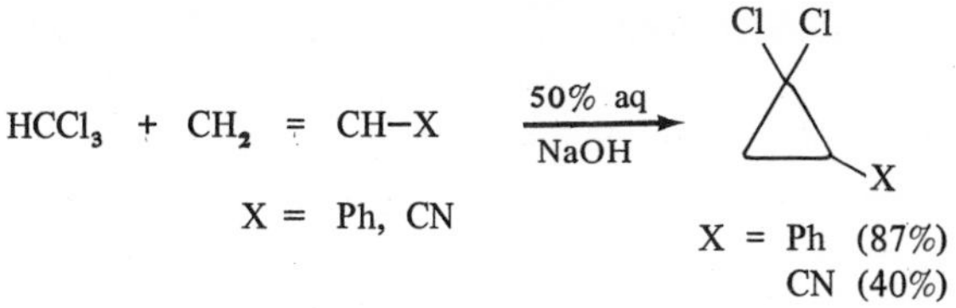

(88)

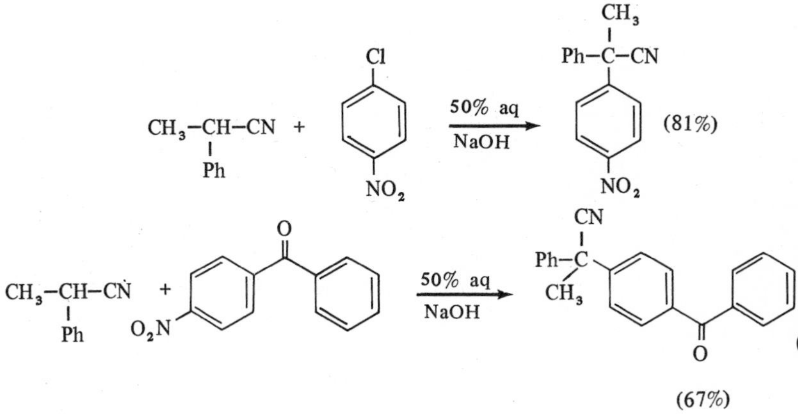

(89)

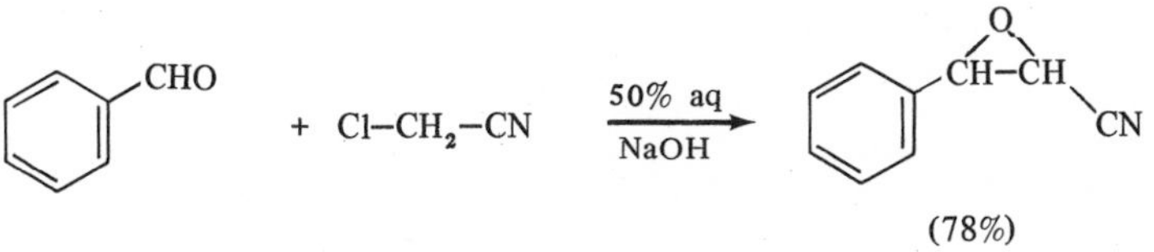

(90)

coefficient and a ratio of oxygen to nitrogen alkylation which was also concentration dependent [Reaction (91)]. Increasing the concentration of the sodium oximate by 10^2 caused a decrease in the second-order rate coefficient by a factor of approximately 8, suggesting that there was an equilibrium between

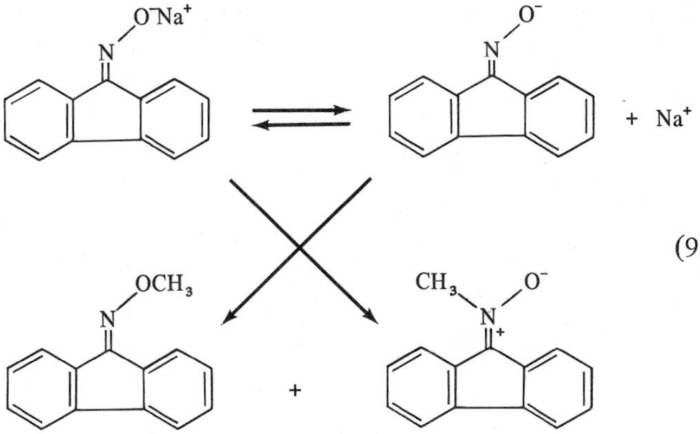

(91)

dissociated and associated species and that the former were more reactive than the latter. Over this same concentration range O-alkylation was found to decrease from 65 to 46%. Support for this interpretation came from the spectroscopic, rate, and product analysis data of Smith and Hansen (1971). Addition of an equivalent amount of (VI) to a solution of the oximate produced a shift in the max from 424 to 470 nm, an observation characteristic of increased ion separation (Wong *et al.*, 1970; Almy *et al.*, 1970; Ford, 1970). The rate and

TABLE LVII

Rate Constants and Products for the Reaction of Sodium Fluorenone Oximate with Methyl Iodide or Methyl Tosylate in 33.5% Acetonitrile-*t*-Butyl Alcohol at 25°

10^2 (Oximate) (*M*)	(VI)	10^2 [$NaBPh_4$) (*M*)	$10^2 k_2$ (liters mol^{-1} sec^{-1})	Oxygen methylation (% yield)
Methyl iodide				
50	51		97 ±5	64
46	48		88 ±9	65
24.3	25.6		105 ±4	61
14.9	15.5		110 ±20	66
7.5	8.2		108 ±5	65
3.6	4.1		99 ±5	65
9.0		7.7	1.37±0.02	46
9.3		10.1	1.44±0.02	43
9.7		10.6	1.4 ±0.1	
9.7		14.0		41
9.1		15.0	1.34	41
9.5		17.0		40
Methyl tosylate				
55.8	56.1		66 ±1	97
53.0	52.5		54 ±3	
25.5	28.0		61 ±1	98
21.4	22.5		69 ±0.1	
17.4	19.0		67 ±4	99
10.0	11.0		67 ±4	99
9.8	10.5		69	
5.3	5.6		69	99
5.2	5.1		60 ±1	95
9.0		7.7	1.5 ±0.1	48
9.3		10.1	1.59±0.04	45
9.7		10.6	1.67±0.01	
9.5		14.2	1.9 ±0.1	
9.1		15.0		42
9.5		17.0		43

product analysis data for the reaction of the sodium oximate with methyl iodide and methyl tosylate (Table LVII) also supported this interpretation. In the presence of an equivalent amount of (VI), the second-order rate coefficient for the reaction with methyl iodide remained at approximately 65%, the previously extrapolated value for the reaction run at infinite dilution in the absence of crown. The presence of $Na^+BPh_4^-$ in place of crown suppressed the ionization of the sodium oximate and allowed the exploration of the properties of the associated species. Under these conditions the rate coefficient for reaction with methyl iodide was depressed by a factor of 70 from that observed in the presence of crown ether and the percentage of O-alkylation was decreased to approximately 42%. Smith and Hansen concluded that the free ion alkylated perferentially at oxygen whereas the ion pair alkylated primarily at nitrogen. The data in Table LVII concerning CH_3OTs as alkylating agent provided further support for the foregoing interpretation and clearly demonstrated that the nature of the leaving group on the alkylating agent influenced the O/N alkylation ratio.

Zaugg *et al.* (1972) have reported a series of kinetic studies involving the alkylation of sodio diethyl *n*-butylmalonate with *n*-butyl bromide in the solvents benzene and cyclohexane in the presence of varying concentrations of the additives THF, DME, DMF, and (V). The results are summarized in Tables LVIII and LIX. The absence of a common-ion rate depression in pure DMF, an

TABLE LVIII

Salt Effects on Alkylation Rates at 25°C in DMF of Diethyl Sodio-*n*-Butylmalonate with *n*-Butyl Bromide

Salt	Concentration (M)	$10^5 k_2$ (M^{-1} sec^{-1})
None	0	323 ± 6
		345 ± 7
		339 ± 5
$NaClO_4$	0.101	338 ± 3
		332 ± 3
$LiClO_4$	0.100	35.0 ± 0.8
	0.101	46.2 ± 0.5

excellent cation-solvating medium, indicated that the alkylation reaction is an ion pair process (Table LVIII). The second-order rate constant increased with increasing concentration of additive (Table LIX). Sedimentation and nmr studies ruled out a sparsely solvated ion pair as the reactive species. In benzene solutions, only 0.036 M in (V), the alkylation rate is already equal to that observed in neat (8.70 M) DMF (Table LIX). Zaugg *et al.* (1972) suggested that a highly solvated species is the reactive nucleophilic agent, since the more donor sites that are accommodated by the cation, the more its charge will tend to

TABLE LIX

Alkylation Rates at 25°C of Diethyl Sodio-*n*-Butylmalonate with *n*-Butyl Bromide in Four Binary Solvent Systems

Concentration (M)	$10^5 k_2$, (M^{-1} sec^{-1})	Concentration (M)	$10^5 k_2$ (M^{-1} sec^{-1})	Concentration (M)	$10^5 k_2$ (M^{-1} sec^{-1})
A. Dicyclohexo-18-crown-6 (V) in benzene		B. DME in cyclohexane		D. Dicyclohexo-18-crown-6 (V) in THF	
0.0116	5.72 ± 0.17	0.301	0.93 ± 0.02	0.01	13.5 ± 0.5
0.0164	11.2 ± 0.8	0.388	1.15 ± 0.01	0.02	19.7 ± 0.4
0.0228	17.0 ± 0.8	0.684	1.48 ± 0.02	0.03	24.5 ± 0.3
0.0249	18.3 ± 1.3	0.954	1.84 ± 0.02	0.04	29.7 ± 0.5
0.0290	21.1 ± 1.1	2.820	9.11 ± 0.22	0.06	41.1 ± 0.6
0.0305	23.0 ± 1.2	4.720	18.7 ± 0.6	0.10	55.3 ± 0.5
0.0313	23.3 ± 2.0	6.710	25.5 ± 0.1	0.20	69.2 ± 0.9
0.0358	30.6 ± 1.2			0.50	84.7 ± 1.2
0.0406	32.7 ± 0.8	C. DMF in cyclohexane			
0.0447	42.1 ± 1.7	0.308	2.91 ± 0.06		
0.0509	43.2 ± 0.5	0.464	4.97 ± 0.09		
0.0741	51.3 ± 1.2	0.681	7.69 ± 0.13		
0.1015	63.6 ± 0.5	1.194	15.96 ± 0.55=		
0.202	95.4 ± 1.6	3.293	53.1 ± 0.6		
0.527	138.1 ± 2.4				

become dispersed, with a resultant increase in the nucleophilicity of the anionic portion of the ion pair. The potent cation-solvating ability of crown was clearly illustrated by its extraordinary capacity to accelerate alkylation rates at concentrations equivalent to that of the substrate; a rate maximum was observed in both benzene and THF. These observations suggested that this multidentate ligand possessed enough donor sites in one molecule to provide maximum solvation and it was the monosolvated ion pair which was the reactive species in the crown-catalyzed reaction. It was observed that small quantities of crown have a greater effect in benzene as compared to THF, reflecting the better cation-solvating ability of THF.

Kurts *et al.* (1974) have reported that addition of (V) to the alkali metal salts of ethyl acetoacetate in dioxane caused the destruction of ionic agglomerates and formation of monomeric ion pairs. In the reaction of this enolate with

TABLE LX

Rate Constants and Ratio of C and O Isomers in Ethylation of the Enolate of Ethyl Acetoacetate with Ethyl Tosylate in Dioxane at 25°C

Cation	$c \times 10^2$ $(M)^a$	(V) c_{en}	$k_{obs} \times 10^5$ (liters mole^{-1} sec^{-1})	C/O ratio
Na^+	1.83	4.3	6.3	2.3
	2.00	2.0	6.27	2.3
	1.95	1.0	2.05	–
	1.95	0.5	0.82	–
K^+	2.02	3.8	27.0	0.70
	1.97	3.0	27.0	0.70
	5.00	3.2	27.0	0.70
	1.35	2.0	26.6	0.70
	0.82	2.0	27.0	0.70
	2.14	1.9	26.0	0.65
	2.18	1.0	20.5	–
	1.83	0.77	14.2	–
	1.90	0.53	10.0	–
	1.88	0.27	5.0	–
	2.06	0.00	<0.01	–
Rb^+	1.96	3.0	44.0	1.10
	2.00	2.0	44.6	1.05
Cs^+	0.96	6.0	93.4	1.00
	1.85	6.5	94.0	1.00
	1.85	4.3	94.1	1.00
	0.98	2.2	69.1	–
Ph_4As^+	1.04	–	700	0.27
	1.02	–	800	0.26
	1.70	–	680	0.27

[a]Initial concentrations of the enolate and the ethyl tosylate.

ethyl tosylate a marked acceleration of rate was observed in the presence of (V) (2000-fold in the presence of equimolar crown). The second-order rate coefficients as a function of crown concentration and the carbon versus oxygen alkylation product ratios are summarized in Table LX. At (V)/enolate ratios greater than two, the C/O ratio remained essentially constant for a given alkali metal. Since the enolate can exist in a number of reactive conformations [(XXXI), (XXXII), and (XXXIII)], the authors also examined the ratio of cis/trans O-alkylated products. The results are summarized in Table LXI. It may be seen that as the radius of the cation increased (Na < K < Rb < Cs), a gradual increase in the proportion of O-alkylated product in the trans configuration was observed.

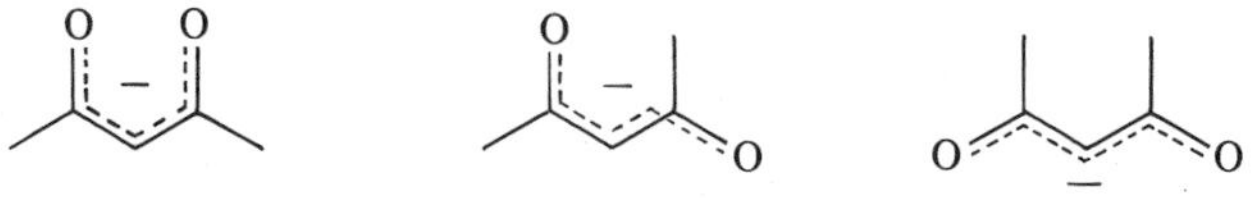

(XXXI) (XXXII) (XXXIII)

TABLE LXI

Values of Rate Constants with Respect to C and O Centers of the Ion Pairs of the Enolates of Ethyl Acetoacetate and the Crystallographic Radii (r) of the Cations

Cation	r (Å)	$k_{obs} \times 10^{5\,a}$	$k_P^C \times 10^{5\,a}$	$k_P^O \times 10^{5\,a}$	O-*cis* / O-*trans*
Na^+	0.97	6.3	4.4	1.9	0.60
K^+	1.33	27	16.0	11.0	0.43
Rb^+	1.48	44	23.0	21.0	0.43
Cs^+	1.69	93	45.5	47.5	0.12
Ph_4As^+	4.30	720	160	560	0.00
Free anion	∞	2600	–	–	–

[a]In liters $mole^{-1}$ sec^{-1}.

Cambellau *et al.* (1976) have investigated the reaction of ethyl tosylate and ethyl iodide with the sodium salt of ethyl acetoacetate in tetrahydrofuran in the presence and absence of (IV) and (IX)c. The results are summarized in Tables LXII and LXIII and in Scheme VI. The results with ethyl tosylate were interpreted as follows. In pure tetrahydrofuran, the sodium enolate existed primarily as ion-paired aggregates. Upon addition of (IV) or (IX)c the aggregates were broken up to form species which were more nucleophilic than the aggregated ions, gave a greater ratio of oxygen to carbon alkylation, and gave a greater ratio of *trans*- to *cis*-oxygen alkylation. Since oxygen is more electronegative than carbon, the naked enolate would have most of its negative charge at the oxygen atoms, thus increasing the extent of attack at these centers. In addition, in

TABLE LXII

Reaction of the Sodium Salt of Ethyl Acetoacetate with Ethyl Tosylate in THF at 50°C[a]

Rate (M^{-1} min^{-1})	Without complexing agent	(IV)/enolate =1	(IV)/enolate =10	(IX)c/enolate =1	(IX)c/enolate =1.5
k_{obs}	0.7×10^{-2}	3.1×10^{-2}	10.5×10^{-2}	1.05	1.65
$k_{obs}/k_{obs\ pure\ THF}$	1	4.41	15	150	235
k_O	0.07×10^{-2}	1.35×10^{-2}	5.2×10^{-2}	0.88	1.39
$k_O/k_{O\ pure\ THF}$	1	19	74	1250	2000
k_C	0.63×10^{-2}	1.75×10^{-2}	5.2×10^{-2}	0.17	0.26
$k_C/k_{C\ pure\ THF}$	1	2.8	8.2	27	41
% O total	9	43	50	84	84
100(O-trans/O-total)	10	70	85	100	100

[a] [Enolate] = 0.02 *M*; [EtOTs] = 0.30 *M*.

TABLE LXIII

Reaction of the Sodium Salt of the Ethyl Acetoacetate with Ehtyl Iodide in THF at 40°C[a]

Rate (M^{-1} min^{-1})	Without complexing agent	(IV)/enolate = 1	(IX)c/enolate = 1
k_{obs}	3×10^{-2}	2.7×10^{-1}	11
k_{obs}/k_{obs} pure THF	1	9	370
% O total	0	Trace (1%)	25

[a] [Enolate] = 0.02 *M*; [EtI] = 0.02-0.30 *M*.

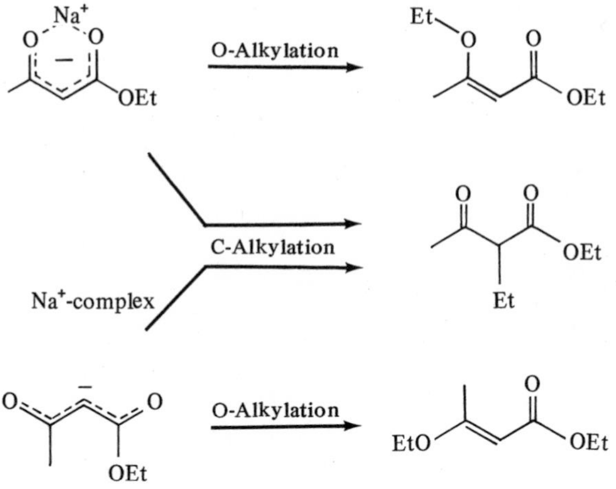

Scheme VI

order to minimize electrostatic repulsions between the two partially negatively charged oxygens, the more extended W-conformation became dominant, thus producing the observed trend in *trans*- to *cis*-oxygen alkylation. It is interesting to note that the cryptate was much more effective in producing these reactive species than the crown. A maximum rate effect was obtained with a crown to enolate ratio of 10, whereas 1 equivalent of cryptate was 10 times as effective. Kurts (1974) had observed that a maximum rate was attained when 2 equivalents of crown were added, but these studies were carried out in the solvent dioxane, which is a poorer coordinating agent for metal cations. In the reaction with ethyl iodide, the enolate produced no O-alkylation product in the absence of complexing agent. In the presence of 1 equivalent of (IV) only a trace of O-alkylation was observed, whereas in the presence of (IX)c 25% O-alkylation was achieved. These results are reminiscent of those of Kurts *et al.*, who observed 13% O-alkylation in HMPA with ethyl iodide and 88% O-alkylation with ethyl tosylate.

N. Alkali Metals

Dye *et al.* (1970) have reported the solubilization of potassium and cesium metals in tetrahydrofuran and ethyl ether using (V). Metal concentrations of approximately 1×10^{-4} *M* were obtained using 5×10^{-3} *M* solutions of crown. Solution of the metals in THF were performed at room temperature, while solution of potassium in ethyl ether was carried out at −78°C. At −78°C, the THF-metal solutions were stable for days. Electron paramagnetic resonance analysis indicated that potassium monomer was present to the extent of 10^{-7} *M*, while the concentration of solvated electron was approximately 10^{-8} *M*. Optical absorption spectra indicated that the major species in the presence of crown was the alkali metal anion, M^-, arising from the solubility equilibrium [Reaction (92)].

$$2M\ (s) \longrightarrow M^+ + M^- \qquad (92)$$

Lok *et al.* (1972) have also reported the solubility enhancement of sodium and potassium metals in secondary amines and in straight and branched chain ethers in the presence of crown ether (V) and cryptate (IX)c. The alkali metal anions, Na^- and K^-, were identified from optical spectra when (V) was used as the solubilizing agent. In contrast to this, the solvated electron was the major species in the presence of (IX)c.

Kormarynsky and Weissman (1975) have reported the preparation of the anion radicals of benzene and toluene in nonpolar solvents and in the parent solvents using (V) in contact with a potassium mirror. Evidence was presented suggesting ion pairing between the radical anion and the counter cation. Kaempf *et al.* (1974) have reported the dissolution of Na, K, Rb, and Cs in benzene and toluene or in cyclic ethers containing these aromatic hydrocarbons in the presence of (V) and (IX)c to produce the corresponding radical anions. The interaction between the complexed cation and the radical anion was shown to be weaker when (IX)c was employed. The anion radicals formed from biphenyl, naphthalene, and anthracene were also reported. Nelson and von Zelewsky (1975) reported the use of (IV) in the formation of the radical anions of mesitylene, toluene, and benzene.

REFERENCES

Adams, R., and Thal, A. F. (1948). *Chem. Rev.* **42**, 107.
Allen, F. H. (1948). *Chem. Rev.* **42**, 156.
Almy, J., Garwood, D. C., and Cram, D. J. (1970). *J. Am. Chem. Soc.* **92**, 4321.
Barton, D. H. R., Lier, E. F., and McGhie, J. F. (1968). *J. Chem. Soc. C* 1031.

Bartsch, R. A., and Wiegers, T. K. E. (1972). *Tetrahedron Lett.* 3819.
Bartsch, R. A. *et al.* (1973). *J. Am. Chem. Soc.* **95**, 6745.
Bartsch, R. A., and Shelly, T. A. (1973). *J. Org. Chem.* **38**, 2911.
Bartsch, R. A., and Kayser, R. H. (1974). *J. Am. Chem. Soc.* **96**, 4346.
Bartsch, R. A., Mintz, E. A., and Parlman, R. M. (1974). *J. Am. Chem. Soc.* **96**, 4249.
Bartsch, R. A. (1975). *Accounts Chem. Res.* **8**, 239.
Biellmann, J. F., and Schmitt, J. L. (1973). *Tetrahedron Lett.* 4615.
Boden, R. M. (1975). *Synthesis* 783.
Bohme, D. K., Machay, G. I., and Payzant, J. D. (1974). *J. Am. Chem. Soc.* **96**, 4027.
Borchardt, J. K., and Saunders, W. H. (1974). *J. Am. Chem. Soc.* **96**, 3912.
Bowers, A. (1961). *J. Org. Chem.* **26**, 2043.
Bowers, M. T., Ave, D. H., Webb, H. M., and McIver, R. T. (1971). *J. Am. Chem. Soc.* **94**, 4314.
Brandstrom, A. *et al.* (1969). *Acta Chem. Scand.* **23**, 2202.
Brauman, J. I., Olmstead, W. N., and Lieder, C. A. (1974). *J. Am. Chem. Soc.* **96**, 4030.
Cambillau, C., Sarthou, P., and Bram, G. (1976). *Tetrahderon Lett.* 281.
Cardillo, G., Orena, M., and Sandri, S. (1976). *J. Chem. Soc. Chem. Commun.* 190.
Christensen, J. J., Hill, J. O., and Izatt, R. M. (1971). *Science* **174**, 459.
Christensen, J. J., Eatough, D. J., and Izatt, R. M. (1974). *Chem. Rev.* **74**, 351.
Cima, F. Del, Biggi, G., and Pietra, F. (1973). *J. Chem. Soc. Perkin Trans.* **2**, 55.
Cinquini, M., Montanari, F., and Tundo, P. (1975). *Chem. Commun.* 393.
Coetzee, J. F., and Sharpe, W. R. (1972), *J. Solut. Chem.* **1**, 77.
Corey, E. J., Nicolaou, K. C., Shibasaki, M., Machida, Y., and Shiner, C. S. (1975). *Tetrahedron Lett.* **37**, 3183.
Cook, F. L., Bowers, C. W., and Liotta, C. L. (1974). *J. Org. Chem.* **39**, 3416.
Cram, D. J., and Cram, J. M. (1974). *Science* **183**, 803.
Curci, R., and Difuria, F. (1975). *Int. J. Chem. Kinet.* **7**, 341.
Dehm, D., and Padwa, A. (1975). *J. Org. Chem.* **40**, 3139.
Dehmlow, E. V. (1974a). *Angew. Chem.* **86**, 187.
Dehmlow, E. V. (1974b). *Angew Chem. Int. Ed. English* **13**, 170.
Dehmlow, E. V. (1975). *Chemtech* 210
Dietrich, B., and Lehn, J. M. (1973). *Tetrahedron Lett.* 1225.
Dietz, R., Forno, A. E. J., Larcombe, B. E., and Peover, M. E. (1970). *J. Chem. Soc. B* 816.
Dockx, J. (1973). *Synthesis* **8**, 441.
Durst, H. D. (1974). *Tetrahedron Lett.* 2421.
Durst, H. D., Zubrick, J. W., and Kieczykowski, G. R. (1974). *Tetrahedron Lett.* 1777.
Durst, H. D., Milano, M., Kikta, E. J., Connelly, S. A., and Grushka, E. (1975). *Anal. Chem.* **47**, 1797.
Dye, J. L., DeBacker, M. G., and Nicely, V. A. (1970). *J. Am. Chem. Soc.* **92**, 5226.
Edwards, J. O. (1954). *J. Am. Chem. Soc.* **76**, 1540.
Edwards, J. O. (1968). *J. Chem. Educ.* **45**, 386.
Engemyr, L. B., and Songstad, J. (1972). *Acta Chem. Scand.* **26**, 4179.
Evans, D. A., and Golob, A. M. (1975). *J. Am. Chem. Soc.* **97**, 4765.
Fiandanese, V., Marchese, G., Naso, F., and Sciacovelli, O. (1973). *J. Chem. Soc. Perkin Trans.* **2**, 1336.
Fishman, J., and Torigoe, M. (1965). *Steroids* **5**, 599.
Ford, W. T. (1970). *J. Am. Chem. Soc.* **92**, 2857.
Fraenkel, G., and Pechhold, E. (1970). *Tetrahedron Lett.* 153.
Freidman, L., and Schechter, J. (1960). *J. Org. Chem.* **25**, 877.
Gokel, G. W., and Durst, H. D. (1976a). *Aldrichchim. Acta* **9**, 3.
Gokel, G. W., and Durst, H. D. (1976b). *Synthesis* 168.

Grimsrud, E. P., and Kratcchvil, B. (1973). *J. Am. Chem. Soc.* **95**, 4477.
Grovenstein, E., and Williamson, R. E. (1975). *J. Am. Chem. Soc.* 97, 646.
Grunwald, E., and Leffler, J. E. (1963). "Rates and Equilibria of Organic Reactions." Wiley, New York.
Grushka, E., Durst, H. D., and Kikta, E. J. (1975). *J. Chromatogr.* **112**, 673.
Handel, H., and Pierre, J. L. (1975). *Tetrahedron* **31**, 997.
Henbest, H. B., and Jackson, W. R. (1967), *J. Chem. Soc. C* 2465.
Herriott, A. W., and Picker, D. (1975). *J. Am. Chem. Soc.* 97, 2345.
House, H. O., Cope, A. C. and Holmes, H. L. (1957). *Org. React.* 9, 107.
House, H. O., and Richey, F. A. (1969). *J. Org. Chem.* **34** (5), 1430.
Hudson, R. F. (1974). *In* "Chemical Reactivity and Reaction Paths" (G. Klopman, ed.), Chapter 5. Wiley, New York.
Hunter, D. H., and Shearing, D. J. (1971). *J. Am. Chem. Soc.* **93**, 2348.
Hunter, D. H., Lee, W., and Sins, S. K. (1974). *J. Chem. Soc., Chem. Comm.* 1018.
Johnson, R. A., and Nidy, E. G. (1975). *J. Org. Chem.* **40**, 1680.
Johnson, W. S., and Keane, J. F. W. (1963). *Tetrahedron Lett.* 193.
Kaempf, B., Raynal, S., Collet, A., Schué, F., Boileau, L., and Lehn, J. M. (1974). *Angew. Chem.* **86**, 670.
Kebarle, P. (1972). Ions and Ion-Pairs in Organic Reactions, *In* "Ions and Ion-Solvent Molecule Interactions in the gas phase" (M. Szwarc, ed.). Wiley (Interscience), New York.
Knöchel, A., and Rudolph G. (1974). *Tetrahedron Lett.* 3739.
Knöchel, A., Rudolph, G., and Thiem, J. (1974). *Tetrahedron Lett.* 551.
Knöchel, A., Oehler, J., and Rudolph, G. (1975). *Tetrahedron Lett.* **36**, 3167.
Kormarynsky, M. A., and Weissman, S. I. (1975). *J. Am. Chem. Soc.* 97, 1589.
Kosower, E. M. (1968). *In* "Physical-Organic Chemistry," pp. 77-81, 337-339, Wiley, New York.
Kostikov, R. R., and Molchanov, A. P. (1975). *Zh. Org. Khim.* **11**, 1767.
Kurts, A. L., Sakembaeva, S. M., Beletskaya, I. P., and Reutov, O. A. (1974). *Zh. Org. Khim.* **10**, 1572.
Landini, D., Montanari, F., and Pirisi, F. M. (1974). *Chem. Commun.* 879.
Lehn, J. M. (1973). *Struct. Bond.* **16**.
Liotta, C. L. and Abidaud, A., unpublished results.
Liotta, C. L., and Harris, H. P. (1974). *J. Am. Chem. Soc.* **96**, 2250.
Liotta, C. L., Harris, H. P., McDermott, M., Gonzalez, T., and Smith, K. (1974). *Tetrahedron Lett.* 2417.
Liotta, C. L., Grisdale, E. E., and Hopkins, H. P. (1975). *Tetrahedron Lett.*, 4205.
Lok, M. T., Tehan, F. J., and Dye, J. L. (1972): *J. Phys. Chem.* **76**, 2975.
Luteri, G. F., and Ford, W. T. private communication.
Mack, M. M., Dehm, D., Boden, R., and Durst, H. D., *Tetrahedron Lett.*
Makosza, M. (1966). *Tetrahedron Lett.* 4621.
Makosza, M., and Ludwikow, M. (1974). *Angew Chem. Int. Ed. English* **13**(10), 665.
Marshall, J. A., and Johnson, W. S. (1962). *J. Am. Chem. Soc.* **84**, 1485.
Martin, J. C., and Bioch, D. R. (1971). *J. Am. Chem. Soc.* **93**, 451.
Marvel, C. S., and McColm, E. M. (1941). "Organic Syntheses," 2nd ed., Collect. Vol. 1, p. 536. Wiley, New York.
Maskornick, M. J. (1972). *Tetrahedron Lett.* 1797.
Matsuda, T., and Koida, K. (1973). *Bull. Chem. Soc. Jpn.* **46**, 2259.
McIver, R. T. (1970). *Rev. Sci. Instrum.* **41**, 555.
Merritt, M. V., and Sawyer, D. T. (1970). *J. Org. Chem.* **35**, 2157.
Meyer, W., and Schnautz, N. (1962). *J. Org. Chem.* **27**, 2011.
Meyer, W., and Wolfe, J. (1964). *J. Org. Chem.* **29**, 170.

Moss, R. A., and Pilkiewicz, F. G. (1974). *J. Am. Chem. Soc.* **96**, 5632.
Moss, R. A., Joyce, M. A., and Pilkiewicz, F. G. (1975). *Tetrahedron Lett.* 2425.
Mowry, D. J. (1948). *Chem. Rev.* **42**, 189.
Nagata, W., Terasawa, T., and Takeda, K. (1960). *Tetrahedron Lett.* **17**, 27.
Nagata, W. (1961). *Tetrahedron* **13**, 278.
Nagata, W., Narisada, M., and Sugasawa, T. (1962). *Tetrahedron Lett.* 1041.
Nagata, W., Kikkawa, K., and Fujimoto, M. (1963a). *Chem. Pharm. Bull.* **11**, 226.
Nagata, W., Sugasawa, T., Narisada, M., Wakabayoshi, T., and Hayase, Y. (1963b). *J. Am. Chem. Soc.* **85**, 2342.
Chem. Soc. **89**, 1483.
Nagata, W., Yoshioka, M., and Hirai, S. (1972a). *J. Am. Chem. Soc.* **94**, 4635.
Nagata, W., Yoshioka, M., and Murakami, M. (1972b). *J. Am. Chem. Soc.* **94**, 4644, 4654.
Nagata, W., Yoshioka, N., and Murakawa, M. (1972c). *J. Am. Chem. Soc.* **94**, 4654.
Nagata, W., Yoshioka, M., and Terasawa, T. (1972d). *J. Am. Chem. Soc.* **94**, 4672.
Nagata, W., Sugasawa, T., Narisada, M., Wakabayashi, T., and Hayase, Y. (1963b). *J. Am.*
Naso, F., and Ronzini, L. (1974). *J. Chem. Soc. Perkin Trans.* **1**, 340.
Nelson, G. V., and von Zelewsky, A. (1975). *J. Am. Chem. Soc.* 97, 6279.
Parker, A. J. (1969). *Chem. Rev.* **69**, 1.
Pearson, R. G., and Songstad, J. (1967). *J. Org. Chem.* **32**, 2899.
Pedersen, C. J. (1967a). *J. Am. Chem. Soc.* **89**, 2495.
Pedersen, C. J. (1967b). *J. Am. Chem. Soc.* **89**, 7017.
Pedersen, C. J. (1968). *Fed. Proc. Fed. Am. Soc. Exp. Biol.* **27**, 1305.
Pedersen, C. J. (1970a). *J. Am. Chem. Soc.* **92**, 386.
Pedersen, C. J. (1970b). *J. Am. Chem. Soc.* **92**, 391.
Pedersen, C. J. (1971a). *Aldrichchim. Acta* **4**, 1.
Pedersen, C. J. (1971b). *J. Org. Chem.* **36**, 254.
Pedersen, C. J. (1972). *In* "Organic Synthesis," Vol. 52, p. 66. Wiley, New York.
Pedersen, C. J., and Frensdorff, H. K., (1972). *Angew. Chem. Int. Ed. English* **11**, 16.
Pierre, J. L., and Handel, H. (1974). *Tetrahedron Lett.* 2317.
Ritchie, C. D. (1974). *Accounts Chem. Res.* **5**, 348.
Ritchie, C. D., and Virtanen, P. O. I. (1972). *J. Am. Chem. Soc.* **94**, 4966.
Rodig, O. R., and Johnston, N. J. (1969). *J. Org. Chem.* **34**(6), 1942 & 1949.
Roitman, J. N., and Cram, D. J. (1971). *J. Am. Chem. Soc.* **93**, 2231.
Sam, D. J., and Simmons, H. E. (1972). *J. Am. Chem. Soc.* **94**, 4024.
Sam, D. J., and Simmons, H. E. (1974). *J. Am. Chem. Soc.* **96**, 2252.
San Filippo, J., Chern, C., and Valentine, J. S. (1975). *J. Org. Chem.* **40**, 1678.
San Filippo, J., Chern, C., and Valentine, J. S. (1976). *J. Org. Chem.* **41**, 1077.
San Filippo, J., Romano, L. J., Chern, C., and Valentine, J. S. (1976). *J. Org. Chem.* **41**, 586.
Saunders, W. H., Bushman, D. G., and Cockerill, A. F. (1968). *J. Am. Chem. Soc.* **90**, 1775.
Sepp, D. T., Scherer, K. V., and Weber, W. P. (1974). *Tetrahedron Lett.* 2983.
Shchori, E., and Jaqur-Grodzinski, J. (1972a). *Israel J. Chem.* **10**, 959.
Shchori, E., and Jaqur-Grodzinski, J. (1927b). *Israel J. Chem.* **10**, 935.
Shchori, E., and Jaqur-Grodzinski, J. (1972c). *J. Am. Chem. Soc.* **94**, 7957.
Sicher, J. (1972). *Angew. Chem. Int. Ed. English* **11**, 200.
Smiley, R. A., and Arnold, C. (1960). *J. Org. Chem.* **25**, 257.
Smith, S. G., and Milligan, D. V. (1968). *J. Am. Chem. Soc.* **90**, 2393.
Smith, S. G., and Hanson, M. P. (1971). *J. Org. Chem.* **36**, 1931.
Staley, S. W., and Erdman, J. P. (1970). *J. Am. Chem. Soc.* **92**, 3832.
Starks, C. M. (1971a). *J. Am. Chem. Soc.* **93**, 195.

Starks, C. M. (1973). *J. Am. Chem. Soc.* **95**, 3613.
Starks, C. M., and Owens, R. W. (1973). *J. Am. Chem. Soc.* **95**, 3613.
Streitweiser, Jr., A. (1962). "Solvolytic Displacement Reactions." McGraw-Hill, New York.
Sugimoto, N. *et al.* (1962). *Chem. Pharm. Bull.* **20**, 427.
Svoboda, M., Hapala, J., and Zavada, J. (1972). *Tetrahedron Lett.* 265.
Swain, C. G., and Scott, C. B. (1953). *J. Am. Chem. Soc.* **75**, 141.
Tabner, B. J., and Walker, T. (1973). *J. Chem. Soc. Perkin Trans. II,* 1201.
Taft, R. W. (1975). Gas Phase Proton Transfer Equilibria, *In* "Proton Transfer Reactions" (E. F. Caldin and V. Gold, eds.). Chapman and Hall, London.
Terrier, F., Ah-Kow, G., Pouet, M., and Simonnin, M. (1976). *Tetrahedron Lett.* 227.
Thomassen, L. M., Ellingsen, T., and Ugelstad, J. (1971). *Acta Chem. Scand.* **25**, 3024.
Traynham, J. G., Stone, D. B., and Couvillion, J. L. (1967). *J. Org. Chem.* **32**, 510.
Valentine, J. S., and Curtis, A. B. (1975). *J. Am. Chem. Soc.* **97**, 224.
Vogtle, F., and Neumann, P. (1973). *Chem. Z.* **97**, 600.
Wenkert, E., and Strike, D. P. (1964). *J. Am. Chem. Soc.* **86**, 2044.
Wieser, K., Poon, L., Jirkovsky, I., and Fishman, M. (1969). *Can. J. Chem.* **27**, 433.
Wong, K. H., Konizer, G., and Smid, J. (1970). *J. Am. Chem. Soc.* **92**, 666.
Yamdagni, R., and Kebarle, P. (1972). *J. Am. Chem. Soc.* **94**, 2940.
Ykman, P., and Hall, Jr., H. K. (1975). *Tetrahedron Lett.* 2429.
Young, L. B., Lee-Ruff, E., and Bohme, D. K. (1973). *J. Chem. Soc. Chem. Commun.* 35.
Zaugg, H. E., Ratajczyk, J. F., Leonard, J. E., and Schaefer, A. D. (1972). *J. Org. Chem.* 37, *2249.*
Zavada, J., Svoboda, M., and Pankova, M. (1972). *Tetrahedron Lett.* 711.
Zubrick, J. W., Dunbar, B. I., and Durst, H. D. (1975). *Tetrahedron Lett.* 71.

4 STRUCTURAL STUDIES OF SYNTHETIC MACROCYCLIC MOLECULES AND THEIR CATION COMPLEXES

N. Kent Dalley

Department of Chemistry
Brigham Young University
Provo, Utah

I. Introduction 207
II. Structural Studies 209
A. Cation Complexes of Synthetic Macrocyclic Molecules 209
B. Uncomplexed Synthetic Macrocyclic Molecules . 236
III. Short C-C Distances in Synthetic Macrocyclic Molecules 240
IV. Conclusions 241
References 241

I. INTRODUCTION

Interest in the structures of synthetic macrocyclic compounds and their cation complexes has increased in recent years. The interest in these compounds has been generated by their ability to surround or enclose many different cations. The reaction with cations, particularly metal ions, usually involves a planar coordination sphere in the case of small cyclic polyethers or a three-dimensional cage in the cases of cryptates and large cyclic polyethers. Information obtained from structural studies of macrocyclic molecules and their complexes makes it possible to investigate some of the factors which determine the type of coordination which occurs. These factors include cation size, type, and charge; ligand size, donor atom type, and substituent; and solvent type.

The kinds of compounds included in this chapter are limited to cryptates, saturated polyethers and their derivatives, and complexes of these two ligand types. Several of these compounds have also been discussed by Truter (1973). Although cryptate complexes are similar in that the cation is always enclosed in the cage, the polyether complexes may be classified into four groups according to the positioning of the cation relative to the ligand donor atoms.

The first group includes the complexes in which the cation fits nicely in the cage or cavity of the molecule. The second group consists of those complexes in which the cation is too large to fit in the ligand cavity. In these cases the cation lies above the cavity of the cyclic polyether. In the third group the cation is smaller than the cavity, usually resulting in the ligand wrapping around the cation or complexing with two cations. In the few cases (e.g., Na^+-dibenzo-18-crown-L*) in which the cation is in the center of the cavity, the ligand is somewhat distorted to accommodate the small ion. In the fourth group the cation does not complex with all the potential donor atoms. This usually occurs when ligands which contain both oxygen and other donor atoms form complexes with certain transition metal ions which have greater affinity for the accompanying anion than for some of the donor atoms of the macrocyclic ligand. Without the ordering influence of a cation there are few common characteristics which can be used to classify cyclic polyether molecules.

The term cavity is used to describe the complexing volume of the cyclic polyethers. This term has validity for complexes of the first two groups and for a few complexes in group three. In these compounds the cation organizes the ligand so that the donor atoms are directed toward the cation and form a highly symmetrical, nearly planar polygon. For group three complexes in which the ligand wraps around the cation or shares it unequally and for group four complexes in which the cation is outside the ligand, the concept of a cavity as just described is not valid.

The conformations of the well-organized ligands of groups one and two are similar. The torsion (sometimes called the dihedral) angles exhibit a low level of torsional strain in these complexes. The O–C–C–O torsion angles are near 60° (Fig. 1) and the C–C–O–C angles are near 180° (Fig. 1). A ligand with these torsion angles resembles the structural formulas of Fig. 2. Deviation of torsion angles from these values, such as is found in groups three and four, indicates significant conformational changes from that of the highly symmetrical compounds.

In all cyclic polyethers whose structural parameters have been determined, the C–C bond distances have been calculated to be significantly shorter than C–C bond lengths observed in other compounds. Although this abnormality does not affect the type of complex formed, it is a common feature of these macrocyclic compounds. More will be said about this characteristic in Section III.

All the studies discussed in this chapter are based on X-ray crystallographic data. It must be remembered that these structures may differ significantly from the structures in solution, although it is likely that there will be similarities between the two conformations, particularly in the case of the cation-ordered complexes.

*The cyclic polyether nomenclature used here is an extension of that proposed by Pedersen (1967; see also this volume, Chapter 1).

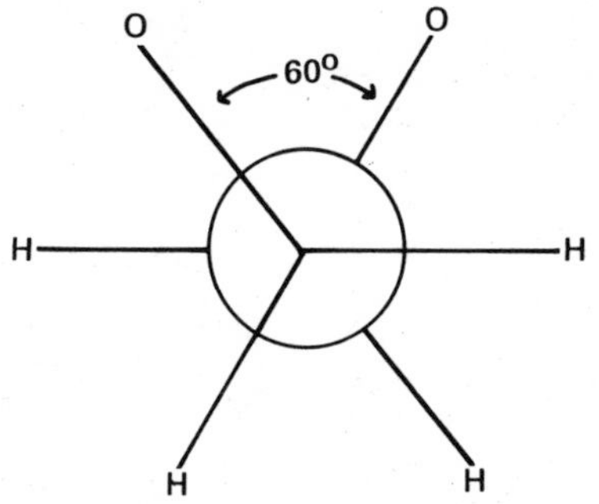

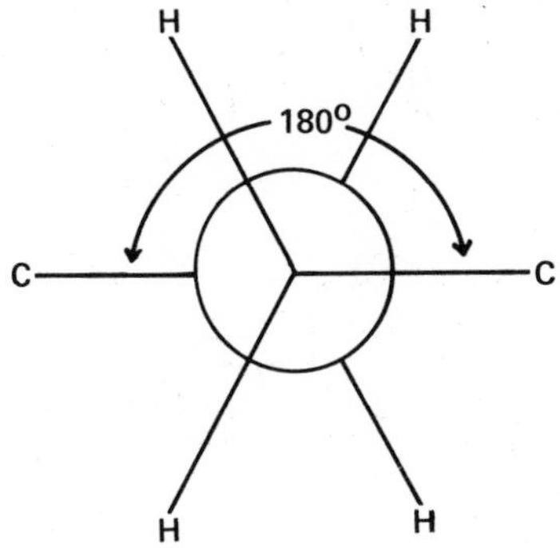

Fig. 1. (a) O–C–C–O torsion angle of 60°.
(b) C–C–O–C torsion angle of 180°.

II. STRUCTURAL STUDIES

A. Cation Complexes of Synthetic Macrocyclic Compounds

1. Cryptates and Other Multicyclic Ligands

The crystal structures of several cryptate complexes have been determined by R. Weiss and co-workers. Studies involving complexes of 222 (I) (see Fig. 2 for formulas) with the cations Na^+ (Moras and Weiss, 1973a), K^+ (Moras *et al.*,

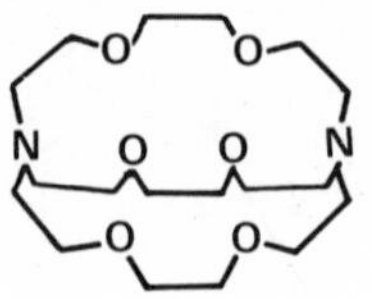

I

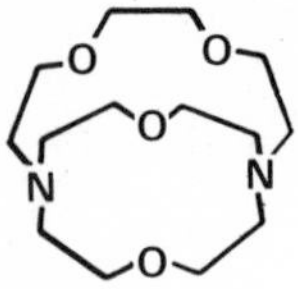

II

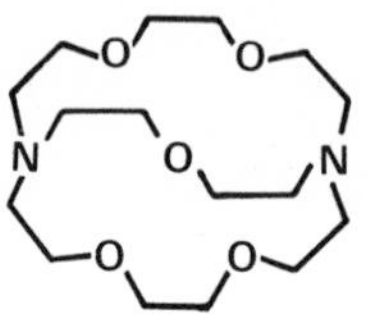

III

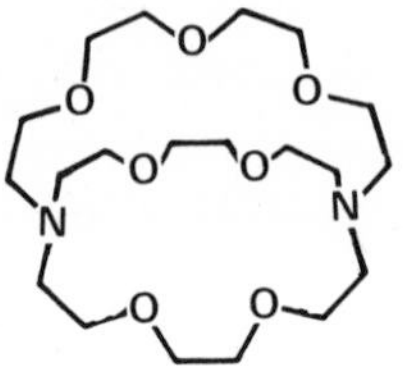

IV

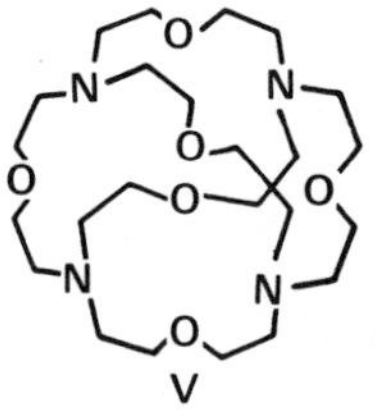

V

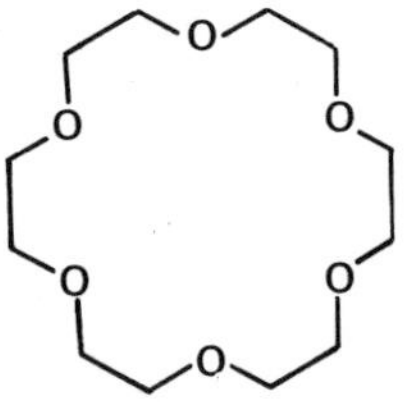

VI

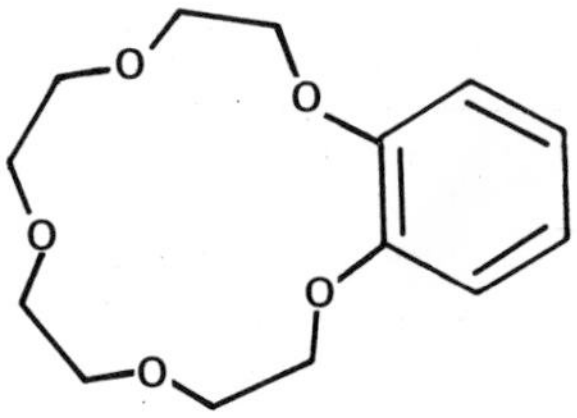

VII

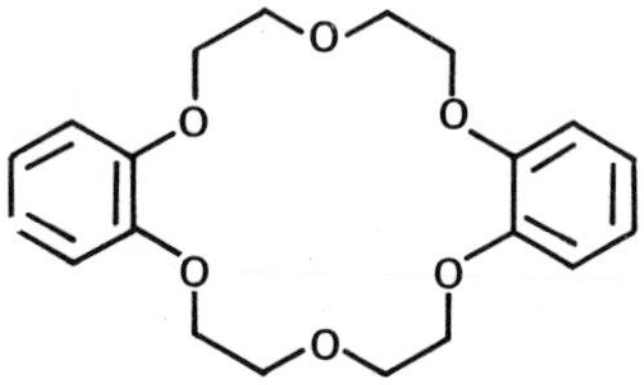

VIII

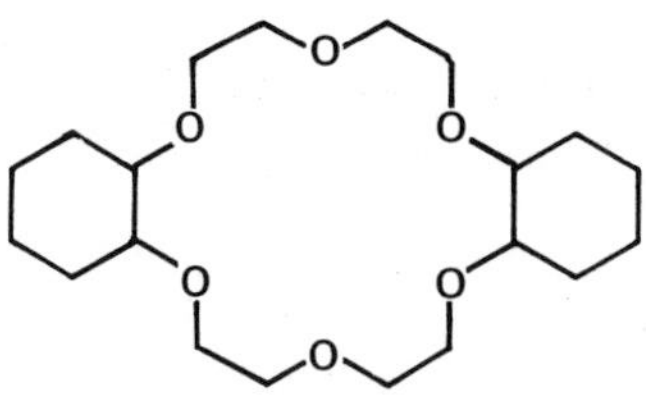

IX

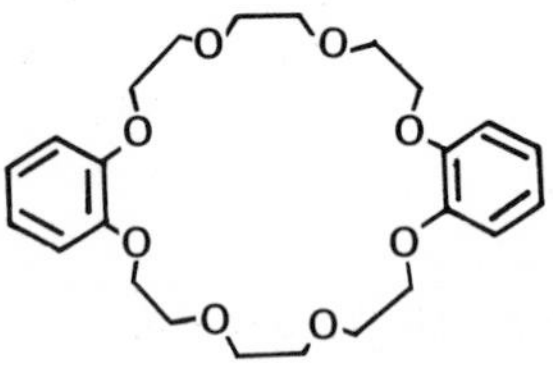

X

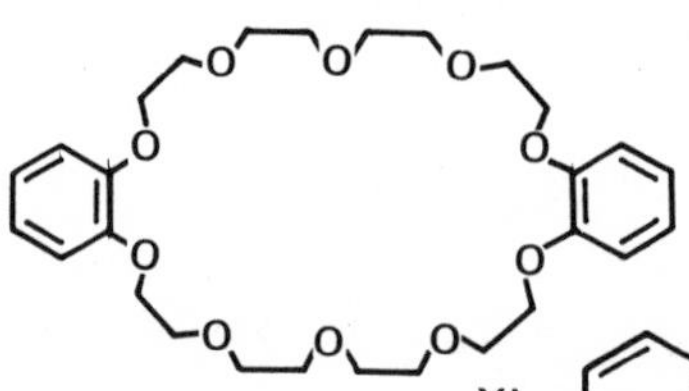

XI

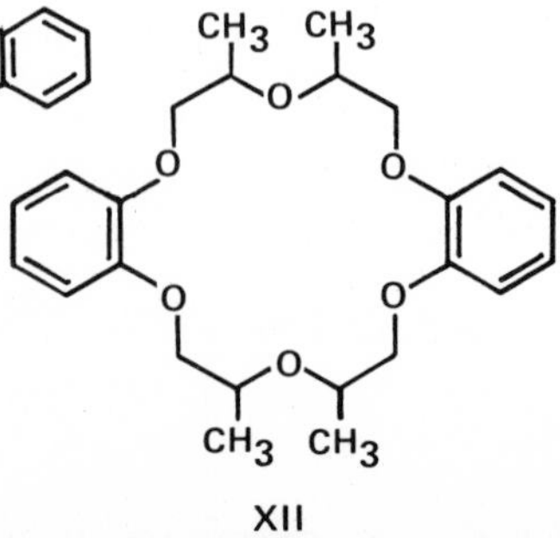

XII

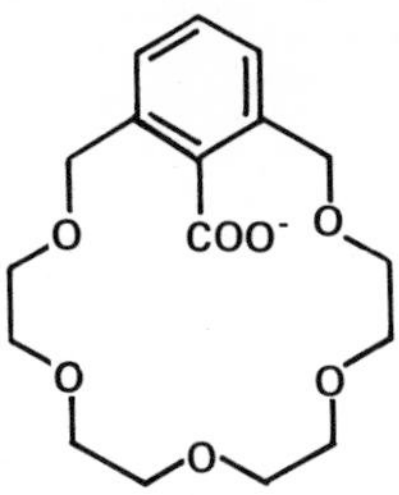

XIII

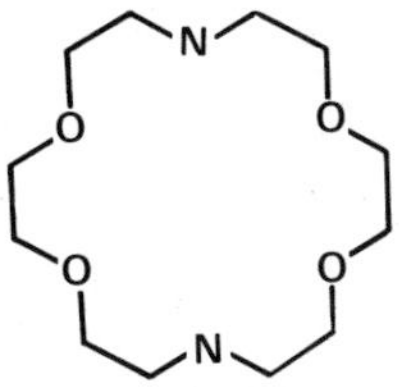

XIV

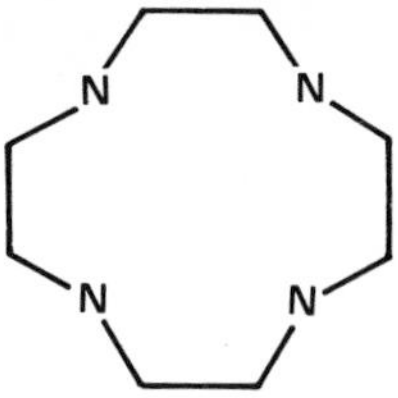

XV

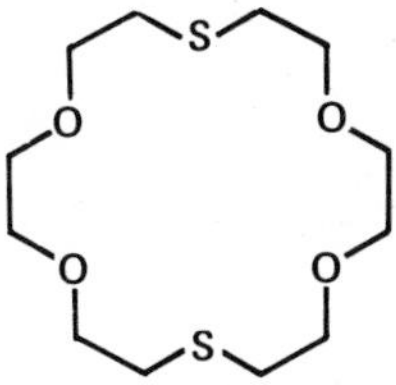

XVI

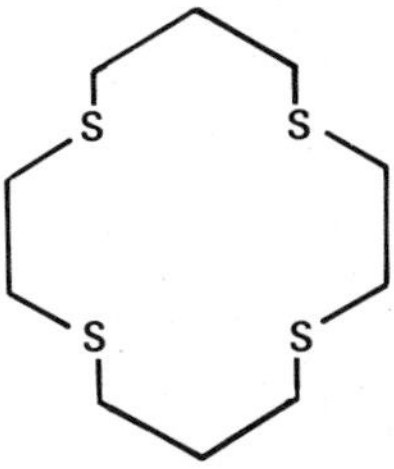

XVII

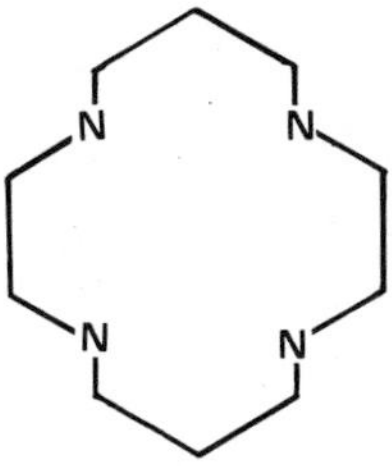

XVIII

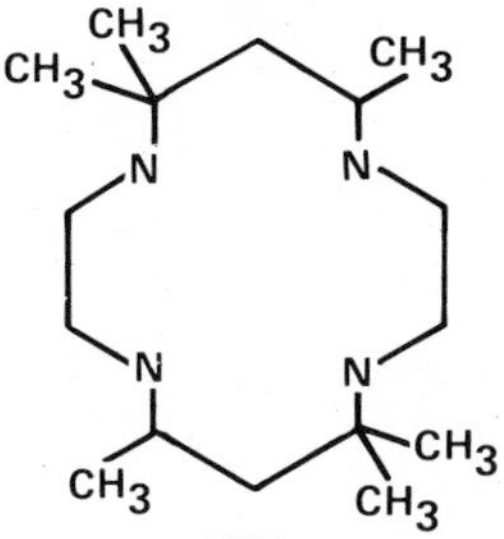

XIX

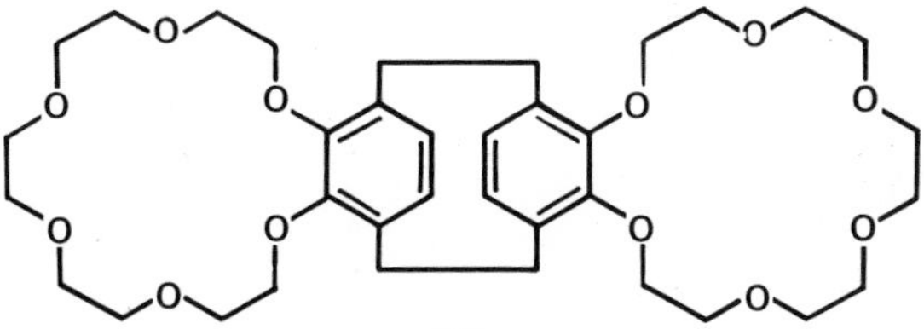

XX

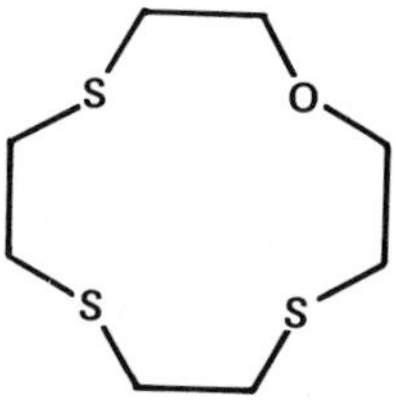

XXI

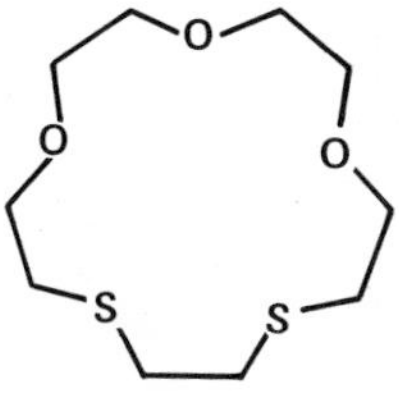

XXII

Fig. 2. Structural formulas of macrocyclic molecules discussed in this chapter.

TABLE I

Multicyclic Complexes

Ligand[a] L	Central ion M^{n+}	Anion A^-	L : M	Torsion angle range[b] O–C–C–X[e]	Torsion angle range[b] C–C–O–C	Torsion angle range[b] C–C–N–C	Distance ranges[c] (Å) M-L	Distance ranges[c] (Å) M-A⁻ M-solvent	Type of coordination	Type of structure	Ref.[d]
I	Na^+	I^-	1 : 1	40.6-58.1°	148.3-163.1°	76.3-87.3° 157.3-160.6°	Na-N 2.722-2.782 Na-O 2.566-2.901	No interaction	Na^+ enclosed by ligand	Ionic $[Na(C_{18}H_{36}N_2O_6)]^+I^-$	*1*
I	K^+	I^-	1 : 1	49.6-57.6°	169.9-179.3°	64.3-65.4° 167.0-169.0°	K-N 2.874 K-O 2.776-2.790	No interaction	K^+ enclosed by ligand	Ionic $[K(C_{18}H_{36}N_2O_6)]^+I^-$	*2*
I[f]	Rb^+	SCN^-	1 : 1	63.9-67.9°	176.9-179.8°	73.0-74.4° 164.0-165.5°	Rb-N 2.993-3.011 Rb-O 2.879-2.928	No interaction	Rb^+ enclosed by ligand	Ionic $[Rb(C_{18}H_{36}N_2O_6)]^+ SCN^-$	*3*
I[f]	Cs^+	SCN^-	1 : 1	68.1-73.4°	174.0-179.9°	73.0-79.6° 160.5-171.5°	Cs-N 3.020-3.048 Cs-O 2.960-2.973	No interaction	Cs^+ enclosed by ligand. Rb^+ and Cs^+ complexes are isomorphous	$[Cs(C_{18}H_{36}N_2O_6)]^+ SCN^-$	*3*
I	Ca^{2+}	Br^-	1 : 1	33-59°	78-178°	59-90° 147-178°	Ca-N 2.720-2.722 Ca-O 2.487-2.551	No interaction with Br^- Ca-OH_2 2.425	Ca^{2+} enclosed by ligand. Coordination sphere also includes H_2O	Ionic $[Ca(C_{18}H_{36}N_2O_6)(H_2O)]^{2+}2Br^-$	*4*

Ligand[a] L	Central ion M^{n+}	Anion A^-	L : M	Torsion andlge range[b] O–C–C–X[e]	C–C–O–C	C–C–N–C	Distance ranges[c] (Å) M-L	M-A^- M-solvent	Type of coordination	Type of structure	Ref.[d]
I[f]	Ba^{2+}	SCN^-	1 : 1	43-69°	127-180°	62-91° 135-177°	Ba-N 2.944-3.002 Ba-O 2.742-2.886	Ba-NCS 2.879-2.911 Ba-OH_2 2.838-2.883	Ba^{2+} enclosed by ligand. Coordination sphere also includes one H_2O and one SCN^-	Ionic $[Ba(C_{18}H_{36}N_2O_6)(SCN)(H_2O)]^+SCN^-$	*5*
I	Tl^+	$HCOO^-$	1 : 1	66.8-70.8°	177.3-179.5°	70.1-71.1° 167.6-168.5°	Tl-N 2.946 Tl-O 2.898-2.913	No interaction	Tl^+ enclosed by ligand	Ionic $[Tl(C_{18}H_{36}N_2O_6)]^+ HCOO^-$	*6*
II	Li^+	I^-	1 : 1	54.9-60.5°	88.8-179.8°	78.3-81.4° 149.8-152.8°	Li-N 2.288 Li-O 2.081-2.173	No interaction	Li^+ enclosed by ligand	Ionic $[Li(C_{14}H_{28}N_2O_4)]^+ I^-$	*7*
III	Co^{2+}	$Co(SCN)_4^{2-}$	1 : 1	36-52°	*g*	*g*	Co-N 2.201-2.243 Co-O 2.096-2.219	No interaction	Co^{2+} enclosed by ligand	$[Co(C_{16}H_{32}N_2O_5)]^{2+} [Co(SCN)_4]^{2-}$	*8*
IV	Ba^{2+}	SCN^-	1 : 1	47-70°	84-177°	59-82° 156-178°	Ba-N 3.084-3.179 Ba-N 2.796-3.092	No interaction with anion Ba-OH_2 2.810-2.874	Ba^{2+} enclosed by ligand also coordinated to 2 water ligands	Ionic $[Ba(C_{20}H_{40}N_2O_7)(H_2O)_2]^{2+}2SCN^-$	*9*
V	NH_4^+	I^-	1 : 1	*g*	*g*	*g*	Av. N-NH_4^+ 3.13 Av. O-NH_4^+ 3.11	No interaction	NH_4^+ enclosed by ligand hydrogen bonding likely	$[NH_4(C_{24}H_{48}N_4O_6)]^+ I^-$	*10*

TABLE I (cont.)

Multicyclic Complexes

Ligand[a] L	Central ion M^{n+}	Anion A^-	L : M	Torsion angle range[b] O–C–C–X[e]	C–C–O–C	C–C–N–C	Distance ranges[c] (Å) M-L	M-A^- iM-solvent	Type of coordination	Type of sturcture	Ref.[d]
VH_4^{4+}	Cl^-	Cl^-	1 : 1	[g]	[g]	[g]	Av. Cl-N 3.09 Av. Cl-O 3.25	[h]	Cl^- enclosed by protonated ligand, held in place by 4 hydrogen bonds	$[C_{24}H_{52}N_4O_6{}^{4+}Cl]^{3+}$ $3Cl^-$	*10*

[a]Numbers refer to compounds listed in Fig. 2.
[b]Range includes smallest to largest absolute value for the torsion angle for the indicated structure.
[c]Range includes smallest to largest central atom-donor atom (ligand, anion, or solvent) distance.
[d]References *1* Moras and Weiss (1973a), *2* Moras *et al.* (1973a), *3* Moras *et al.* (1973b), *4* Metz *et al.* (1973a), *5* Metz *et al.* (1973b), *6* Moras and Weiss (1973c), *7* Moras and Weiss (1973b), *8* Mathieu and Weiss (1973), *9* Metz *et al.* (1973c), *10* Metz *et al.* (1976)
[e]X represents N or O. No distinction is made in listing torsion angles.
[f]Unit cell contains two crystallographically independent molecules.
[g]Data not available.
[h]Does not apply.

1973a), Rb^+, Cs^+ (Moras *et al.*, 1973b), Ca^{2+} (Metz *et al.*, 1973a), Ba^{2+} (Metz *et al.*, 1973b), and Tl^+ (Moras and Weiss, 1973c) have been done. In addition, these workers determined the structures of the Li^+-211 (II), Co^{2+}-221 (III), and Ba^{2+}-322 (IV) complexes (Moras and Weiss, 1973b; Mathieu and Weiss, 1973; Metz *et al.*, 1973c). The type of coordination between the cation and macrocyclic ligand is similar in all these compounds. The metal ion is completely enclosed in the bicyclic ligand with the cation coordinated to all the nitrogen and oxygen atoms. In all these compounds the nitrogen atoms are pointing into the cavity (referred to as the in-in conformation) and the result is a football-shaped molecule that expands or contracts slightly to accommodate larger or smaller cations. The changes in shape are reflected in the deviation of the torsion angles from the normal 60 or 180° (Table I). In the well-ordered cryptate complexes the torsion angles observed are those of low torsional strain. This is similar to the situation found for the polyethers. The O—C—C—O torsion angles are near 60°, the C—C—O—C torsion angles are near 180°, and the C—N—C—C torsion angles fall into two groups, one near 60° and one near 180°. The range of these angles is listed in Table I. The M-O interatomic distances in Table I are also of interest. For example, in the cesium complex the M-O distance is somewhat shorter than the sum of the cesium cation and the oxygen van der Waals radii (Table II).

TABLE II

Cation Radius[a] and Atom van der Waals Radii

Cation	Radius[a] (Å)	Atom	Radius[b] (Å)
Na^+	1.02	O	1.40
K^+	1.38	N	1.5
Rb^+	1.49	S	1.85
Cs^+	1.70	Cl	1.80
Ca^{2+}	1.00		
Ba^{2+}	1.36		
La^{3+}	1.061		

[a] Radius given for cations in oxides with a coordination number of 6 (Shannon and Prewitt, 1969).

[b] Values taken from Pauling (1960).

This would indicate a rather strained molecule, and the low stability of the Cs complex (Table III) bears this out. In all the univalent cation complexes the coordination sphere of the cation includes only the heteroatoms of the ligand. The compounds are ionic and consist of a large complex cation with the metal ion several angstroms from the anion. In the bivalent cation complexes the cation has a similar binding to the ligand and, with the exception of the

TABLE III

Log *K* for M^+-222 Cryptates[a]

Metal ion	Log *K*
Li^+	<2
Na^+	3.90
K^+	5.40
Rb^+	4.35
Cs^+	<2

[a]Data taken from Lehn and Sauvage (1971).

Co^{2+}-221 compound, also to additional solvent molecules, anions, or both. The Ca^{2+} complex is an ionic compound in which one water molecule is also coordinated to the metal ion. In the Ba^{2+} complex again a complexed cation is present but it consists of $[Ba(222)(SCN)(H_2O)]^+$ with one SCN^- group and one water ligand included in the coordination sphere of the Ba^{2+}. In the Ba^{2+}-322 complex two water molecules are included in the coordination sphere of the barium, while the SCN^- groups interact with the cation only by hydrogen bonding through the solvent molecule. The Co^{2+} complex is similar to the univalent cation-222 complexes in that the metal ion is coordinated only by the cryptate ligand. The unusual feature of this compound is that the anion consists of $[Co(SCN)_4]^{2-}$.

Recently, Metz *et al.* (1976) reported two complexes in which the ligand is a tricyclic molecule of formula $C_{24}H_{48}M_4O_6$ (V) and contains four bridgehead nitrogens, all in the in-in conformation. One complex consists of an NH_4^+ in the molecular cavity which is probably held in place by weak hydrogen bonds, as indicated by the average N-N and N-O distances in NH•••N and NH•••O of 3.13 and 3.11 Å, respectively. The other complex is unusual in that the central ion is Cl^- and the four nitrogen atoms are protonated, so the Cl^- interacts with the ligand through hydrogen bonding. In this case, hydrogen bonding of the type Cl•••NH is substantiated by the short average Cl-N distance of 3.09 Å, which is less than the sum of the van der Waals' radii of the two atoms (Table II).

2. Macrocyclic Polyether Complexes

Dunitz and his co-workers examined the effect of cation size on structure for a given ligand. They determined the structure of 18-crown-6 (Fig. 2, VI), and the complexes of this ligand with Na^+, K^+, Rb^+, Cs^+, and Ca^{2+} (Dobler *et al.*, 1974; Seiler *et al.*, 1974; Dobler and Phizackerley, 1974a; Dobler and Phizackerley, 1974b; Dunitz and Seiler, 1974b). It is interesting to examine the results of their work, keeping in mind the size relationships of cations (Table II) and cavities (Table IV). Structural data for these and other crown ether complexes are given in Table V.

TABLE IV

Cavity Sizes of Macrocyclic Polyethers Estimated from Atomic Models[a] and X-Ray Diffraction Data

Atomic Model Estimates

Molecule	Diameter (Å)
12-crown-4	1.2[b]-1.5[c]
15-crown-5	1.7-2.2
18-crown-6	2.6-3.2
21-crown-7	3.4-4.3

X-Ray Diffraction Data

Molecule	Ref.[d]	Diameter (Å)	Adjusted diameter[e] (Å)
Na^+-VII	*1*	4.52[f]	1.72
K^+-VII	*2*	4.64	1.84
K^+-VI	*3*	5.54-5.66[g]	2.74-2.86
Rb^+-VI	*4*	5.47-5.64	2.67-2.84
Cs^+-VI	*5*	5.53-5.65	2.73-2.85
DAC[h]-VI	*6*	5.45-5.94	2.65-3.14
Ba^{2+}-IX(A)	*7*	5.49-5.62	2.69-2.82
Na^+-VIII[i]	*8*	5.32-5.55	2.52-2.75
Na^+-IX(B)	*9*	5.35-5.93	2.55-3.13
Cs^+-XII(F)	*10*	5.33-5.48	2.53-2.68

[a]Data taken from Pedersen, this volume, Chapter 1, Table VII.

[b]According to Corey-Pauling-Koltar atomic models.

[c]According to Fisher-Hirschfelder-Taylor atomic models.

[d]References: *1* Bush and Truter (1972a); *2* Mallinson and Truter (1972); *3* Seiler *et al.* (1974); *4* Dobler and Phizackerley (1974a); *5* Dobler and Phizackerley (1974b); *6* Goldberg (1975a); 7 Dalley *et al.* (1972); Smith (1972); *8* Bush and Truter (1971); *9* Mercer and Truter (1973a); *10* Mallinson (1975a).

[e]Adjusted diameter is the result of subtracting van der Waals radius of atoms from X-ray distance. It is the distance from edge of atom.

[f]Radius of crown-5 ligands obtained by assuming the five oxygen atoms form a regular pentagon, locating its center, and calculating the distance from center to atom. The average pentagon-center to atom-center distance was doubled to give diameter. Distances calculated by author from data in reference.

[g]Diameter of crown-6 ligands obtained by calculating the interatomic distances between opposite oxygens in the ligand. Shortest and longest distance listed. All except K^+-VI and DAC-VI calculated by author from data in reference.

[h]DAC represents dimethyl acetylene dicarboxylate.

[i]Data for two separate molecules in unit cell.

TABLE V

Crown Ether Complexes

Ligand[a] L	Cation M^{n+}	Anion A^-	L : M	Torsion angle range[b] O–C–C–O	Torsion angle range[b] C–C–O–C	Distance ranges[c] (Å) M-L	Distance ranges[c] (Å) M-A^- or M-solvent	Type of coordination	Type of structure	Ref.[d]
VI	Na^+	SCN^-	1 : 1	46.6-63.4°	70.5-177.4°	Na-O 2.452-2.623	M-OH_2 2.321	Na^+ coordinated to 6 oxygens of ligand and to H_2O oxygen	Ionic $[Na(C_{12}H_{24}O_6)(H_2O)]^+SCN^-$	*1*
VI	K^+	SCN^-	1 : 1	65.2-70.0°	170.8-178.9°	K-O 2.770-2.833	Weak interaction with disordered SCN^- K-A 3.19	K^+ coordinated to 6 planar oxygens of ligand, weakly to anion	Ion pair	*2*
VI	Rb^+	SCN^-	1 : 1	59.6-67.3°	168.8-179.1°	Rb-O 2.929-3.146	SCN^- disordered Rb-N 3.226-3.314	Rb^+ coordinated to 6 planar oxygens of ligand, weakly to anion	Ion pair, molecules bridged by two SCN^-	*3*
VI	Cs^+	SCN^-	! : 1	60.9-68.0°	172.2-180.0°	Cs-O 3.035-3.274	Cs-N 3.300-3.315	Isostructural with Rb^+ compound	Ion pair, molecules bridged by two SCN^-	*4*
VI	Ca^{2+}	SCN^-	1 : 1	*e*	*e*	Ca-O 2.56-2.74	Ca-N 2.35	Ca^{2+} in cavity, also coordinated to 2 SCN^- through N	Ion pair	*5*
VII	Na^+	I^-	1 : 1	56-62° 3°[f]	147-178°	Na-O 2.35-2.43	No interaction with anion Na-OH_2 2.29	Na^+ coordinated to 5 oxygens of ether ligand	Ionic $[Na(C_{14}H_{20}O_5)(H_2O)]^+I^-$	*6*

Ligand[a] L	Cation M^{n+}	Anion A^-	L : M	Torsion angle range[b] O–C–C–O	Torsion angle range[b] C–C–O–C	Distance ranges[c] (Å) M-L		Distance ranges[c] (Å) M-A^- or M-solvent	Type of coordination	Type of structure	Ref.[d]
									and 1 water molecule		
VIII[g] Molecule 1	Na^+	Br^-	1 : 1	61.1-65.5° 1.0-1.4°[f]	166.6-180.0°	Na-O	2.63-2.82	Na-OH_2 2.27-2.31 No interaction with Br^-	Na^+ coordinated to 6 oxygens of ether ligand and 2 water ligands	Ionic $[Na(C_{20}H_{24}O_6)(H_2O)]^+Br^-$	*7*
Molecule 2			1 : 1	59.3-68.1° 0.0-1.5°[f]	169.6-179.3°	Na-O	2.54-2.89	Na-Br 2.82 Na-OH_2 2.35	Na^+ coordinated to 6 oxygens of ether ligand, 1 bromo ligand, and 1 water	Ion pair	*7*
IX Isomer B	Na^+	Br^-	1 : 1	59.5-70.1°	159.8-178.3°	Na-O	2.676-2.967	Na-OH_2 2.346	Na^+ coordinated by 6 oxygens of ether ligand and 2 water oxygens	Ionic $[Na(C_{20}H_{36}O_6)(H_2O)_2]^+Br^-$	*8*
X	Na^+	*o*-Nitrophenolate	1 : 2	60-139° 2°[f]	157-180°	Na-O	2.468-2.615	Na-ON 2.399 Na-O phenolate 2.296-2.320	Each Na^+ coordinated to 3 oxygens of ether, 1 nitro oxygen, and shares 2 phenolate oxygens	Ion pair	*9*
VII	K^+	I^-	2 : 1	22-61° 1°[f]	89-179°	K-O	2.777-2.955	No interaction	K^+ sandwiched between 2 ether ligands coordinated to	Ionic $[K(C_{14}H_{20}O_5)_2]^+ I^-$	*10*

(cont.)

TABLE V (cont.)

Crown Ether Complexes

Ligand[a] L	Cation M^{n+}	Anion A^-	L : M	Torsion angle range[b] O–C–C–O	Torsion angle range[b] C–C–O–C	Distance ranges[c] (Å) M-L	Distance ranges[c] (Å) $M-A^-$ or M-solvent	Type of coordination	Type of structure	Ref.[d]
								all 10 oxygen atoms		
X	K^+	SCN^-	1 : 2	61-68° 0°[f]	84-175°	K-O nonshared 2.732-2.825 K-O shared 2.898-2.979	K-N 2.87-2.88	Each K^+ coordinated to 3 oxygens of ring and share 2 others, share anion and benzene ring of neighboring molecule	Ion pair	*11*
XI	K^+	I^-	1 : 1	44-69° 4°[f]	62-178°	K-O 2.850-2.931	No interaction	K^+ coordinated by 10 oxygen atoms of ether ligand	Ionic $[K(C_{28}H_{40}O_{10})]^+I^-$	*12*
VIII[h]	Na^+	SCN^-	1 : 1	65.9-69.4°[f] 1.8-4.5°[f]	172.9-179.8°[i]	Na-O 2.74-2.89	Na-N 3.32	Na^+ coordinated to 6 oxygens of the ether	Ionic $[Na(C_{20}H_{24}O_6)]^+SCN^-$	*13*
	Rb^+	SCN^-	1 : 1	65.9-69.4°	172.9-179.8°	Rb-O 2.861-2.94	Rb-N 2.94	Rb^+ coordinated to 6 oxygens of ether and 1 SCN^- ligand	Ion pair	
XII Isomer	Cs^+	SCN^-	1 : 1	61-72°	162-180°	Cs-O 3.07-3.34	Cs-N 3.19-3.25	Cs^+ coordinated to 6 oxygens of	Ion pairs molecules bridged by 2 SCN^-	*14*

Ligand[a] L	Cation M^{n+}	Anion A^-	L : M	Torsion angle range[b] O–C–C–O	C–C–O–C	Distance ranges[c] (Å) M-L	M-A^- or M-solvent	Type of coordination	Type of structure	Ref.[d]
F				1-2°[f]			Cs-arylcarbon 3.80	ether ligand and **shares 2** SCN^-		
XII Isomer G	Cs^+	SCN^-	2 : 1	61-69° 3°[f]	143-178°	Cs-O 3.12-3.36	**Not given,** SCN^- are disordered	Sandwich compound	Ionic $[Cs(C_{24}H_{32}O_6)_2]^+SCN^-$	*14*
IX Isomer A	Ba^{2+}	SCN^-	1 : 1	55-66°[j]	134-180°[j]	Ba-O 2.80-2.91	Ba-N 2.88 Ba-OH_2 2.80	Ba^{2+} coordinated to 6 oxygens of ligand, 2 SCN^-, and 1 H_2O ligand	Ion pair	*15* *16*
IX Isomer A	La^{3+}	NO_3^-	1 : 1	*e*	*e*	2.61-2.92	La-ONO_2 2.63-2.71	La^{3+} coordinated to 6 ether oxygens and 6 oxygens from the 3 nitrato ligands	Ion pair	*17*
VII H_2O solvent	Ca^{2+}	SCN^-	1 : 1	Some differ considerably from 60 or 180°		Av. Ca-O 2.53	Av. Ca-O solvent 2.39 Av. Ca-NCS 2.44	Ca^{2+} coordinated to 5 oxygens of ether, 1 solvent O, and 2 SCN^-	Ion pair	*18*
VII MeOH Solvent	Ca^{2+}	SCN^-	1 : 1	Same as above		Same as above	Same as above	Same as above	Ion pair	*18*
VI	DAC[k,l]		1 : 1	69.6-75.6°	169.2-179.9°	*m*	*m*	All oxygens of ether ligand participate in	Guest-host	*19*

(cont.)

TABLE V (cont.)

Crown Ether Complexes

Ligand[a] L	Cation M^{n+}	Anion A^-	L : M	Torsion angle range[b] O–C–C–O	Torsion angle range[b] C–C–O–C	Distance ranges[c] (Å) M-L	Distance ranges[c] (Å) M-A^- or M-solvent	Type of coordination	Type of structure	Ref.[d]
								ion-dipole interactions with molecules; some H bonding		
XIII	*Tert*-butyl amine	Ligand has anion function	1 : 1	In ether portion 71.4-72.3°	In ether portion 83.5-175°	*m*	*m*	Some hydrogen bonding of ligand oxygen to amine nitrogen	Guest-host	*20*
VI	Uranylnitrate[l] dihydrate		1 : 1	Gauche ~60°	~180°	*m*	*m*	$UO_2(NO_3)_2(H_2O)_2$ interacts with ether only through H bonds from H_2O and ether	Similar to guest-host	*21*

(cont).

[a]Numbers refer to compounds listed in Fig. 2.

[b]Range includes smallest to largest absolute value for the torsion angle for the indicated structure.

[c]Range includes smallest to largest central atom-donor atom (ligand, anion, or solvent) distance.

[d]References *1* Dobler *et al.* (1974), *2* Seiler *et al.* (1974), *3* Dobler and Phizackerley (1974a), *4* Dobler and Phizackerley (1974b), *5* Dunitz and Seiler (1974b), *6* Bush and Truter (1972a), *7* Bush and Truter (1971), *8* Mercer and Truter (1973a), *9* Hughes (1975), *10* Mallinson and Truter (1972), *11* Mercer and Truter (1973b), *12* Bush and Truter (1972b), *13* Bright and Truter (1970), *14* Mallinson (1975a), *15* Dalley *et al.* (1972), *16* Smith (1972), *17* Harman *et al.* (1976), *18* Owen and Wingfield (1976), *19* Goldberg (1975a), *20* Goldberg (1975b), *21* Bombieri *et al.* (1976).

[e]Data not available.

[f]Torsion angle (O—C—C—O) in ether ring in which carbon segment is also part of benzene ring.

[g]Unit cell contains two crystallographically independent molecules.

[h]Population of M^+ site, 45% Na^+, 55% Rb^+.

[i]Torsion angles obtained from Bush and Truter (1971).

[j]Torsion angles calculated by author from data of Smith (1972).

[k]DAC represents dimethyl acetylene dicarboxylate.

[l]Complexed species is molecule.

[m]Complexed species is molecular, therefore does not apply.

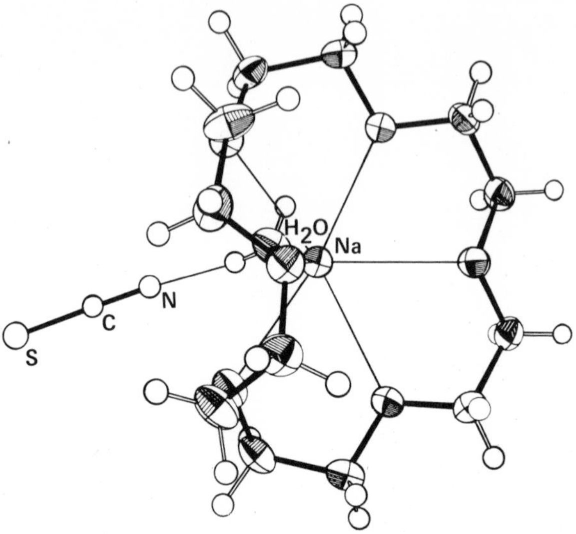

Fig. 3. [$NaSCNH_2O$-18-crown-6] with Na^+ complexed by the ether ligand, a SCN^-, and a water molecule (Dobler *et al.*, 1974). The vibrational ellipsoids are drawn at the 50% probability level (Johnson, 1965).

Examination of the cation and ligand size parameters (Tables II and IV) indicates that Na^+ is too small, K^+ is just right, and Rb^+ and Cs^+ are too large to fit into the cavity of an 18-crown-6 ligand. This is substantiated by the structural studies. The Na^+ is coordinated to all six oxygen atoms, five of which lie approximately in a plane, the sixth being located above the plane. The ligand (Fig. 3) appears to be wrapped about the Na^+ as is typical for a group three complex. The Na^+ is also coordinated to one water molecule. In the K^+ complex the conformation expected for a group one complex is found with the K^+ lying in the cavity of the ligand on a crystallographic center of inversion. The central ion is surrounded by a nearly planar hexagon of oxygen atoms (Fig. 4). The values found for torsion angles in the complex are as expected and are listed in Table V. The K^+ interacts very weakly with the disordered SCN^-. In both the Rb^+ and the Cs^+ complexes the cations are too large for the cavity and therefore lie above the plane of the oxygen atoms of the ligand as expected for group two complexes. The Rb^+ and Cs^+ ions lie 1.2 and 1.44 Å, respectively, above the plane of the oxygen atoms. In each complex unit the metal ion interacts weakly with the nitrogen atoms of two SCN^- anions linking the complex cations together in a chain (Fig. 5). The Ca^{2+} complex is disordered but it appears that the cation is in the cavity of the ligand.

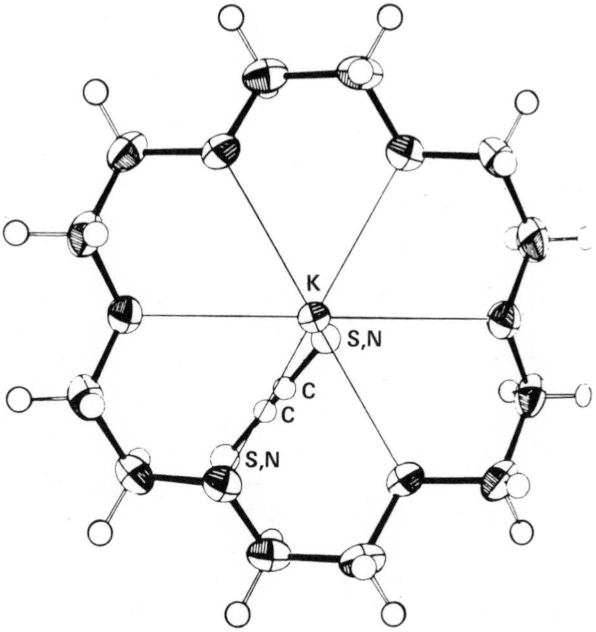

Fig. 4. [KSCN-18-crown-6] shown in a direction normal to the ether plane (Seiler *et al.*, 1974). The vibrational ellipsoids are drawn at the 50% probability level (Johnson, 1965).

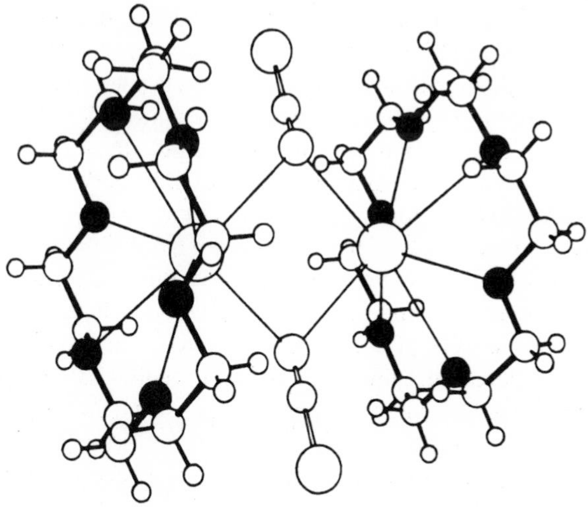

Fig. 5. [CsSCN-18-crown-6] units bridged by SCN ligands. The Cs^+ do not fit into the ether ligands (Dunitz *et al.*, 1974).

Truter and her group have studied several systems involving a given cation with crown ethers of varying numbers of ligand donor atoms. The Na^+ complexes include those with benzo-15-crown-5 (Fig. 2, VII), dibenzo-18-crown-6 (VIII), dicyclohexo-18-crown-6 (IX), and dibenzo 24-crown-8 (X). In the NaI-benzo-15-crown-5 complex (Bush and Truter, 1972a) the cation lies about 0.75 Å above the plane of the oxygen atoms. The structural results and the cation size (Table II) versus cavity size calculated from X-ray data (Table IV) both show that this complex belongs in group two. It is of interest that the cavity diameters for crown-5 ethers calculated from X-ray data agree well with that (1.7 Å) based on Corey-Pauling-Koltun atomic models, but are considerably smaller than that (2.2 Å) based on Fisher-Hirschfelder-Taylor models, (Pedersen, this volume, Chapter I, Table VII). The Na^+ is also coordinated to a water molecule, the entire structure being a pentagonal pyramid. The I^- does not interact with the Na^+. The structures of the NaBr complexes with dibenzo-18-crown-6 (Bush and Truter, 1971) and the cis-anti-cis (B) isomer of dicyclohexo-18-crown-6 (Mercer and Truter, 1973a) have been determined by X-ray diffraction. Based on size relationships, both complexes would be expected to belong to group three. However, in these complexes the ring substituents evidently give added rigidity to the ether ring and the Na^+ lies in the planar cavity. This positioning of the Na^+ is in contrast to that discussed earlier for the Na^+-18-crown-6 complex. In both complexes the normal torsion angles are seen, with the exception that in the dibenzo derivative the O—C—C—O torsion angles about the C—C bonds common to both the benzene ring and the ether ring are near 0°. The Na-O distances (Table V) in both compounds have a wider range of values than was found for the K-O distances in the K^+-18-crown-6 complex, indicating unequal coordination by Na^+ to the oxygen atoms. In the dibenzo compound there are two crystallographically independent molecules. One molecule exists as an ion pair, with one H_2O and one Br^- ligand being included in the coordination sphere of the Na^+. The other is an ionic compound, $[Na(crown)(H_2O)_2]^+Br^-$. The Na^+ in both molecules has hexagonal bypyramidal coordination. The latter type of coordination is also found in the complex with isomer B. In the Na^+-*o*-nitrophenolate compound with dibenzo-24-crown-8 a different type of complex is found (Hughes, 1975). The hole of this ligand is considerably larger than those of the 18-crown-6 ligands, and, in fact, is large enough so that two Na^+ fit inside it. This is another type of coordination found in complexes of group three. Each Na^+ has a coordination number of six, and is coordinated to three oxygen atoms of the ether ligand, leaving two oxygen atoms of the ligand not coordinated to a cation (Fig. 6). The *o*-nitrophenolate anion bridges the two Na^+ and completes the coordination of the cation. The torsion angles differ somewhat from the expected values (Table V).

Truter and co-workers have also reported the structures of several K^+-crown complexes where the number of ligand donor atoms varies from 5 to 10. The

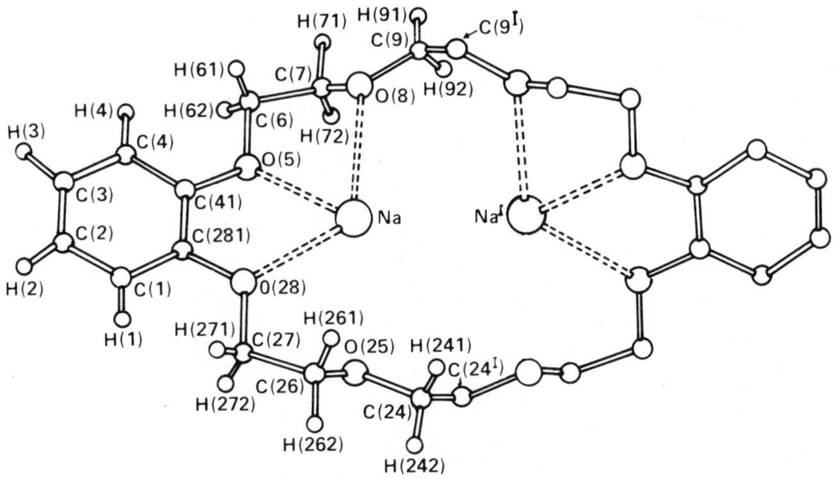

Fig. 6. [Na_2-*o*-nitrophenolate-dibenzo-24-crown-8] with two Na^+ complexed by the ether ligand (Hughes, 1975).

K^+-benzo-15-crown-5 complex (Mallinson and Truter, 1972) is a sandwich type molecule (Fig. 7). The cation does not fit in the small cavity and so it is sandwiched between two crown ethers and is coordinated to all 10 oxygen atoms of the ligands. This type of structure is not unexpected for a group two complex. One of the C—C—O—C torsion angles is much smaller than the expected value (Table V). Truter's group has not studied any K^+-crown-6 compounds, but it has been shown by Seiler *et al.* (1974) that the K^+ fits nicely into the cavity of a ligand of this size (Fig. 4). The K^+ complex with dibenzo-24-crown-8 (Mercer and Truter, 1973b) is similar to the Na^+ complex with the same ligand in that there are two metal ions in the cavity of the ligand. However, in this compound all the oxygen atoms of the ether are involved in coordination. In the K^+-dibenzo-30-crown-10 complex (Bush and Truter, 1972b) the ligand (Fig. 2, XI) folds and wraps about a single K^+ (Fig. 8). The cation is coordinated to all ten oxygen atoms of the ligand, but not to any other anion or solvent molecule.

The first reported structure of a crown complex was that of a 45% Na^+, 55% Rb^+-dibenzo-18-crown-6 (Fig. 2, VIII) crystal (Bright and Truter, 1970). The unit cell also includes an uncomplexed crown molecule. The Na^+ comes closer to fitting into the cavity than the Rb^+, as would be expected. The metal-anion distances suggest that the Na^+ complex exists as an ionic compound while the Rb^+ complex is an ion pair. Mallinson (1975a) investigated the structures of Cs^+ complexes of the cis-anti-cis (F) and trans-anti-trans (G) isomers of tetramethyl-dibenzo-18-crown-6 (Fig. 2, XII). As would be expected, the large Cs^+ did not fit into the crown-6 ring. Two different types of structures are observed. The complex with

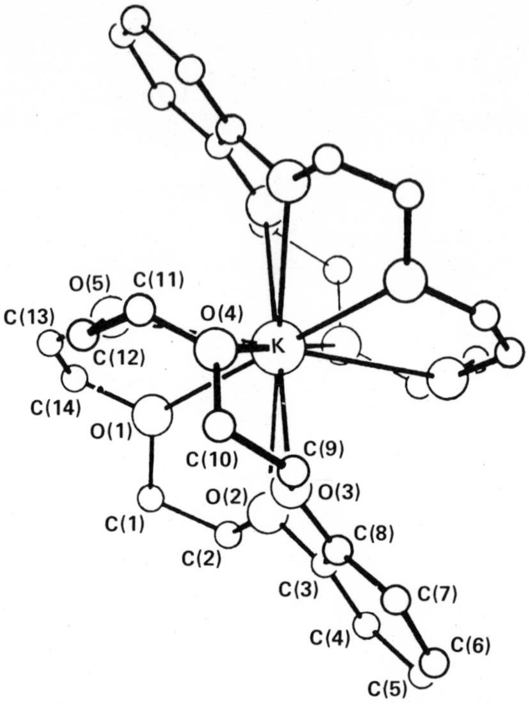

Fig. 7. [K-15-crown-5]$^+$ with the K^+ sandwiched between two ether ligands. (Mallinson and Truter, 1972).

isomer F consists of a Cs^+ coordinated to six oxygen atoms, but lying above the plane formed by these atoms. The SCN^- form bridges between molecules in a manner similar to that found in the Rb^+ and Cs^+ complexes of 18-crown-6. The Cs^+-isomer G crystal is somewhat disordered but a 2:1 ligand-to-cation sandwich

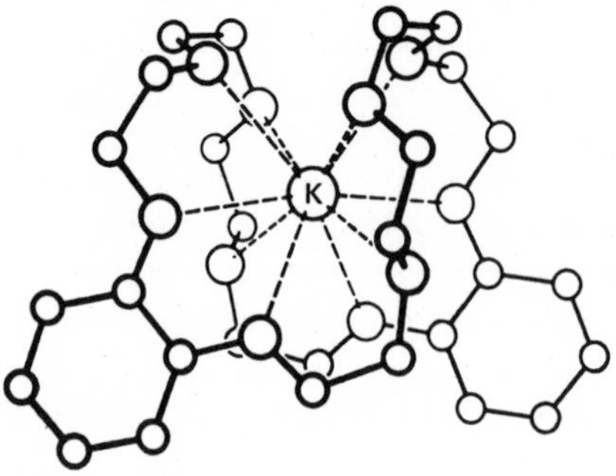

Fig. 8. [K-30-crown-10]$^+$ the ligand wraps around the cation (Bush and Truter, 1972b)

complex is formed. The Cs^+ is coordinated to all 12 oxygen atoms of the two ligands in a manner similar to that seen for the K^+-benzo-15-crown-5 complex.

Dalley *et al.* (1972) determined the structure of the $Ba(SCN)_2$ complex with isomer A of dicyclohexo-18-crown-6 (Fig. 2, IX). This study established that isomer A was the cis-syn-cis isomer. The Ba^{2+}, located on a twofold axis, fits nicely in the cavity of the ligand. The cyclic polyether molecule has a symmetrical conformation, being located about the twofold axis. The structure in the solid state exists as an ion pair, since the metal ion is coordinated to both anions as well as to a water ligand. Isomer B was later shown to be the cis-anti-cis isomer in the study of its Na^+ complex (Mercer and Truter 1973a) which was discussed previously.

Recently, Harman *et al.* (1976) reported the structure of the first tripositive cation-crown compound consisting of a complex formed between $La(NO_3)_3$ and the cis-syn-cis isomer of dicyclohexo-18-crown-6 (Fig. 2, IX). Each La^{3+} is coordinated to 12 oxygen atoms, 6 ether oxygens, and 6 oxygen atoms from the three bidentate nitrato ligands. The ether oxygens are not quite coplanar. The ligand has a pseudo twofold axis, which would be expected for the cis-syn-cis isomer with a cation of this size located in the cavity. The metal-ether oxygen distances are typical La-O distances (about 2.6 Å) except for two which are 2.89 and 2.92 Å (Table V).

Owen and Wingfield (1976) report two structures of $Ca(SCN)_2$-benzo-15-crown-5 (Fig. 2, VII) complexes in which the solvent was methanol and water, respectively. The coordination of the metal ion in the two structures is similar. The Ca^{2+} has a coordination number of 8, being coordinated in each solvent to the five oxygen atoms of the ether, two nitrogen atoms of the SCN^- ligands, and an oxygen atom of the solvent, with the Ca^{2+} positioned between the polyether ligand and the monodentate ligands. The Ca^{2+}-ether oxygen coordination is rather similar to that found in the Na^+ complex with the same ligand but the ligand molecule is more irregular. The solvent does not make an important difference in these structures, probably because of the similarities between methanol and water. The study is significant in that it is the first in which the effect of changing solvents was investigated.

Goldberg has initiated structural studies of an extremely interesting class of compounds which have been called guest-host compounds (Cram and Cram, 1974). These are compounds in which ligands complex organic polar molecules and cations rather than metal ions. He has reported the structures of the 18-crown-6 (VI) complex with dimethyl acetylene-dicarboxylate (Goldberg, 1975a) and of another somewhat similar ligand with *tert*-butylamine (Goldberg, 1975b) at 113 and 120°K, respectively. In the latter compound, the ligand (Fig. 2, XIII), which is systematically named 3,6,9,12,15-pentaoxa-21-carboxybicyclo-[15,3,1]heneicosa-1(21),17,19-triene, but is also referred to as 2,6-dimethyl-benzoic acid-18-crown-5, has five ether oxygens separated by the usual two-

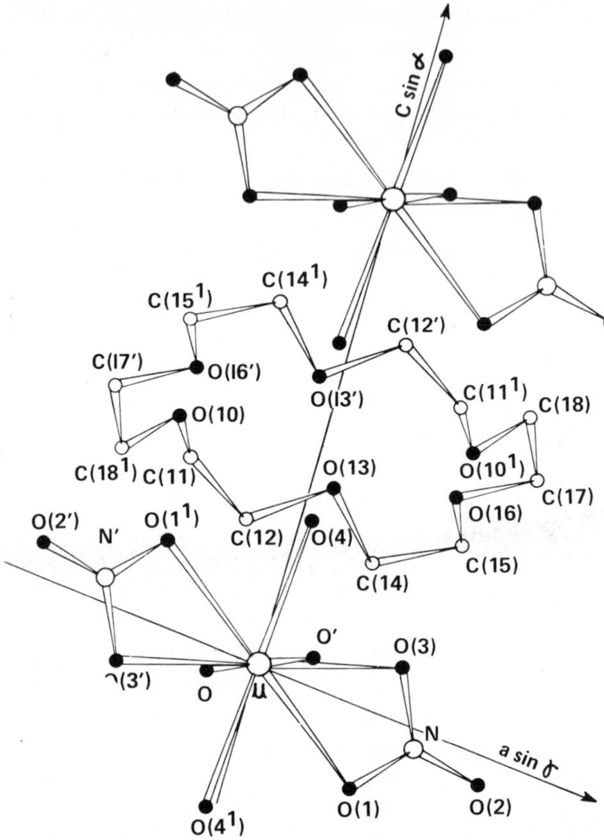

Fig. 9. $[UO_2(NO_3)(H_2O)_2]$ interacts with 18-crown-6 only through hydrogen bonding between the H_2O ligand and the ether (Bombieri *et al.*, 1976).

carbon bridge with an ionized benzoic acid group inserted into the ring. In the 18-crown-6 complex, the molecule is bound to the ligand mainly by dipole interactions and by hydrogen bonding. The ligand is fairly well organized, as illustrated by the torsion angles of Table V. In the other complex, which exists as an ion pair, the cation is bound to the ligand by electrostatic interaction and by rather strong hydrogen bonding. The five oxygen atoms of the ether and one of the oxygen atoms of the acid group are nearly coplanar, with alternate oxygen atoms above and below the plane, a conformation similar to that found in the ordered 18-crown-6 complexes. The torsion angles in that portion of the molecule not connected to the benzene ring are normal for complexes in groups one and two. It is also interesting to note that the carbon-carbon bond distances in the ether portion of both molecules are shorter than expected, a situation similar to that found in other cyclic polyethers.

The structure of the compound formed between uranyl nitrate dihydrate and 18-crown-6 (Fig. 2, VI) (Bombieri *et al.*, 1976) is not that of a normal complex, as might have been expected, but rather a compound that resembles the guest-host type compounds. There is no interaction between UO_2^{2+} and the cyclic polyether but the short water oxygen-ether oxygen distances of 2.98 and 3.03 Å suggest that intermolecular hydrogen bonding occurs between the hydrate water of the uranyl compound and the polyether (Fig. 9). The conformation of the polyether resembles that of the ligands of the ordered group one and two complexes.

3. Substituted Macrocyclic Polyether Complexes

a. Substituted Complexes. Several crown complexes have been reported (Table VI) in which some or all of the ether oxygens have been replaced by other atoms, particularly nitrogen and sulfur. This replacement affects not only the size of the potential cavity, especially in the case of replacement by sulfur, but also the type of coordination and the selectivity. For example, the 18-crown-6 molecule with two nitrogen atoms replacing oxygen atoms opposite each other forms much more stable complexes with silver than the all-oxygen ligand (Frensdorff, 1971).

Structures of the K^+, Pb^{2+}, and Cu^{2+} complexes with 1,10-diaza-18-crown-6 (Fig. 2, XIV) have been reported (Moras *et al.*, 1972; Metz and Weiss, 1973; Herceg and Weiss, 1973). Both K^+ and Pb^{2+} lie in the cavity of the ligand and interact with all the heteroatoms. The torsion angles and the metal-ether oxygen or nitrogen distances indicate that these compounds are group one complexes (Table VI). In the K^+ compound there is weak interaction between K^+ and the SCN^- group, but in the Pb^{2+} complex the Pb^{2+} is bonded to the SCN^- ligand in addition to the ligand. The Cu^{2+} compound is a group four complex. The metal ion is coordinated to both nitrogen atoms and to two of the four oxygen donor atoms. The remaining two coordination sites are occupied by the two chloro ligands. The preference of Cu^{2+} for Cl^- over ether oxygens is probably the cause for this type of coordination. The distortion of the ligand in this complex is evident from the torsion angle data (Table VI). If the counterion had been an anion that does not form strong complexes with metal ions (e.g., ClO_4^-), Cu^{2+} might have coordinated to all the hetero atoms of the ligand, and as a result, have been located more nearly in the cavity.

The structure of dinitro (1,4,7,10-tetraazacyclododecane) cobalt (III) chloride has been reported (Iitako *et al.*, 1974). The four nitrogen atoms of the macrocyclic ligand 1,4,7,10-tetraazacyclododecane (Fig. 2, XV), commonly called cyclen, are coordinated to the metal ion, as are the nitrogen atoms of the two nitro ligands. The result is an octahedral cation $[Co(cyclen)(NO_2)_2]^+$ with the nitro groups cis to one another. There is no interaction between Co^{2+} and the

TABLE VI

Substituted Crown Ether Complexes

Ligand[a]	Cation	Anion	L : M	Torsion angle range[b] X–C–C–X[e]	Torsion angle range[b] C–C–X[e]–C	Distance ranges[c] (Å) M-L	Distance ranges[c] (Å) M-anion or M-solvent	Type of coordination	Type of structure	Ref.[d]
XIV	K^+	SCN^-	1 : 1	66-72°	X=O 175-178° X=N 178-179°	K-O 2.825-2.836 K-N 2.856	K-NCS 3.33	K^+ coordinated to all donor atoms of ligand and SCN^- ligands which bridge complex	Ion pair	*1*
XIV	Pb^{2+}	SCN^-	1 : 1	60.5-67.1°	X=O 173.8-178.1° X=N 179.1-179.5°	Pb-O 2.787-2.879 Pb-N 2.751	Pb-SCN 2.894	Pb^{2+} coordinated to all donor atoms of ligand and to S of 2 SCN^-	Ion pair	*2*
XIV	Cu^{2+}	Cl^-	1 : 1	59.3-71.9°	X=O 92.6-177.2° X=N 168.3-178.0°	Cu-O bonding 2.709-2.754 Nonbonding 3.634-3.663 Cu-N 2.030-2.039	Cu-Cl 2.284-2.337	Cu^{2+} coordinated to 2 oxygens, 2 nitrogens of ligand, also to 2 chloro ligands	Ion pair	*3*
XV	Co^{3+}	Cl^-	1 : 1	34.9-55.5°	*f*	Co-N 1.947-1.976	$Co\text{-}NO_2$ 1.923-1.931 No interaction with Cl^- or H_2O	Co^{3+} coordinated to 4 nitrogens of ligand, 2 nitrogens from NO_2 groups	Ionic $[Co(cyclen)(NO_2)]^+$ $Cl^- \cdot H_2O$	*4*
XVI	Pd^{2+}	Cl^-	1 : 1	50-58°	X=O 63-178° X=S 165-177°	Pd-S 2.302-2.305 Nonbonding Pd-O 3.94-4.16	Pd-Cl 2.313-2.316	Pd^{2+} coordinated to 2 sulfurs of ether and 2 chloro ligands	Ion pair	*5*

(cont.)

				Torsion angle range[b]		Distacne ranges[c] (Å)				
Ligand[a]	Cation	Anion	L : M	X–C–C–X[e]	C–C–X[e]–C	M-L	M-anion or M-solvent	Type of coordination	Type of structure	Ref.[d]
XVII	Cu^{2+}	ClO_4^-	1 : 1	67-77°	X=S 153-180°	Cu-S 2.297-2.308	Cu-O 2.652	Cu^{2+} coordinated to 4 sulfurs in thiaether ligand and to 1 oxygen of each ClO_4^-	Ion pair	*6*
XVII	Ni^{2+}	BF_4^-	1 : 1	61-75°[g]	X=S 157-180°[g]	Ni-S 2.175-2.177	No interaction	Ni coordinated to 4 sulfur atoms of ligand	Ionic $[Ni(C_{10}H_{20}S_4)]^{2+}$ $2BF_4$	*7*
XVII	Nb	Cl	1 : 2	169-177°[h]	X=S 82-132°[h]	Nb-S 2.713	Nb-Cl opposite S 2.252; others 2.30-2.33	Each Nb coordinated to a sulfur of ligand and 5 chloro ligands	Ligand in exo form is bridge between $NbCl_5$ units	*8,9*

[a]Numbers refer to compounds listed in Fig. 2.

[b]Range includes smallest to largest absolute value for the torsion angle of the indicated structure.

[c]Range includes smallest to largest central atom-donor atom (ligand, anion, or solvent) distance.

[d]References *1* Moras *et al.* (1972), *2* Metz and Weiss (1973), *3* Herceg and Weiss (1973), *4* Iitaka *et al.* (1974), *5* Metz *et al.* (1974), *6* Glick *et al.* (1976). *8* DeSimone and Glick (1975), *9* DeSimone and Glick (1976b), *7* Davis *et al.* (1975).

[e]The X represents O, S, C, or N. In the X–C–C–X torsion angles all combinations are included. In the C–X–C–C torsion angle X is defined.

[f]Data not available.

[g]Torsion angles calculated by Glick *et al.* (1976) from data of Davis *et al.* (1975).

[h]Torsion angles calculated by DeSimone and Glick (1976a).

Cl^- or with the H_2O of the crystal structure. The cation is a group frou complex in which all of the C—N—C—C torsion angles are considerably smaller than those found in group one or two complexes (Table VI).

The $PdCl_2$ complex of 1,10-dithia-18-crown-6 (Fig. 2, XVI) has been studied by Metz *et al.* (1974) and found to be square planar (Fig. 10). The Pd is coordinated to the two sulfur atoms of the polyether ligand and to the two Cl^-.

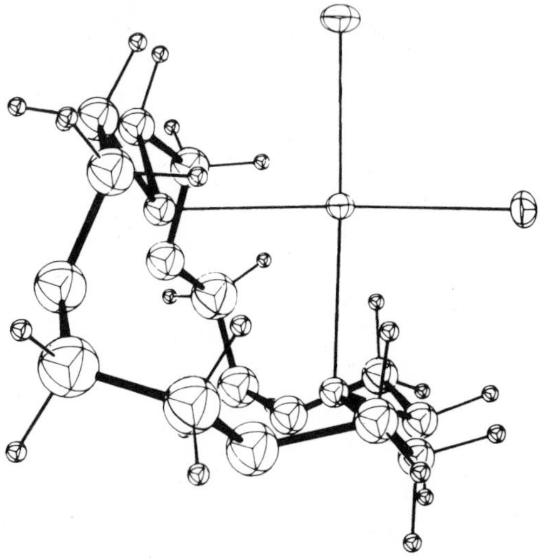

Fig. 10. In [$PdCl_2$-dithia-18-crown-6] the Pd has square planar coordination, being coordinated to two chloro ligands and two sulfur atoms of the thia ether (Metz *et al.*, 1974).

The cation does not interact with any of the oxygen atoms of the cyclic molecule. The important factor in determining the structure of this group four complex appears to be the presence of the two Cl^-, which are known to interact strongly with Pd^{2+}.

b. Substituted Ether Complexes with More than Two Bridging Carbon Atoms. In the macrocyclic molecules and complexes already discussed there have always been bridges consisting of two carbon atoms between the donor atoms. Some related compounds have been studied in which four donor atoms are separated by alternating two and three carbon atom segments. These compounds do not contain oxygen atoms but they can be considered derivatives of the crown ethers in which all the oxygen atoms have been replaced by sulfur or nitrogen atoms. The sulfur-containing compound XVII (Fig. 2) is systematically named 1,4,8,11-tetrathiacyclotetradecane, while the nitrogen-containing compound XVIII is 1,4,8,11-tetraazacyclotetradecane but is commonly named cyclam. In some of the nitrogen complexes, the ligands are cyclam derivatives in which some of the hydrogen atoms of the nitrogen or carbon atoms have been replaced by methyl groups.

Ligand XVII forms complexes with transition metal ions (Glick *et al.,* 1976; Davis *et al.,* 1975; DeSimone and Glick, 1975; DeSimone and Glick, 1976b). A group one complex is formed with $Cu(ClO_4)_2$ in which the metal ion is located in the cavity, with the four sulfur atoms of the multidentate ligand and the two ClO_4^- ligands completing the coordination sphere. The structure of the $Ni(BF_4)_2$ complex with the same ligand is similar in that the Ni^{2+} lies in the cavity; however, the complex exists as the ionic $[Ni(ligand)]^{2+}(BF_4)_2^-$. The complex cation has square planar geometry. The range of the torsion angles listed in Table VI for both structures is similar to those found in other complexes of group one. Another type of coordination is found in the group four $NbCl_5$ complex (Fig. 11) in which the ligand acts as a bridge between the two $NbCl_5$ groups. Each Nb atom is coordinated to one sulfur of the cyclic ligand and to five chloro ligands. The conformation of the ligand is similar to that of the β form (see Section II.B.2) of the noncomplexed ligand (DeSimone and Glick, 1976a).

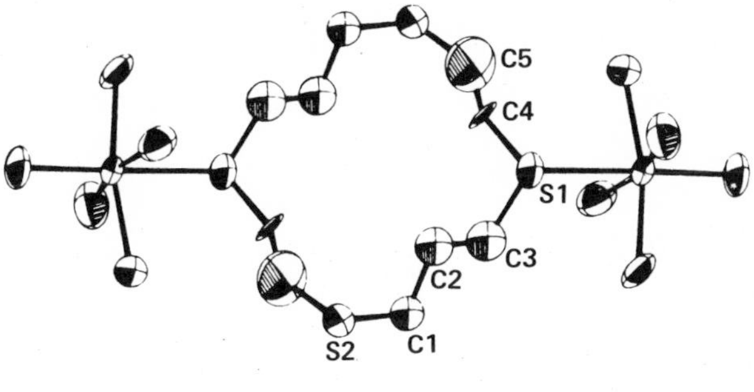

Fig. 11. The four sulfur macrocyclic ligand bridges the two $NbCl_5$ groups (DeSimone and Glick, 1975).

Most of the complexes of the derivatives of cyclam (Fig. 2, XVIII) involve Ni^{2+}, although a Cu^{2+} complex has been reported. The role of the anion is illustrated in these complexes. If the anion is ClO_4^-, BF_4^-, or some other anion that does not form strong complexes with metal ions, then square planar complexes usually result. For example, the Ni^{2+} complex with C-*dl*-5,5,7,12,12,14-hexamethyl-1,4,8,11-tetraazaclotetradecane (tetb) (Fig. 2, XIX) in each of its three configurations is square planar (Curtis *et al.*, 1973b). The anions in these complexes were ClO_4^-, BF_4^-, and $ZnCl_4^{2-}$. On the other hand, anions that form complexes with the metal ion gives rise to octahedral complexes. Examples of this type of coordination are found in the following complexes: $NiCl_2$-cyclam

(Bosnich *et al.*, 1965), $Ni(N_3)_2$-*N*-tetramethylcyclam (Wagner *et al.*, 1974), acetylacetonato-C-*meso*-(5,5,7,12,12,14-hexamethyl-1,4,8,11-tetraazacyclotetradecane) nickel (II) perchlorate (Curtis *et al.*, 1973a) and acetato-C-*dl*-(5,7,7,12,14,14-hexamethyl-1,4,8,11-tetraazacyclotetradecane) nickel (II) perchlorate (Whimp *et al.*, 1970). In the latter two compounds the organic bidentate ligands occupy coordination sites. The ClO_4^- ions are outside the coordination sphere. One other type of complex with a tetraamine ligand has been studied. In the [Cu(tetb)Cl] ClO_4 complex (Bauer *et al.*, 1973) the chloro ligand is in the coordination sphere of two Cu^{2+} and bridges the two Cu atoms (Fig. 12).

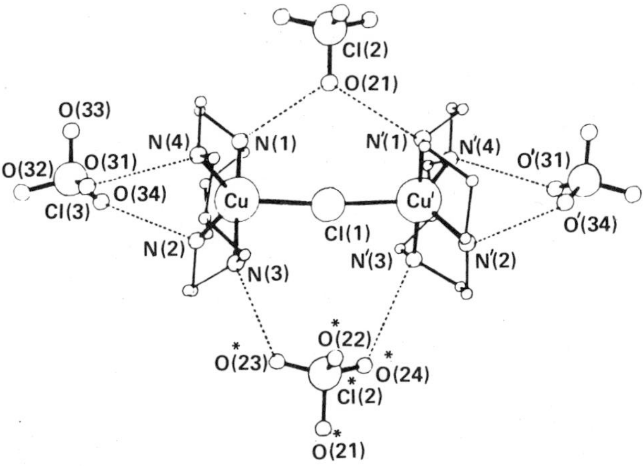

Fig. 12. Two $[Cu(tetb)]^{2+}$ complexes bridged by a Cl^-. The ClO_4^- are hydrogen bonded to the ligand, but do not interact with the cation (Bauer *et al.*, 1973).

B. Uncomplexed Synthetic Macrocyclic Molecules

1. Cryptate and Macrocyclic Polyether Molecules

The structure of 222 (Fig. 2, I) has been studied by Weiss *et al.* (1970). The nitrogen atoms are in the in-in configuration but it appears that several of the torsion angles in the ether segments have changed significantly. The N-N distance was found to be 6.87 Å, which is somewhat longer than the corresponding distance in the various complexes (e.g., about 6.0 Å in the Rb^+ complex). Also, the distances from the center of the N...N line to the oxygen atoms range from 2.41 to 3.50 Å (Truter and Pedersen 1971). This is in contrast to the Rb-O distances in the Rb^+-222 complex, which range from 2.879 to 2.928 Å (Moras and Weiss, 1973b).

A few noncomplexed cyclic polyether molecules have been studied. These include 18-crown-6 (Fig. 2, VI) (Dunitz and Seiler, 1974a), dibenzo-18-crown-6

(VIII) (Bright and Truter., 1970), the cis-anti-cis isomer of tetramethyl dibenzo-18-crown-6 (XII) (Mallinson, 1975b), the cis-syn-cis and cis-anti-cis isomers of dicyclohexo-18-crown-6 (IX) (Dalley *et al.*, 1975a), and dibenzo-30-crown-10 (XI) (Bush and Truter, 1972b). Two unusual polyether molecules have also been studied. Goldberg (1976) reported the structure of 2,6-dimethyl-ylbenzoic acid-18-crown-5 (XIII), the host of a guest-host compound discussed in Section II.A.2, and Parker *et al.* (1973) determined the structure of [2.2] para-cyclophane-bis-18-crown-6 (XX), a molecule containing two polyether rings. The number of possible structures of noncomplexed molecules that can be studied using X-ray diffraction techniques is limited because many of the compounds have low melting points.

A comparison of the structures of the noncomplexed molecules with the same molecules in complexes suggests types of conformational changes which may occur during complexation. The reported structures of cyclic polyether molecules have some common characteristics. None of them have the ordered conformations found in the complexes of groups one and two. The rings are rather elliptical (Table VII) and in all reported structures some of the oxygen atoms point outward. As a result of the irregular conformations of the rings, some of the torsion angles differ considerably from those found in the organized metal complexes. Even though the molecules do not have highly ordered structures, there are several cases in which they are located about centers of inversion. This is the case for 18-crown-6, the cis-anti-cis isomer of dicyclohexo-18-crown-6, dibenzo-18-crown-6, dibenzo-30-crown-10, the cis-syn-cis isomer of tetramethyl dibenzo-18-crown-6, and [2.2] paracyclophane-bis-18-crown-6. Only the cis-syn-cis isomer of dicyclohexo-18-crown-6 and 2,6-dimethylyl-benzoic acid-18-crown-5 do not lie about a center of inversion.

Without the presence of organizing metal ions, and because energy differences between some conformations are small, the structures determined for these molecules in the solid state are greatly affected by packing energies. Infrared spectral evidence (Dale and Karstiansen, 1972) suggests however that the solution conformation of the 18-crown-6 molecule resembles that found in the solid state (Dunitz and Seiler, 1974).

2. Substituted Macrocyclic Polyether Molecules

The noncomplexed substituted cyclic polyethers that have been studied are 1,10-diaza-18-crown-6 (Fig. 2, XIV) (Herceg and Weiss, 1972), 1,10-dithia-18-crown-6 (XVI), 1,4,7-trithia-12-crown-4 (XXI), 1,4 dithia-15-crown-5 (XXII) (Dalley *et al.*, 1975b), and the α, β_1, and β_2 forms of 1,4,8,11-tetrathiacyclo-tetradecane (XVII) (DeSimone and Glick 1976a). Of all the noncomplexed cyclic polyethers studied, including oxygen ethers and the substituted ethers, only the aza ether molecule resembles the ordered ligand observed in complexes of groups one and two. The only significant structural difference between the

TABLE VII

Noncomplexed Crown Ether and Substituted Crown Molecules

Molecule[a]	Torsion angle range[b] X–C–C–X[e]	Torsion angle range[b] C–C–X–C	Distance across ring[c] (Å)	Comments	Ref.[d]
VI	67.6-174.7°	X=O 79.7-175.5°	4.267-6.974[f]	On center of inversion, cavity elliptical in shape	*1*
VIII	77-164°[g] 5°[g,h]	X=O 74-176°[g]	4.72-6.58[f]	On center of inversion, cavity elliptical in shape	*2*
XII	69-170° 5°[h]	X=O 84-177°	4.40-6.89[f]	On center of inversion, cavity elliptical in shape	*3*
IX A	58-178[i]	X=O 81-179°[i]	4.01 to >7.00	Cavity elliptical in shape	*4*
IX B	60-174°[i]	X=O 78-176°[i]	4.11 to >7.00	On center of inversion, cavity elliptical in shape	*4*
XI	63-72° 2°[h]	X=O 95-180°	*j*	On center of inversion	*5*
XIII	49.7-69.7°[k]	X=O 71.8-180°	From acid oxygen to ether oxygen 2.71 diagonally opposite ether oxygens 6.05 and 6.27	Ether elliptical in shape, 1 acid oxygen intramolecularly hydrogen bonded to ether oxygen	*6*
XX	*j*	*j*	*j*	On center of inversion	*7*
XIV	64.2-73.3°	X=O 176.4-177.9° X=N 175.9-178.9°	O-O 5.61-5.670 N-N 5.841	On center of inversion, cavity nearly circular	*8*
XXI	69-176°[i]	X=O 174-175°[i] X=S 61-74°[g]	Shortest distance S-O across ring 4.46	Heart-shaped molecule, all sulfurs point out of cavity	*9*
XXII	60-171°[i]	X=O 89-180°[i] X=S 74-89°[g]	4.83-6.44	Cavity elliptical in shape, all sulfurs point out of cavity	*9*

Molecule[a]	Torsion angle range[b] X–C–C–X[e]	C–C–X–C	Distance across ring[c] (Å)	Comments	Ref.[d]
XVI	68-176°[i]	X=O 173-174°[i] X=S 65-72°[g]	Shortest distance between opposite heteroatoms 4.6	On center of inversion, molecule elliptical in shape, all sulfurs point out of cavity	*9*
XVII α form	176-178°	X=S 60-67°	[j]	On center of inversion, exo conformation, sulfurs at corner of rectangle, point out of cavity	*10*
XVII β_1 form	175-180°	X=S 60-66°	[j]	On center of inversion, exo conformation, resembles α conformation	*10*
XVII β_2 form	177-179°	X=S 74-129°	[j]	On center of inversion; less favorable energetic form in exo conformation. Resembles conformation in Nb complex	*10*

[a] Numbers refer to compounds listed in Fig. 2.
[b] Range includes smallest to largest absolute value for the torsion angle for indicated structure.
[c] Distance across ring given includes shortest and longest distance.
[d] References *1* Dunitz and Seiler (1974a), *2* Bright and Truter (1970), *3* Mallinson (1975b), *4* Dalley *et al.* (1975a), *5* Bush and Truter (1972b), *6* Goldberg (1976), *7* Parker *et al.* (1973), *8* Herceg and Weiss (1972), *9* Dalley *et al.* (1975b), *10* DeSimone and Glick (1976a),
[e] The X represents O, S, C, or N. In the X–C–C–X torsion angles all combinations are included; in the C–X–C–C torsion angle X is defined.
[f] Distances or angles calculated by author of chapter from data given in references.
[g] Torsion angles from Mallinson (1975b).
[h] Torsion angle (X–C–C–X) in ether ring in which carbon segment is also part of benzene ring. X in these cases always represents oxygen.
[i] Torsion angles calculated by author.
[j] Data not available.
[k] Includes only torsion angles in ether part of ring.

noncomplexed aza molecule and its potassium complex (see Tables VI and VII) is the inversion about the nitrogen atoms. The nitrogen lone pair electrons are directed out of the cavity in the noncomplexed molecule but into the cavity in the complex. The sulfur-containing molecules are elliptical in shape, with the sulfur atoms always directed out of the cavity.

III. SHORT C-C DISTANCES IN SYNTHETIC MACROCYCLIC MOLECULES

Short C–C bonds are observed in all cyclic polyethers and their cation complexes whose structures have been determined. Average bond lengths in typical compounds are reported in Table VIII, where it is seen that the bond lengths (~1.50 Å) are significantly shorter than the expected value of 1.537 (Table, 1960) or 1.523 Å which Davis and Hassel (1963) reported for gas phase 1,4-dioxane. Several investigators have commented on the reality of these short

TABLE VIII

Average C–C Bond Lengths in Crown Compounds

Compound	C-C distance	Reference[a]
K^+-benzo-15-crown-5	1.45[b,c]	*1*
18-Crown-6	1.507	*2*
K^+-18-crown-6	1.504	*3*
Dibenzo-30-crown-10	1.492[b,d]	*4*
K^+-dibenzo-30-crown-10	1.48[b,d]	*4*
Na^+-dibenzo-18-crown-6	1.49[b]	*5*
1,10-Diaza-18-crown-6	1.50[d]	*6*
DAC[e]-18-crown-6	1.497[d,f], 1.501[d,f,g]	*7*
2,6-Dimethylbenzoic acid-18-crown-5	1.500[d,f,h], 1.505[d,f,g,h]	*8*
1,10-Dithia-18-crown-6	1.50	*9*
TTP[i], α form	1.509	*10*
$[Ni(TTP)]^{2+}(BF_4)_2$	1.509[d]	*11*

[a]References *1* Mallinson and Truter (1972), *2* Dunitz and Seiler (1974a), *3* Seiler *et al.* (1974), *4* Bush and Truter (1972b), *5* Bush and Truter (1971), *6* Herceg and Weiss (1972), *7* Goldberg (1975a), *8* Goldberg (1976), *9* Dalley *et al.* (1975b), *10* DeSimone and Glick (1976a), *11* Davis *et al.* (1975).

[b]Omitting aromatic C–C bonds.

[c]Omits C–C bond involving disordered carbon atom.

[d]Average bond lengths calculated by author from data in reference.

[e]DAC represents dimethyl acetylene dicarboxylate.

[f]Data taken at –160°C.

[g]Bond lengths calculated from parameters obtained using modified weights in refinement in order to more adequately treat thermal motion.

[h]Includes only C–C bonds from ether portion of the molecule.

[i]TTP represents 1,4,8,11-tetrathiacyclotetradecane.

values. Dunitz *et al.* (1974) attribute the shorter than expected C–C bonds to inadequate treatment of the thermal motion in refinement of positional parameters. Truter, who earlier also attributed the abnormality to thermal motion (Bush and Truter, 1971, 1972a), recently concluded that the short bonds may be real (Mercer and Truter, 1973a,b). More recently, Goldberg (1975a, 1975b, 1976) has studied cyclic polyethers and similar molecules at low temperatures to reduce thermal motion effects. The C-C distances determined by him (Table VIII) are not significantly different from those found at room temperature. Goldberg also used weighing schemes in the refinement of his data to attempt to correct for the thermal motion and continued to calculate short C–C bond lengths. Short C-C distances are also observed in cryptates. For example, the average C–C bond in the K^{+}-222 complex is 1.45 Å (Moras and Weiss, 1973a), while in the Ba^{2+}-322 complex it is 1.47 Å (Metz *et al.*, 1973c). These averages were calculated by the author from the data in the references. At present no satisfactory explanation has been given for the presence of shorter than expected C–C bond distances in these macrocyclic compounds.

IV. CONCLUSIONS

The foregoing structural studies lead to several conclusions, some of which are well documented while others suggest areas for further study.

(1) The cation complexes of the cyclic polyethers can be arranged into four groups based on the ratio of the cation diameter (Table II) to the cavity diameter (Table IV).

(2) For complexes in which the metal-ligand bonding is covalent, the salt anion and the solvent have considerable effect in determining the type of coordination.

(3) Shorter than usual C–C bonds are observed in all cyclic polyethers, cyclic polyether complexes, and cryptate complexes studied to date.

The known structures of cation-macrocyclic compound complexes have led to interesting correlations. However, the available data suggest several areas which should be pursued in a systematic manner. These include the effect on structure of transition and post-transition metal ions, and of anion, solvent, and ligand substitution.

REFERENCES

Bauer, R. A., Robinson, W. R., and Margerum, D. W. (1973). *J. Chem. Soc. Chem. Commun.* 289-290.

Bombieri, G., DePaoli, G., Cassol, A., and Immirzi, A. (1976). *Inorg. Chim. Acta* **18**, L23-L24.

Bosnich, B., Mason, R., Pauling, P. J., Robertson, G. B., and Tobe, M. L. (1965). *Chem. Commun.* 97-98.

Bright, D., and Truter, M. R. (1970). *J. Chem. Soc. (B)* 1544-1550.
Bush, M. A., and Truter, M. R. (1971). *J. Chem. Soc. (B)* 1440-1446.
Bush, M. A., and Truter, M. R. (1972a). *J. Chem. Soc. Perkin Trans.* **2**, 341-344.
Bush, M. A., and Truter, M. R. (1972b). *J. Chem. Soc. Perkin Trans.* **2**, 345-350.
Cram, D. J., and Cram, J. M. (1974). *Science* **183**, No. 4127, 803-809.
Curtis, N. F., Swann, D. A., and Waters, T. N. (1973a). *J. Chem. Soc. Dalton Trans.* 1408-1413.
Curtis, N. F., Swann, D. A., and Waters, T. N. (1973b). *J. Chem. Soc. Dalton Trans.* 1963-1974.
Dale, J., and Kristiansen, P. O. (1972). *Acta Chem. Scand.* **26**, 1471-1478.
Dalley, N. K., Smith, D. E., Izatt, R. M., and Christensen, J. J. (1972). *J. Chem. Soc. Chem. Commun.* 90-91.
Dalley, N. K., Smith, J. S., Larson, S. B., Christensen, J. J., and Izatt, R. M. (1975a). *J. Chem. Soc. Chem. Commun.* 43-44.
Dalley, N. K. Smith, J. S., Larson, S. B., Matheson, K. L., Christensen, J. J., and Izatt, R. M. (1975b). *J. Chem. Soc. Chem. Commun.* 84-85.
Davis, M., and Hassel, O. (1963). *Acta Chem. Scand.* **17**, 1181.
Davis, P. H., White, L. K., and Belford, R. L. (1975). *Inorg. Chem.* **14**, 8, 1753-1757.
DeSimone, R. E., and Glick, M. D. (1975). *J. Am. Chem. Soc.* **97:4**, 942-943.
DeSimone, R. E., and Glick, M. D. (1976a). *J. Am. Chem. Soc.* **98:3**, 762-767.
DeSimone, R. E., and Glick, M. D. (1976b). Private communication.
Dobler, M., and Phizackerley, R. P. (1974a). *Acta Crystallogr.* **B30**, 2746-2748.
Dobler, M., and Phizackerley, R. P. (1974b). *Acta Crystallogr.* **B30**, 2748-2750.
Dobler, M., Dunitz, J. D., and Seiler, P. (1974). *Acta Crystallogr.* **B30**, 2741-2743.
Dunitz, J. D., and Seiler, P. (1974a). *Acta Crystallogr.* **B30**, 2739-2741.
Dunitz, J. D., and Seiler, P. (1974b). *Acta Crystallogr.* **B30**, 2750.
Dunitz, J. D., Dobler, M., Seiler, P., and Phizackerley, R. P. (1974). *Acta Crystallogr.* **B30**, 2733-2738.
Frensdorff, H. K. (1971). *J. Am. Chem. Soc.* **93:3**, 600-606.
Glick, M. D., Gavel, D. P., Diaddario, L. L., and Rorabacher, D. B. (1976). *Inorg. Chem.*, **15**, 5, 1190-1193.
Goldberg, I. (1975a). *Acta Crystallogr.* **B31**, 754-762.
Goldberg, I. (1975b). *Acta Crystallogr.* **B31**, 2592-2600.
Goldberg, I. (1976). *Acta Crystallogr.* **B32**, 41-46.
Harman, M. E., Hart, F. A., Hursthouse, M. B., Moss, G. P., and Raithby, P. R. (1976). *J. Chem. Soc. Chem. Commun.* 396-397.
Herceg, M., and Weiss, R. (1972). *Bull. Soc. Chim. Fr.* 1972 **2**, 549-551.
Herceg, M., and Weiss, R. (1973). *Acta Crystallogr.* **B29**, 542-547.
Hughes, D. L. (1975). *J. Chem. Soc. Dalton Trans.* 2374-2378.
Iitaka, Y., Shina, M., and Kimura, E. (1974). *Inorg. Chem.* **13**, 12, 2886-2891.
Johnson, C. K. (1965). ORTEP. Rep. ORNL-3794, Oak Ridge Nat. Lab. Oak Ridge Tennessee.
Lehn, J. M., and Sauvage, J. P. (1971). *Chem. Commun.* 440-441.
Mallinson, P. R. (1975a). *J. Chem. Soc. Perkin Trans.* **2**, 261-266.
Mallinson, P. R. (1975b). *J. Chem. Soc. Perkin Trans.* **2**, 266-269.
Mallinson, P. R., and Truter, M. R. (1972). *J. Chem. Soc. Perkin Trans.* **2**, 1818-1823.
Mathieu, F., and Weiss, R. (1973). *J. Chem. Soc. Chem. Commun.* 816.
Mercer, M., and Truter, M. R. (1973a). *J. Chem. Soc. Dalton Trans.* 2215-2220.
Mercer, M., and Truter, M. R. (1973b). *J. Chem. Soc. Dalton Trans.* 2469,2473.
Metz, B., and Weiss, R. (1973). *Acta Crystallogr.* **B29**, 1088-1093.

Metz, B., Moras, D., and Weiss, R. (1973a). *Acta Crystallogr.* **B29**, 1377-1381.
Metz, B., Moras, D., and Weiss, R. (1973b). *Acta Crystallogr.* **B29**, 1382-1387.
Metz, B., Moras, D., and Weiss, R. (1973c). *Acta Crystallogr.* **B29**, 1388-1393.
Metz, B., Moras, D., and Weiss, R. (1974). *Inorg. Nucl. Chem.* **36**, 785-790.
Metz, B., Rosalky, J., and Weiss, R. (1976). *J. Chem. Soc. Chem. Commun.* 533-534.
Moras, D., and Weiss, R. (1973a). *Acta Crystallogr.* **B29**, 396-399.
Moras, D., and Weiss, R. (1973b). *Acta Crystallogr.* **B29**, 400-403.
Moras, D., and Weiss, R. (1973c). *Acta Crystallogr.* **B29**, 1059-1063.
Moras, D., Metz, B., Herceg, M., and Weiss, R. (1972). *Bull. Soc. Chim. Fr.* 1972, **2**, 551-555.
Moras, D., Metz, B., and Weiss, R. (1973a). *Acta Crystallogr.* **B29**, 383-388.
Moras, D., Metz, B., and Weiss, R. (1973b). *Acta Crystallogr.* **B29**, 388-395.
Owen, J. D., and Wingfield, J. N. (1976). *J. Chem. Soc. Chem. Commun.* 318-319.
Parker, K., Helgeson, R. C., Maverick, E., and Trueblood, K. N. (1973). *Am. Cryst. Assoc. Abstr.* Winter Meeting, Florida, paper E8.
Pauling, L. (1960). "The Nature of the Chemical Bond," 3rd ed. p. 260. Cornell Univ. Press, Ithaca, New York.
Pedersen, C. J. (1967). *J. Amer. Chem. Soc.* **89:26**, 7017-7034.
Seiler, P., Dobler, M., and Dunitz, J. D. (1974). *Acta Crystallogr.* **B30**, 2744-2745.
Shannon, R. D., and Prewitt, C. T. (1969). *Acta Crystallogr.* **B25**, 925-946.
Smith, D. E. (1972). Ph.D. Dissertation, Brigham Young Univ., Provo, Utah.
Tables of Interatomic Distances and Configuration in Molecules and Ions (1960). Spec. Publ. No. 18, London, The Chemical Society.
Truter, M. R. (1973). *Structure and Bonding* **16**, 71-111.
Truter, M. R., and Pedersen, C. J. (1971). *Endeavour* **30** (III), 142-146.
Wagner, F., Mocella, M. T., D'Aniello, Jr. M. J., Wang, A. H. -J, and Barefield, E.K. (1974). *J. Am. Chem. Soc.* **96:8**, 2625-2627.
Weiss, R., Metz, B., and Moras, D. (1970). *Proc. Int. Conf. Co-ord. Chem., 13th* **2**, 85-86.
Whimp, P. O., Bailey, M. F., and Curtis, N. F. (1970). *J. Chem. Soc. (A)* 1956-1963.

5 KINETIC STUDIES OF SYNTHETIC MULTIDENTATE MACROCYCLIC COMPOUNDS

Gerard W. Liesegang and Edward M. Eyring

Department of Chemistry
University of Utah
Salt Lake City, Utah

I. Introduction 245
A. Other Surveys of Alkali Metal Complexation Kinetics 246
B. Valinomycin Complexation Kinetics 246
II. Macrocyclic Polyether Complexation Kinetics . . 253
III. Macrobicyclic Ligand Complexation Kinetics . . 260
IV. Macrocycles Containing Nitrogen 273
V. Macrocycles Containing Sulfur 280
VI. Host-Guest Chemistry 284
References 286

I. INTRODUCTION

The ramifications of the crown ethers for the broad field of reaction kinetics are already so numerous that a comprehensive survey is not practicable. We will therefore concentrate our attention here on what is known about the rates of the reactions and the mechanisms by which crown ethers and related ligands complex substrates in solutions. This work will also lead us to a discussion of probable uses of crown ethers in catalytically active sites of synthetic macromolecules that begin to simulate the specificity and efficiency of enzymes.

While the equilibrium studies of complexation reactions involving crown ethers are well advanced and kinetic studies of biological macromolecules with some of the same complexation properties as the crown ethers have also been extensively reported, complexation kinetic studies of the crown ethers have gotten off to a late start. Kinetic studies of alkali metal complexation in general are impeded by several factors: complexes are weak and therefore must be studied at high metal ion concentrations; the reaction rates are intrinsically high and the experimental difficulties are further compounded by the requirement that high concentrations be used; and finally, the complexes are usually colorless, so that spectrophotometric measurements of rates are rarely possible.

To whatever degree any chapter on a scientific subject can be definitive, the present chapter on crown ether kinetics is certainly preliminary. It is hoped that readers will catch the vision of the work left to be done in this area and will speed the obsolescence of the following remarks.

A. Other Surveys of Alkali Metal Complexation Kinetics

Although we have not found any previous surveys of crown ether rate studies, there are several reviews of alkali and alkaline earth metal ion complexation kinetics to which we have referred in preparing the present report. For instance, Manfred Eigen, Ruthild Winkler, their collaborators, and others have published several lengthy review chapters (Diebler *et al.*, 1969; Eigen and Winkler, 1970; Grell *et al*;. 1975; Laprade *et al.*, 1975) that address themselves primarily to the kinetic properties of biological carriers of these metal ions across membranes. Finally, in a broader review of the field, Lehn *et al.* (1973) devote three pages to the kinetics of complexation of these metal ions by a macrobicyclic ligand ("cryptand"), valinomycin, etc.

B. Valinomycin Complexation Kinetics

Valinomycin, an antibiotic first isolated from extracts of *Streptomyces fulvissimus* in 1955, is depicted in Fig. 1. Valinomycin is a 36-membered ring

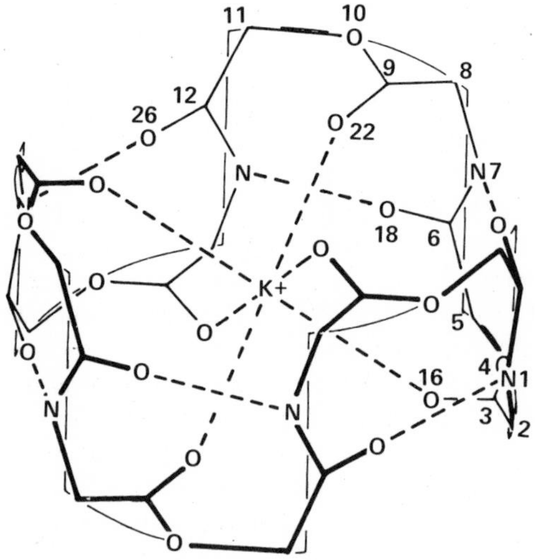

Fig. 1. Schematic of the structure of the K^+ complex of valinomycin showing S_6 symmetry of the ring skeleton, coordination, and hydrogen bonding. (Taken with permission from Neupert-Laves and Dobler, 1975.)

with a molecular weight of 1111.36 that makes it smaller than even the smallest proteins. This antibiotic apparently typifies those transmembrane carriers of metal ions that do not themselves extend across the entire thickness of the biological membrane (typically about 100 Å). Valinomycin is called an oligodepsipeptide because it is built up of alternating amino acids and oxyacids.

Polar moieties of valinomycin are deployed at the interior of this ball-like molecule when it is thrust into a nonaqueous environment, such as the lipid interior of a biological membrane. The hydrophobic exterior of the valinomycin molecule makes it readily soluble in the lipid, and the polar interior forms a cavity across which as many as six hydrogen bonds may form in the absence of a metal ion, or into which a fully desolvated alkali metal cation may be introduced that is fully coordinated by the ester carbonyl groups of the valinomycin at six first coordination sphere sites. The formation of the complex in methanolic solution is thought to proceed through the following successive equilibria (Grell *et al.*, 1975): First a diffusion-controlled bimolecular collision occurs between one of the open forms of the valinomycin molecule and the fully solvated cation. In this first step one of the ester carbonyl groups of the valinomycin is substituted for one of the solvent molecules coordinating the cation. Displacements of additional coordinated methanol molecules by single ester carbonyl groups of the antibiotic rapidly follow until any further exchange of ligands by the cation is limited by the stereogeometry of the incompletely folded valinomycin. This multistep process can be thought of, in principle, as a fast preequilibrium forming an intermediate complex (I). The rate-determining step in the subsequent formation of the final complex (II) is a concerted conformational change of the valinomycin to produce the compact form that characterizes the final structure of the K^+-valinomycin complex ion. Concurrently, there is a rapid displacement of the remaining solvent molecules until all six valinomycin ester carbonyl groups coordinate the central cation. In a first approximation we may write the familiar mechanism

$$\underset{}{M^+\text{(solvated)}} + L\text{(solvated)} \underset{k_{21}}{\overset{k_{12}}{\rightleftharpoons}} \underset{\text{(I)}}{M^+\text{(solvent)}L} \underset{k_{32}}{\overset{k_{23}}{\rightleftharpoons}} \underset{\text{(II)}}{ML^+} \tag{1}$$

which may be written somewhat more explicitly in this special case as

$$M^+ + V \overset{k_{12}}{\rightleftharpoons} \underset{\text{(I)}}{[M^+ - V]} \overset{k_{23}}{\rightleftharpoons} \underset{\text{(II)}}{MV^+} \tag{2}$$

where V denotes the valinomycin.

The true complexity of this process is better appreciated if we note that Grell and Funck (1973) observed at least two concentration-independent ultrasonic absorption maxima in an *n*-hexane solution of pure valinomycin. This indicates that at least three predominant conformations of uncomplexed valinomycin coexist in this nonpolar solvent. Addition of small amounts of ethanol causes a third ultrasonic absorption maximum to appear, implying the coexistence of at least four valinomycin conformations in this more polar mixed solvent. Finally, in highly polar pure methanol, a nearly continuous ultrasonic absorption spectrum is recorded that can be explained in terms of the coexistence of no fewer than five valinomycin conformations in rapid equilibrium with one another. These data conclusively disprove the early notion (Shemyakin *et al.*, 1969) of a one-step valinomycin conformational equilibrium between a closed conformation with six intramolecular hydrogen bonds and a half-open conformation with three intramolecular hydrogen bonds. Thus, the only justification for as gross an oversimplification as Eq. (1) is that a more complicated mechanism is not required to account for two experimentally measured relaxation times in a given sample solution containing metal ions and ligand. Rapid conformational equilibria must indeed be present, but with very small relaxation amplitudes in the time range of interest, so that they are revealed only by the rate-determining first-order process

$$(\mathrm{I}) \underset{}{\overset{k_{23}}{\rightleftharpoons}} (\mathrm{II}).$$

Actually, even greater simplification is frequently necessary in analyzing kinetic data for the formation of complex ions. Often the first reaction step in Eq. (1) is so rapid under the experimental conditions of interest that $k_{21} \gg k_{23}$. In this circumstance we have the simplifying relations

$$k_{\mathrm{f}} = \frac{k_{12}k_{23}}{k_{21}} \quad \text{and} \quad k_{\mathrm{d}} = k_{32} \tag{3}$$

where the numerical subscripts are from Eq. (1), the subscript f denotes formation, and the d dissociation. In a case where the formation of the "outer sphere" complex (I) in Eq. (1) is experimentally too fast to measure, it may still be possible to calculate a value of $K_{\mathrm{os}} = k_{12}/k_{21}$ (where the subscript os denotes outer sphere) from a theoretical equation derived by Fuoss (1958). There are many instructive examples of its use (Hammes and Steinfeld, 1962; Hemmes, 1972). Thus, we frequently encounter discussions of metal ion complexation in terms of the first-order specific rate k_{23} or sometimes k_{sub} obtained from the relation

$$k_{\mathrm{f}} = K_{\mathrm{os}}k_{\mathrm{sub}} \tag{4}$$

where sub denotes solvent molecule substitution.

Some representative valinomycin kinetic data are shown in Table I. An interesting challenge to the kineticist is to explain mechanistically differences

TABLE I

Kinetic Parameters for the Formation of Valinomycin Cation Complexes in Methanol at 25°[a]

Cation	K'^{c} (M^{-1})	k_{12} (M^{-1} sec^{-1})	k_{21} (sec^{-1})	$K_{12}=k_{12}/k_{21}$ (M^{-1})	k_{23} (sec^{-1})	k_{32} (sec^{-1})	$K_{23}=k_{23}/k_{32}$	Exp. method
NH_4^+	47	1×10^9	1.5×10^8	6.5	2×10^6	2.5×10^5	8	US[e]
Na^+	4.7	7×10^7	2×10^7	3.5	4×10^6	2×10^6	2	US
K^{+b}	3×10^4	4×10^8	1×10^8	4.0	1×10^7	1.3×10^3	7.7×10^3	US,TJ[f]
Rb^{+b}	6.5×10^4	k_{on}, (M^{-1} sec^{-1})[d]:	5.5×10^7		k_{off}, (sec^{-1})[d]:	7.5×10^2		TJ
Cs^{+b}	8×10^3		2×10^7			2.2×10^3		TJ

[a] Data taken from Grell *et al.* (1975).
[b] Apparent stability constants were determined in the presence of 0.1 *M* tetra-*n*-butylammonium perchlorate.
[c] The apparent stability constants (K') were determined by spectrophotometric titrations.
[d] k_{on} and k_{off} are the overall rate constants for forward and reverse reaction.
[e] US denotes ultrasonic absorption spectrometry.
[f] TJ denotes temperature jump relaxation spectrometry carried out in methanol containing 0.1 *M* tetra-*n*-butylammonium perchlorate.

in a given rate constant, let us say k_f, as various metal ions are reacted with the same ligand in the same solvent. Linear variations of log k_f or log k_{sub} with ionic radius are known (Eigen and Wilkins, 1965) for many ligands reacting with a sequence of metal ions in water, all having the same charge but different ionic radii. We must remember that the charge density of a metal ion (and its consequent ability to hold a water molecule in its first coordination sphere) is proportional to ionic charge divided by the ionic radius. Thus the now familiar explanation for the linearity of a log k_{sub} versus r^{-1} plot is that the rate-determining step in such a complexation process is the dissociation of a first coordination sphere water molecule rather than the approach or entry into the first coordination sphere of the new ligand. If the cation has so low a charge density that the rate-determining step in the loss of a water molecule from the first coordination sphere of an ion is the diffusion of the water molecule into the neighboring bulk solvent, the first-order rate constant k_{sub} for the water loss would be $\sim 10^9$-10^{10} sec^{-1}. As we see from Table II, in an ultrasonic absorption

TABLE II

Kinetic Parameters for the Formation of Complex Ions with Uramildiacetate in Water at 20°[a] and with Murexide in Methanol at 25°[b]

	Uramildiacetate			Murexide		
	k_{sub} (sec^{-1})	k_{diss} (sec^{-1})	log K_{stab}	k_f (M^{-1} sec^{-1})	k_{diss} (sec^{-1})	log K_{stab}
Li^+	1×10^8	4×10^2	5.4	$(4 \pm 1) \times 10^9$	5×10^6	2.9
Na^+	5×10^8	2.5×10^5	3.3	1.4×10^{10}	5.6×10^6	3.4
K^+	1×10^9	1×10^7	~2.0	$>1.2 \times 10^{10}$	$>10^7$	3.1

[a] Data taken from Eigen and Maass (1966).
[b] Data taken from Diebler *et al.* (1969).

kinetic study (Eigen and Maass, 1966) of the complexation of three alkali metal ions by uramildiacetate, a ligand with the formula (I), the values of k_{sub} are

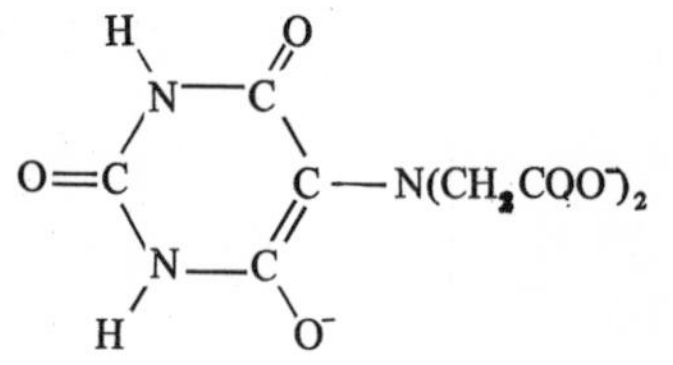

(I)

almost as large as the theoretical diffusion-controlled limit and the same linear increase in the specific rate of substitution with increasing ionic radius occurs as has been observed with so many other metal ions. Also shown in Table II are the corresponding kinetic data for the complexation of the same metal ions by murexide (II) (the ammonium salt of purpuric acid) in methanol measured by

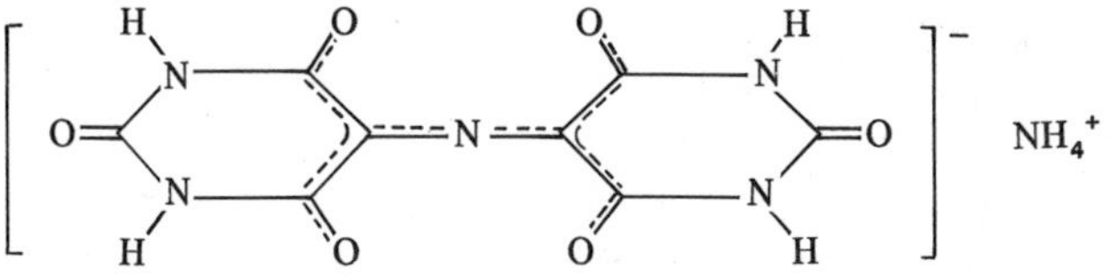

(II)

the spectrophotometric electric field jump relaxation method (Diebler *et al.*, 1969). The most significant features of Table II are that the values of these first-order alkali metal ion rate constants more closely approach the diffusion-controlled value than do those of more highly charged metal ions and the increase in k_{sub} for a given increase in ionic radius is much less pronounced in the case of these alkali metal ions than for more highly charged ions.

Returning now to the valinomycin data of Table I, we find that a plot (Fig. 2) of the logarithm of either the formation or dissociation rate constants versus ionic radii is very nonlinear, contrary to the expectations raised by the preceding discussion. Diebler *et al.* (1969) pointed out that a variety of interaction terms might play a role in making the complexation of alkali metal ions by a complex ligand such as valinomycin more complicated than the simple radius-charge density considerations noted earlier would lead us to believe. They mentioned specifically "metal ion-ligand attraction, including electrostatic (ion-ion or ion-multipole) interaction; polarization and ligand field effects; (electrostatic ligand-ligand repulsion and van der Waals terms)." In this and subsequent papers (e.g., Eigen and Winkler, 1970), Eigen and co-workers emphasized that an ion size specificity can be accounted for by the superposition of solvation and chelation effects, provided that proper account is taken of the fact that a polydentate ligand may "freeze" into fixed positions (because of steric hindrances) below a certain optimal cavity size with little or no gain in binding energy with decreasing metal ion radius to compensate for an increasing solvation energy. Of late, this has led (see, e.g., Diamond, 1975) to a typecasting of Eigen and Winkler (as well as Simon and Morf, 1973) in the role of "heavies" who incorrectly (?) surmised that ion selectivity has a primarily steric rather than electrostatic basis.

The experimental arguments mustered by Diamond (1975) for the greater importance of electrostatic effects are interesting: Valinomycin and hexade-

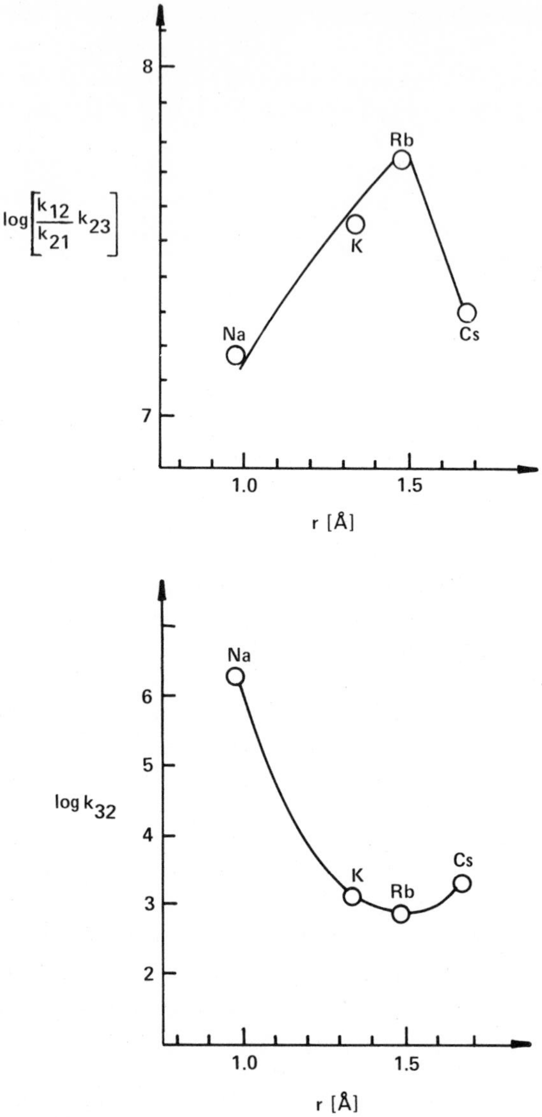

Fig. 2. Characteristic kinetic parameters for the formation ($[k_{12}/k_{21}]k_{23}$) and dissociation (k_{32}) of valinomycin alkali ion complexes as a function of the ionic radius. (Taken with permission from Grell *et al.,* 1975.)

cavalinomycin, which have relatively rigid cavities of differing sizes but with identical ligands, exhibit very similar selectivity patterns for alkali metal cations (Moreno and Diamond, 1974). Substituent group effects on the selectivity of actins (which have easily deformed cavities) can be successfully explained on the basis of inductive effects on the charge of the ligands lining the cavity (Szabo *et al.*, 1969). Finally, Phillies *et al.* (1975) have made a laser Raman spectroscopic study of nonactin and its complexes with Na^+, K^+, Rb^+, Cs^+, Tl^+, $NH_4{}^+$, NH_4OH^+, and $(NH_2)_2CNH_2^+$ in a methanol-chloroform solvent and obtained linear correlations of carbonyl (ligand) stretching frequencies with electrostatic interaction energies but not with ionic radii. Phillies *et al.* conclude their communication by noting that their work supports the *electrostatic* models of Diebler *et al.* (1969) and Krasne and Eisenman (1973), thus rehabilitating Eigen and co-workers, at least temporarily.

In an equally recent paper by Cahen and Popov (1975), infrared and Raman spectroscopic data on lithium and sodium cryptates (macrobicyclic ligands studied extensively by Lehn *et al.* (1973) in nonaqueous solvents lead these authors to conclude that "the cation-ligand interaction is predominately electrostatic in nature." These authors incidentally cite another far-infrared study (Tsatsas *et al.*, 1972) of Na^+- and K^+-dibenzo-18-crown-6 complexes in nonaqueous solvents that concludes that "it is ion solvation, rather than steric or electronic properties of the enclosing cavity, that dominates ion selectivity in solution." As we will see in what follows, the state of disarray on the subject of factors responsible for ion selectivity is no worse than the confusion regarding complexation rates.

Since the ring sizes of the most common crown ethers are substantially smaller than that of valinomycin, we may expect that electrostatic effects will predominate over steric effects in determining ion selectivity patterns for the crown ethers. Talekar (1975) has cataloged incidentally a considerable number of effects that may be important in accounting for the ion selectivity of the crown ethers and ionophores in general.

II. MACROCYCLIC POLYETHER COMPLEXATION KINETICS

In an early ^{23}Na nmr study of the complexation of Na^+ by dibenzo-18-crown-6 (denoted hereafter by DBC6) (III) in dimethylformamide (DMF)

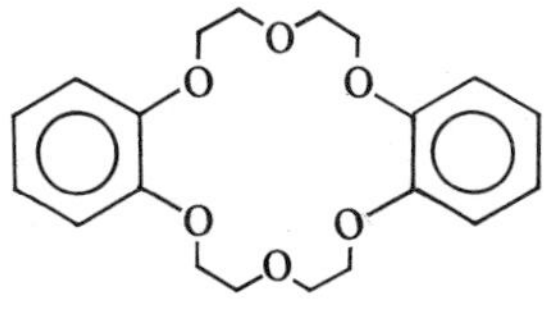

(III)

Shchori *et al.* (1971) drew two interesting conclusions: A strong ionic strength dependence of the kinetics indicates that the primary reaction being observed is the complexation

$$\mathrm{Na^+,\ }n(\mathrm{DMF}) + \mathrm{DBC6} \underset{k_\mathrm{d}}{\overset{k_\mathrm{d}}{\rightleftharpoons}} \mathrm{Na^+,\ DBC6} + n(\mathrm{DMF})$$

and not the simple exchange equilibrium

$$\mathrm{{}^*Na^+,\ }n(\mathrm{DMF}) + \mathrm{Na^+,\ DBC6} \rightleftharpoons \mathrm{{}^*Na^+,\ DBC6} + \mathrm{Na^+,\ }n(\mathrm{DMF})$$

Also, their rate constant $k_\mathrm{f} \sim 6 \times 10^8\ M^{-1}\ \mathrm{sec}^{-1}$ (see Table III) closely resembles a value of $k_\mathrm{f} = 3 \times 10^8\ M^{-1}\ \mathrm{sec}^{-1}$ reported by Diebler *et al.* (1969) for the complexation of Na^+ by the macrocyclic ligand monactin in methanol, but both rate constants are substantially larger than those reported by Lehn *et al.* (1970) for the complexation of Na^+ by a macrobicyclic ligand (cryptand) in water.

One of the interesting features of this paper by Shchori *et al.* (1971) is the use of LiSCN to maintain constant ionic strength. This second salt does affect the observed rate, but because the stability constant for Na^+ complexation is more than 100 times greater than that for Li^+ with DBC6, the authors concluded that changes in activity or ionic pairing and not a competing chemical reaction had to be responsible for the observed effect. Conductivity measurements ruled out ion pairing. The use of tetrabutylammonium ion instead of Li^+ would have made the case for changes in activity being important more persuasive. However, this study clearly puts one on notice that the activity coefficient term in the transition state expression

$$k_\mathrm{f} = K^{\neq}\left(\frac{kT}{h}\right)\left[\frac{\gamma^n_\mathrm{DMF}\gamma_\mathrm{NaDBC6}}{\gamma^{\neq}}\right]$$

must be reckoned with.

In chronological order, the next major contribution to our present understanding of crown ether complexation kinetics was made by Chock (1972). Using a spectrophotometric, Joule heating temperature jump apparatus at a near-ultraviolet wavelength (280-285.7 nm), he determined the rate constants for complexation of several monovalent cations by dibenzo-30-crown-10 in methanol (see Table III). In all experiments, except those with thallium, ionic strength was maintained with LiCl at 0.15 *M*. From his relaxation amplitudes, Chock inferred the existence of a fast crown ether conformational transition preceding the complexation reaction. The Joule heating temperature jump apparatus he utilized did not permit measurement of relaxation times under a microsecond, so that Chock was unable to measure the kinetics of this fast preequilibrium. The least complicated reaction scheme that provided a fit of

TABLE III

Experimentally Determined Rate Constants for the Reaction $M^+ + \text{Ligand} \underset{k_d}{\overset{k_f}{\rightleftharpoons}} \text{Complex Ion}^+$

Ligand	Cation	Solvent	Temp. (°C)	μ (M)	k_f (M^{-1} sec^{-1})	k_d (sec^{-1})	K_{stab}	Kinetic Exp. method	Ref.[a]
Dibenzo-18-crown-6 (DBC6)	Na^+	DMF	25	→0	~6 × 10^7	~10^5	~600 M^{-1}	^{23}Na nmr	*1*
			−13			800			
Dibenzo-30-crown-10 (DBC6)	Na^+	MeOH	25	0.15	>1.6 × 10^7	>1.3 × 10^5	(1.3±0.2) × 10^2	TJ	*2*
	K^+				(6±2) × 10^8	(1.6±0.5) × 10^4	(3.7±0.4) × 10^4		
	Rb^+				(8±2) × 10^8	(1.8±0.4) × 10^4	(4.4±0.5) × 10^4		
	Cs^+				(8±1) × 10^8	(4.7±0.6) × 10^4	(1.7±0.2) × 10^4		
	NH_4^+				>3 × 10^7	>1.1 × 10^5	(2.7±0.3) × 10^2		
	Tl^+				(8±1) × 10^8	(2.5±0.3) × 10^4	(3.2±0.4) × 10^4		
Antamanide	Na^+	MeOH	26.3	0.1	1.1 × 10^5	2.1 × 10^2	5 × 10^2	TJ	*3*
		CH_3CN			7.7 × 10^5	2.6 × 10^1	3 × 10^4		
	Ca^{2+}	MeOH			4.9 × 10^3	1.9 × 10^2	2.5 × 10^1		
Monactin	Na^+	MeOH	30	*b*	3 × 10^8	6 × 10^5	5 × 10^2	EJ, US	*4*
Valinomycin	Na^+	MeOH	25	0.4	1.2 × 10^7	2 × 10^6	4.7	US	*5*
cis-4,4′-Dinitrodibenzo-18-crown-6 (NDBC6)	Na^+	DMF	25	→0	2.3 × 10^7	2.0 × 10^5	115	^{23}Na nmr	*6*
cis-4,4′-Diaminodibenzo-18-crown-6 (AmDBC6)		DMF			1.2 × 10^8	1.9 × 10^5	615		
DBC6		MeOH			3.2 × 10^8	1.4 × 10^4	2.3 × 10^4		
Dicyclohexyl-18-crown-6 (isomer IB) (DCC6IB)		MeOH			2.6 × 10^8	5.2 × 10^4	4.8 × 10^3		
DBC6	K^+	MeOH	−34	0.5		610		^{39}K nmr	7
	Rb^+		−50	0.64		>10^4		^{87}Rb nmr	

[a] References *1* Shchori *et al.* (1971), *2* Chock (1972), *3* Burgermeister *et al.* (1974), *4* Diebler *et al.* (1969), *5* Funck *et al.* (1972), *6* Shchori *et al.* (1973), *7* Shporer and Luz (1975).

[b] Ionic strength not reported.

Chock's relaxation amplitudes and relaxation times between 40 and 5 μsec is

$$\begin{matrix} CR_1 \\ \text{fast} \upharpoonleft\downharpoonright \\ CR_2 \end{matrix} + M^+ \underset{k_d}{\overset{k_f}{\rightleftharpoons}} MCR^+$$

in which CR_1 and CR_2 denote different conformations of the uncomplexed crown ether. This scheme differs in an interesting way from the valinomycin complexation scheme mentioned earlier that was suggested by Grell *et al.* (1975). The latter authors assumed that a rate-determining conformational transition of the valinomycin occurred after the formation of a first bond between the valinomycin and the cation. Neither publication provides detailed kinetic data that could be used to prove the distinguishability of these two mechanisms.

Since Chock's rate constants (k_f and k_d) for Na^+ are only limiting values, it is not possible to make a plot of his data similar to Fig. 2. However, it is clear from his rate constants for K^+, Rb^+, and Cs^+ that the maximum in the plot of log k_f and the minimum in the plot of log k_d versus ionic radius for dibenzo-30-crown-10 would be somewhat flatter than for valinomycin. It is also clear that the specific rates in both directions are significantly faster for dibenzo-30-crown-10 than for valinomycin. Chock also makes an argument for the dissimilarity of ammonium ion kinetics and complex ion stability from those of the alkali cations based on the distribution of partial positive charge over the tetrahedral ammonium ion structure requiring a more specialized ligand configuration that dibenzo-30-crown-10 cannot assume. Thallium (I) in reacting with dibenzo-30-crown-10, on the other hand, behaves like an alkali metal cation in spite of a different electronic configuration.

The rapid, conformational flexibility of dibenzo-30-crown-10 that is hinted at by Chock's temperature jump amplitude measurements is a vital feature of his picture of the complexation reaction mechanism. The nearly diffusion-controlled rate constants k_f are only possible if a low activation energy barrier is maintained by the immediate substitution of a crown moiety for a solvent molecule as the cation is simultaneously desolvated and complexed. A less flexible ligand would require the much more energetically expensive total desolvation of the cation before complexation occurred, with a consequent substantial reduction in reaction rates.

Rodriguez *et al.* (1977) have preliminary ultrasonic absorption kinetic data on aqueous solutions of 15-crown-5 that cast some doubt on the foregoing supposition. A concentration-independent relaxation occurs in ~1 *M* solutions of 15-crown-5 (with no added alkali metal salts) at a frequency of 22 ± 3 MHz; that is, $\tau^{-1} = 2\pi f_r \cong 1.4 \times 10^8$ sec^{-1}. This is distinctly slower than a similar concentration-independent relaxation noted at f_r = 101 ± 3 MHz (Liesegang *et al.*, 1976) in aqueous solutions of 18-crown-6 to which no alkali metal salt has been added. Since the smaller 15-crown-5 ring is somewhat less flexible than that of 18-crown-6, a longer relaxation time for a conformational equilibrium of 15-crown-5 is consistent with Chock's conceptions. However, preliminary data with aqueous solutions of 15-crown-5 and potassium chloride suggest that the concentration-dependent complexation reaction is as fast as that between K^+ and 18-crown-6 (Liesegang *et al.*, 1976). One conceivable explanation for this apparent contradiction would be that the observed 15-crown-5 conformational equilibrium does not involve conformations that participate in complexation.

In a recent study (Burgermeister *et al.*, 1974) of the complexation of Na^+ and Ca^{2+} by the cyclodecapeptide antamanide in methanol, conformational transitions of this ligand as envisioned by Chock were actually measured ultrasonically in the 0.5- to 100-MHz frequency range. The complexation rate constants for this ligand are shown in Table III. The k_f and k_d values for antamanide are both substantially smaller than those for dibenzo-30-crown-10, monactin, and valinomycin (also shown in Table III). This is surprising, since all four macrocyclic ligands are uncharged and have approximately the same ring size. Burgermeister *et al.* attribute this difference to less flexibility in the antamanide ring, with the consequence that substantial desolvation of the cation, rather than a ligand conformational change, is rate limiting for antamanide complexation. They attribute this comparative lack of flexibility to ten peptide bonds in this cyclodecapeptide that show no cis-trans isomerism as interconversion between two globular conformations of antamanide occurs (in the absence of cations), as well as to four rigid C—N bonds belonging to the four proline residues in the ring.

A substantial effort has been made by Shchori *et al.* (1973) to clarify the effect of solvents and aromatic ring substituents on the kinetics of complexation of sodium ion by dibenzo-18-crown-6 (DBC6). They used ^{23}Na nmr spectroscopy and worked with *N,N*-dimethylformamide, methanol, and dimethoxyethane solvents. Rate constants are shown in Table III. Although the presence of electrophilic nitro groups on the benzene rings of *cis*-4,4′-dinitrodibenzeno-

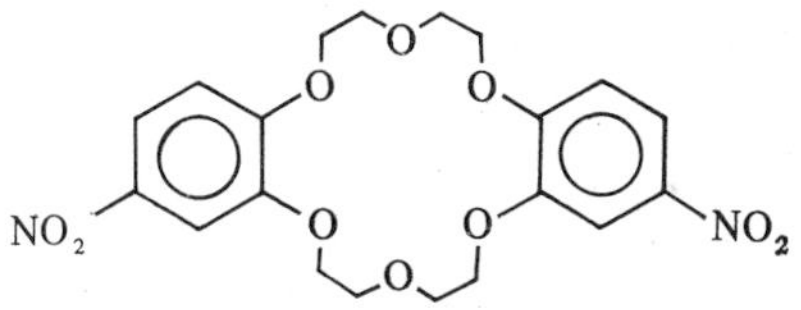

(IV)

18-crown-6 (IV) does slow down complex formation and destabilizes the complex ion, substitution of electron-donating amino groups in these same benzene ring positions does not significantly alter complex ion stability from that of DBC6. Shchori *et al.* hypothesize that changes in the hydrogen-bonding interactions of the ring involving the amino groups may alter the conformation of the ring.

On the basis of their observation that the activation energy for decomplexation of Na^+ by DBC6 is the same in all solvents (~12.6 kcal/mole) but is substantially less (8.3 kcal/mole) for dicyclohexyl-18-crown-6 in methanol, Shchori *et al.* (1973) speculated that the major barrier to removal of Na^+ from DBC6 and its derivatives is just the energy required to effect a conformational rearrangement. The lower E_a for the decomplexation of Na^+ by dicyclohexyl-18-crown-6 would then be simply the consequence of the greater flexibility of this ligand.

Shporer *et al.* (1974) have used ^{23}Na nmr to confirm the rate constant $k_d = 1 \times 10^6$ to 3×10^6 sec^{-1} (at 25° in methanol) for the decomplexation of sodium ion by valinomycin. Shporer and Luz (1975) in a similar communication have used ^{39}K and ^{87}Rb nmr to determine rate constants for K^+ and Rb^+ decomplexation, respectively, by dibenzo-18-crown-6 in methanol at low temperatures (see Table III). The much faster decomplexation of Rb^+ than of K^+ (or Na^+) provides an interesting contrast with Chock's (1972) dibenzo-30-crown-10 study (also in methanol), wherein the k_d values at 25° for K^+ and Rb^+ are very similar to one another (Table III) and k_d for Na^+ is somewhat larger. If the potassium ion decomplexation, $k_d = 610$ sec^{-1} at −34°, of Shporer and Luz (1975) is extrapolated with their activation energy of 12.6 kcal/mole to 25°, a $k_d = 1.1 \times 10^5$ sec^{-1} results. This is a power of ten larger than Chock's $k_d = (1.6 \pm 0.5) \times 10^4$ sec^{-1} for K^+-dibenzo-30-crown-10. When combined with Frensdorff's (1971) stability constant $K_{23} = 10^{5.00}$ at 25°, a $k_d = 1.1 \times 10^5$ sec^{-1} yields a $k_f = 1.1 \times 10^{10}$ M^{-1} sec^{-1}, which equals the theoretical value for a diffusion-controlled reaction (Liesegang *et al.*, 1976). The reasonable conclusion to draw from this large value for k_f is that the ^{39}K nmr k_d and E_a values may be good in the neighborhood of −34° but do not lend themselves to so long an extrapolation as that to room temperature. In fairness to Shporer and Luz (1975) it should be said that in their communication they did not attempt this extrapolation.

In the previously noted ultrasonic absorption kinetic study (Liesegang *et al.*, 1976, 1977) of the complexation of several ions by 18-crown-6 in water (see Table IV) values of k_f have been obtained that are still subject to a common multiplicative correction factor (i.e., the conformational equilibrium constant). Thus a comparison of these k_f values with those of other authors in Table III is premature but the internal trends of Table IV are nonetheless instructive.

TABLE IV

Rate Constants for the Reaction $M^+ + \text{18-crown-6} \underset{k_d}{\overset{k_f}{\rightleftharpoons}} \text{Complex Ion}^+$

Ligand	Cation	Solvent	Temp. (°C)	μ (M)	k_f (M^{-1} sec^{-1})	k_d (sec^{-1})	K_{stab}	Kinetic exp. method	Ref.[a]
18-Crown-6	Na^+	H_2O	25	~0.3	$\sim 2.2 \times 10^8$	3.4×10^7	6.3	Ultrasound	*1*
	K^+				$\sim 4.3 \times 10^8$	3.7×10^6	115.0		
	Rb^+				$\sim 4.4 \times 10^8$	1.2×10^7	36.3		
	Cs^+				$\sim 4.3 \times 10^8$	4.4×10^7	9.8		
	NH_4^+				$\sim 5.6 \times 10^8$	4.4×10^7	12.6		
	Ag^+				$\sim 11.2 \times 10^8$	3.5×10^7	31.6		
	$T\ell^+$				$\sim 9.0 \times 10^8$	4.8×10^6	186.0		

[a]Reference *1* Liesegang *et al.* (1977).

III. MACROBICYCLIC LIGAND COMPLEXATION KINETICS

Macrobicyclic ligands, as their name suggests, are two-ringed macrocycles (see structures (V) through (VIII) as examples). A number of macrobicyclic ligands containing both nitrogen and oxygen atoms, with bridgehead nitrogen atoms, were synthesized by Dietrich *et al.* (1969a, b). These same workers observed that these ligands form very stable complexes with various metal cations, in which the cation is contained within the central molecular cavity of the macrobicyclic ligand. Thus it is interesting to review the complexation-decomplexation kinetics of these macrobicyclic ligands and to compare their kinetic behavior to that of the previously discussed macrocyclic polyethers.

Several macrobicyclic ligands of the general type (V) completely encapsulate metal and halide ions [Parks *et al.*, 1970; Park and Simmons 1968]. Simmons and Park (1968), in proton magnetic resonance (pmr) experiments with the macrobicyclic diamines (V), observed that these compounds undergo conformational

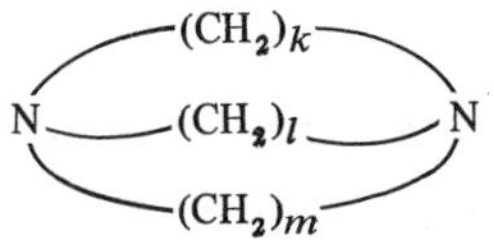

(V)

changes in solution (thus following the trend noted earlier for the antibiotics and crown ethers). For the macrobicyclic diamine, where $k = l = m = 8$, the pmr spectra in $CFCl_3$ solvent indicated a nitrogen inversion with an activation energy of 7.7 kcal/mole and a rate constant of 1.4×10^7 sec^{-1} at 25°.

These same workers reported that three stereoisomers of the bis ammonium ions of (V) can exist when $k, l, m \geqslant 6$. These stereoisomers are designated as *out-out, out-in,* and *in-in* [see (Va)-(Vc)].

(*out-out* [o^+o^+]) (*out-in* [o^+i^+]) (*in-in* [i^+i^+])

(Va) (Vb) (Vc)

An equilibrium exists between the o^+o^+ and i^+i^+ isomers for various amines, whereas the o^+i^+ isomers are apparently of higher free energy than either the o^+o^+ or i^+i^+, because of a nonbonded repulsion effect. In aqueous solution these workers observed that for the macrobicyclic diamine where $k = l = m = 8$ at 25°, an equilibrium between o^+o^+ and i^+i^+ was slowly established. An eight-step mechanism was postulated for the interconversion of the o^+o^+ and i^+i^+ ions in aqueous solvents. It is clear from the above-noted rate constant (1.4×10^7 sec^{-1}) that this conformational change is slower and possibly more complex than those discussed previously in valinomycin, dibenzo-30-crown-10, and 18-crown-6.

In the first major study of the complexation kinetics of the cryptands, Lehn *et al.* (1970) observed temperature-dependent changes in the pmr spectra when potassium, sodium, rubidium, or thallium salts were added to aqueous cryptand (VI), which Lehn would denote as [2.2.2]. These spectral changes were attributed

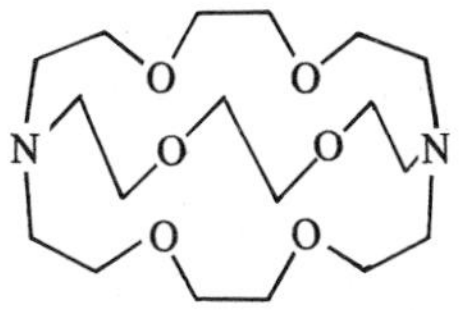

(VI)

to variations in the exchange rates of the cryptated cation with temperature.

As in the case of (V), Lehn *et al.* suggested that the cryptand (VI) and cryptate exist as an equilibrium mixture of three conformations (out-out, out-in, in-in), and exchange may occur in any of these forms. Unlike Simmons and Park (1968), Lehn *et al.* did not observe conformational changes of cryptand (VI).

The exact mechanism of the exchange process may be quite complex when one considers the foregoing conformational changes, together with the possibility of different exchange mechanisms being present. For instance, either of the following may be observed:

(1) A complexation-decomplexation process

$$(M^{n+},\ mH_2O) + (VI)_{aq} \underset{k_{-1}}{\overset{k_1}{\rightleftharpoons}} (M^{n+},\ VI)_{aq} + mH_2O \qquad (5)$$

or

(2) a simple bimolecular exchange process

$$(M^{n+},\ mH_2O)^* + (M^{n+},\ VI)_{aq} \underset{k_{-2}}{\overset{k_2}{\rightleftharpoons}} (M^{n+},\ VI)^*_{aq} + (M^{n+},\ mH_2O) \qquad (6)$$

Their experimental results led Lehn *et al.* to favor process (5), where they were observing the decomplexation step (k_{-1}). The decomplexation of the K^+-cryptate (VI) is unaffected by various anions (Cl^-, F^-, and Br^-; see Table V). This points out that, at least in this specific case, ion pairing does not affect the rate of decomplexation of the cryptate.

Tables V and VI list the exchange rates, stability constants, free energies of

TABLE V

Kinetic Parameters for the Macrobicyclic Ligands

Complex	Solvent	Temp. (°C)	Complexation rate ($k_c \times M^{-1}$ sec^{-1})	Decomplexation rate ($k_d \times sec^{-1}$)	Method	Ref.[a]
[VI,Na^+] Cl^-	Water	3	2×10^5 [b]	27	^1H nmr	1
[VI,K^+] F^-	Water	36	7.5×10^6 [b]	38	^1H nmr	1
[VI,K^+] Br^-	Water	35	8×10^6 [b]	42	^1H nmr	1
[VI,K^+] Cl^-	Water	36	7.5×10^6 [b]	38	^1H nmr	1
[VI,Rb^+] Cl^-	Water	9	7.5×10^5 [b]	38	^1H nmr	1
[VI,Tl^+] Cl^-	Water	40	2.5×10^8 [b]	60	^1H nmr	1
[VI,Tl^+] NO_3^-	Water	−6	2×10^8 [b]	51	^1H nmr	1
[VI,Ca^{2+}] $(Cl^-)_2$	Water	25	$>10^3$ [b]	0.1	P[d]	1
[VI,Sr^{2+}] $(Br^-)_2$	Water	25	6×10^3 [b]	10^{-4}	P	1
[VI,Ba^{2+}] $(Cl^-)_2$	Water	25	3×10^4 [b]	10^{-5}	P	1
[VI,Na^+] Br^-	Ethylenediamine	34	[c]	3×10^2	^{23}Na nmr	2
[VII,Li^+] ClO_4^-	Pyridine	25	[c]	0.12×10^{-3}	^7Li nmr	3
[VII,Li^+] ClO_4^-	Water	25	0.98×10^3	4.9×10^{-3}	^7Li nmr	3
[VII,Li^+] ClO_4^-	Dimethylsulfoxide	25	[c]	2.32×10^{-2}	^7Li nmr	3
[VII,Li^+] ClO_4^-	Dimethylformamide	25	[c]	1.3×10^{-2}	^7Li nmr	3
[VII,Li^+] ClO_4^-	Formamide	25	[c]	7.4×10^{-3}	^7Li nmr	3
[VIII,Li^+] ClO_4^-	Pyridine	25	[c]	1.23	^7Li nmr	3
[VI,Ca^{2+}] $(Cl^-)_2$	Water	25	6.6×10^3	0.26	SF[e]	4
[VIII,Ca^{2+}] $(Cl^-)_2$	Water	25	1.2×10^4	1.9×10^{-3}	SF[e]	4
[VII, Ca^{2+}] $(Cl^-)_2$	Water	25	1.6×10^2	0.1	SF[e]	4
[V(*k*=*l*=*m*=*8*)]	$CFCl_3$	25	[c]	1.4×10^7 [g]	^1H nmr	5
[$Ca^{2+}\subset$IX] $(Cl^-)_2$	D_2O	30	[c]	58[h]	^{13}C nmr FT[f]	6

$[SR^{2+} \subset IX]$ $(Cl^-)_2$	D_2O	3	*c*	171[h]	^{13}C nmr FT[f]	*6*
$[Ba^{2+} \subset IX]$ $(Cl^-)_2$	D_2O	3	*c*	155[h]	^{13}C nmr FT[f]	*6*
$[La^{3+} \subset IX]$ $(NO_3^-)_3$	D_2O	93	*c*	58[h]	^{13}C nmr FT[f]	*6*

[a]References *1* Lehn *et al.* (1973), *2* Ceraso *et al.* (1975), *3* Cahen *et al.* (1973), *4* Loyola *et al.* (1975), *5* Simmons *et al.* (1968), *6* Lehn and Stubbs (1974).

[b]Complexation rates are approximate, since they were calculated at the temperatures indicated and the stability constant at 25°.

[c]No value for the stability constant.

[d]P denotes potentiometric method.

[e]SF denotes stopped-flow method.

[f]FT denotes Fourier transform spectroscopy.

[g]Rate constant for the conformational change.

[h]Intramolecular exchange rates.

TABLE VI

Activation Parameters for the Dissociation Process of the Cryptates

Complex	Solvent	Log K_s[a]	$\Delta H^{\neq}_T$ (kcal/mole)	$\Delta S^{\neq}_T$ cal $(mol\ ^\circ K)^{-1}$	$\Delta G^{\neq}_T$ (kcal/mole)	$\Delta E^{\neq}_a$ (kcal/mole)	Temp. (°C)	Ref.[f]
[VI, Na^+] Cl^-	H_2O	3.9	[c]	[d]	14.2	[e]	27	1
[VI, K^+] F^-	H_2O	5.4	[c]	[d]	15.8	[e]	36	1
[VI, K^+] Br^-	H_2O	5.4	[c]	[d]	15.7	[e]	35	1
[VI, K^+] Cl^-	H_2O	5.4	[c]	[d]	15.8	[e]	36	1
[VI, Rb^+] Cl^-	H_2O	4.35	[c]	[d]	14.4	[e]	9	1
[VI, Tl^+] Cl^-	H_2O	6.3	[c]	[d]	15.8	[e]	40	1
[VI, Tl^+] NO_3^-	H_2O	6.3	[c]	[d]	13.5	[e]	−6	1
[VI, Ca^{2+}] $(Cl^-)_2$	H_2O	4.4	[c]	[d]	>17.0	[e]	25	1
[VI, Sr^{2+}] $(Br^-)_2$	H_2O	8.0	[c]	[d]	>17.0	[e]	25	1
[VI, Ba^{2+}] $(Cl^-)_2$	H_2O	9.5	[c]	[d]	>17.0	[e]	25	1
[VI, Na^+] (Br^-)	EDA	[f]	[c]	[d]	[g]	12.2	25	2
[VII, Li^+] ClO_4^-	Pyridine	[f]	19.0	−12.5	22.7	19.6	25	3
[VII, Li^+] ClO_4^-	DMS	[f]	15.5	−13.8	19.7	16.1	25	3
[VII, Li^+] ClO_4^-	DMS	[f]	15.4	−15.5	20.0	16.0	25	3
[VII, Li^+] ClO_4^-	Formamide	[f]	13.5	−22.8	20.8	14.1	25	3
[VII, Li^+] ClO_4^-	H_2O	5.3	20.7	+0.4	20.6	21.3	25	3
[VIII, Li^+] ClO_4^-	Pyridine	[f]	12.9	−14.9	17.9	13.5	25	3

Complex	Solvent	Log K_s[a]	$\Delta H^{\neq}_T$ (kcal/mole)	$\Delta S^{\neq}_T$ cal (mol °K)$^{-1}$	$\Delta G^{\neq}_T$ (kcal/mole)	$\Delta E^{\neq}_a$ (kcal/mole)	Temp. (°C)	Ref.[f]
[VI, Ca^{2+}] $(Cl^-)_2$	H_2O	4.4	8.2 (7.5)[h]	−33 (−15)[h]	g	e	25	*4*
[VIII, Ca^{2+}] $(Cl^-)_2$	H_2O	6.8	17.8 (10.6)[h]	−11 (−4)[h]	g	e	25	*4*
[VII, Ca^{2+}] $(Cl^-)_2$	H_2O	3.2	2.7 (7.8)[h]	−54 (−22)[h]	g	e	25	*4*
[V(k=l=m=8)]	$CFCl_3$	i	c	d	g	7.7[i]	25	*5*
[Ca^{2+}⊂VIII] $(Cl^-)_2$	j	f	c	d	15.3	e	30	*6*
[Sr^{2+}⊂IX] $(Cl^-)_2$	j	f	c	d	14.5	e	27	*6*
[Ba^{2+}⊂IX] $(Cl^-)_2$	j	f	c	d	<13.3	e	<3	*6*
[La^{3+}⊂IX] $(NO_3^-)_3$	j	f	c	d	>18.6	e	>93	*6*

[a] K_s denotes the stability constants.

[b] References *1* Lehn *et al.* (1973), *2* Ceraso and Dye (1973), *3* Cahen *et al.* (1975), *4* Loyola *et al.* (1975), *5* Simons and Park (1968), *6* Lehn and Stubbs (1974).

[c] No reported enthalpy of activation.

[d] No reported entropy of activation.

[e] No reported activation energies.

[f] No reported stability constants.

[g] No reported free energy of activation.

[h] Bracketed quantities are for the complexation process.

[i] Conformational changes

[j] Solvent composition, if any, not reported.

activation, and so on, for the cryptates investigated. The dissociation rate becomes slower as the stability of the cryptates increases. This slowing of decomplexation with growing stability constants is also observed in the solvent effect study of sodium-DBC6 decomplexation by Shchori *et al.* (1971), and in the studies of 18-crown-6 complexation of monovalent cations by Liesegang *et al.* (1976, 1977) (see Table VI). For macrocyclic polyether complexation, a decrease in the rate of decomplexation with increasing association constant is not surprising, since the observed complexation rate constants are quite close to the diffusion-controlled limit. In the cryptates, rough estimates of the complexation rates (Lehn *et al.*, 1973) are $\sim 10^6\ M^{-1}\ sec^{-1}$ (for the alkali metals). The slowing of decomplexation with increases in the association constant, in the cryptates, is not balanced. That is to say, in ligand (VI) the stability constant increases by approximately a factor of 10 while the decomplexation rate changes by only a factor of 1.4 in going from Na^+ to K^+ in water. Thus, in the case of the cryptates, the differences in the stability constants do not seem to be reflected solely in variations in the decomplexation rate but additionally in differences in the complexation rate. At this time, we may also point out that the dissociation rates of the aqueous alkali metal cryptates, studied by Lehn *et al.*, are a full factor of 10^6 slower than the dissociation rates of the same ions with aqueous 18-crown-6 (Tables IV and V). It is also interesting to note that the stability constants, on the average, are a factor of $\sim 10^3$ larger (Lehn *et al.*, 1970) for the alkali metal cryptates of type (VI) as compared to the alkali metal 18-crown-6 complexes in water.

An investigation of the exchange rates for sodium cryptates [type (VI)] in ethylenediamine (EDA) was reported by Ceraso and Dye (1973). Their preliminary measurements indicated that the exchange rate in water is similar to that in EDA. This close correspondence between exchange rates, in two different solvents, supports the conclusion that the rate-limiting step is the dissociation of the complex. A similar conclusion was reached by Shchori *et al.* (1971) in the complexation of Na^+ with DBC6 in DMF. The kinetic results of Ceraso and Dye (1973) and Shchori *et al.* (1971) appear in Tables III and V.

In an informative review article, Lehn *et al.* (1973) summarized the macrobicyclic complexation kinetics reported to that time:

(1) Dissociation rates of cryptate (VI) are a factor of $\sim 10^3$ slower than those of complexes with the macrocyclic polyethers. (Compare Tables III and V.)

As Lehn points out, this is a comparison between cryptates in an aqueous medium and DBC6 in methanol. A better comparison between dissociation rates may now be made for aqueous 18-crown-6 (Table IV) and aqueous (VI) (Table V). The factor is more nearly 10^6.

(2) Rates and free energies of activation for dissociation follow the same sequence as stability constants, whereas the rates for exchange of water

molecules in the hydration shell of the cations increase in the reverse order (Diebler *et al.*, 1969).

(3) All association rates for the cryptand (VI) are much lower than those for a diffusion-controlled process ($\sim 10^{10}$ M^{-1} sec^{-1}). The rates and free energies of activation for the association process seem to vary less from one system to another than the corresponding quantities for the dissociation process.

We shall attempt to expand upon this last statement. Table 13 in the article by Lehn *et al.* (1973) (similar to our Tables III, V, and VI) lists the macrobicyclic ligand (VI), the macrocyclic polyether DBC6, and the antibiotics monactin and valinomycin. In comparing the association and dissociation rates of the alkali metals with these various ligands, it is apparent that the most drastic changes occur in the rates of dissociation rather than in the rates of association. However, if one looks at just the decomplexation of ligand (VI) with the alkali metal cations, one observes that the dissociation rates are more nearly constant, while the rates of the association seem to vary by an order of magnitude or so. In the case of the macrocyclic polyethers and the antibiotics, the reverse of this variation seems to be true.

Another interesting feature of Lehn's Table 13, or our Table VI, is the free energy of activation of 4.5 kcal/mole for the complexation of $T\ell^+$ with ligand (VI). The corresponding free energy of activation for the alkali metal cations is 8.6 kcal/mole. The specific rate of association for $T\ell^+$ with (VI) is a factor of 10^2 larger than that for the alkali metal cations. Although the difference is not as pronounced, $T\ell^+$ complexes faster with aqueous 18-crown-6 than do any of the alkali metal cations (see Table IV). It is also observed that Ag^+ complexes very rapidly with aqueous 18-crown-6. Thus the orbitals of the metal ions appear to play as decisive a role in the complexation as do dissimilarities in the macrocyclic ligands. Perhaps an associative addition of ligand to $T\ell^+$ occurs without loss of solvent that provides an anchor for the ligand which facilitates a conformational change (Purdie, 1975). Chock (1972) measured a complexation rate for $T\ell^+$ by dibenzo-30-crown-10 in methanol that differs insignificantly from those measured for the alkali metal cations (see Table III). This suggests that no advantage accrues to $T\ell^+$ when the macrocyclic ligand is quite flexible.

Finally, the dissociation and association rates of cryptand (VI) are quite different for alkali as compared to alkaline earth metal cations. The alkaline earth metals complex and dissociate much less rapidly than the alkali metals (see Table V). There has been no published kinetic study of the complexation or decomplexation of the alkaline earth metals with the macrocyclic polyethers. Complexation of Ba^{2+} by aqueous 18-crown-6 is too slow to measure by conventional ultrasonic absorption techniques, which fail below $\sim$5 MHz.

Returning now to Lehn's generalizations regarding the cryptands, we have the following:

(4) Since the association rates differ less than the dissociation rates, the transition state may more closely resemble the reactants than the complex. The following effects upon the transition state may account for the slowness of association:

(a) incomplete compensation for the attraction between cation and the entering binding sites, or removal of several solvent molecules simultaneously; and
(b) a conformational change in the ligand.
Lehn remarks that in valinomycin (b) occurs, whereas process (a) is more probably applicable for a rigid ligand such as cryptand (VI).

We should point out here that if the cryptand undergoes a conformational rearrangement prior to complexation, then a reaction scheme similar to that in the macrocyclic polyethers obtains.

$$\text{out-out} \underset{k_{21}}{\overset{k_{12}}{\rightleftharpoons}} \text{in-in}, \qquad \text{in-in} + M^{n+} \underset{k_{32}}{\overset{k_{23}}{\rightleftharpoons}} [M(\text{in-in})]^{n+}$$

If the conformational rearrangement is fast compared to the complexation step, than the relaxation times are written as

$$\tau_{I}^{-1} = k_{12} + k_{21}$$

$$\tau_{II}^{-1} = \frac{k_{23}}{1+K_{21}} \left[(M^{n+}) + (\text{in-in}) + (\text{out-out}) + K_{T}^{-1}\right]$$

where K_T is the thermodynamic equilibrium constant

$$\left(= \frac{[M(\text{in-in})]^{n+}}{[(\text{in-in})+(\text{out-out})]+[M^{n+}]]} \right)$$

and K_{21} is the equilibrium constant for the conformational change. Thus the reported complexation rate constants are only observed complexation rate constants (= $k_{23}/(1+K_{21})$ being attenuated by K_{21}. If K_{21} is large (~10^5), then the true complexation rate constant k_{23} may be near diffusion controlled while the observed rate constant is quite small (~10^4). Since there are no reported equilibrium constants (K_{21}) for the cryptands (VI)-(VIII), a knowledge of the true complexation rate constant is not possible. This point is raised by Loyola *et al.* (1975).

(5) Since the stability constant increases from water to methanol, the specific rate of dissociation most probably decreases in going from water to methanol.

Cahen *et al.* (1975) reported the effect of various solvents upon the lithium ion-cryptate exchange rates, using a lithium-7 nmr method. The primary cryptand investigated was (VII), denoted by Lehn as (2.1.1). The exchange rates and

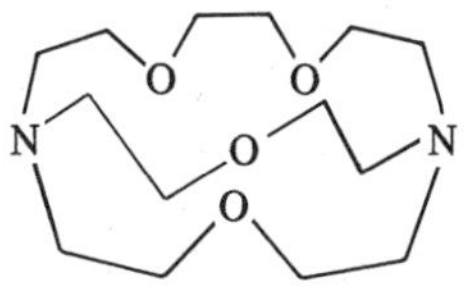

(VII)

thermodynamic parameters, in various solvents, are shown in Tables V and VI. As in the previous cases, Cahen *et al.* observed a decomplexation process similar to that in Eq. (5). Unfortunately, the association constants in the various solvents have not been reported. If these association constants were known, we would be able to verify the foregoing comment (5) by Lehn with direct experimental evidence. In addition, it is difficult to say, without the stability constants, whether the total variation in the decomplexation rates equals the total variation in the stability constants for various metals, and if not, then what percent is accounted for by the complexation rate.

As seen from inspection of Table V, the decomplexation rate of the lithium-cryptate (VII) in water is much slower (a factor of 10^3) than for K^+ and Na^+ cryptates of (VI). Since cryptand (VII) is selective for lithium, whereas cryptand (VI) is selective for potassium, it is difficult to explain this decrease in terms of steric interactions in the fit of the metal ion in the cavity. It would appear rather to be a function of the difference in the two cryptands. Cryptand (VII) is much less flexible than (VI) and so its cryptate should require more energy to dissociate. The activation energy for decomplexation of Li^+-(VII) in water is 21.3 kcal/mole, whereas for K^+-(VI) in water it is only 15.8 kcal/mole.

Cahen *et al.* also investigated the decomplexation rate for different cryptands [(VII) and (VIII)] with Li^+ in pyridine. The rate of decomplexation for the more flexible Li^+-(VIII) complex is 10^4 times as rapid as for the Li^+-(VII) complex, and the activation energy increases from 13.5 kcal/mole to 19.6 kcal/mole. This increase in the rate of decomplexation may be attributable to the looser fit of the Li^+ in (VIII) compared to its fit in (VII), to the increase in the flexibility of the ligand (VIII), or to a combination of these factors.

Since the rates of complexation and dissociation of the cryptates previously discussed are many orders of magnitude lower than the diffusion-controlled limit for a cation reacting with a neutral ligand ($\sim 10^{10}\ M^{-1}\ sec^{-1}$), Loyola *et al.* (1975) were able to investigate the kinetics of these cryptates with a stopped-flow apparatus. They used murexide as an indicator, since cryptates are nonabsorbing in the visible and near ultraviolet. They investigated the complexation of Ca^{2+} by cryptates (VI), (VII), and (VIII) and observed a small dependence on ionic strength for both the forward and reverse rates for which they offered no explanation. Loyola *et al.* also reported that the larger the stability constant for

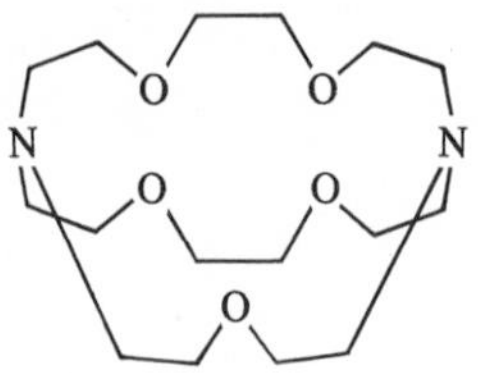

(VIII)

the Ca^{2+}-complexes, the slower the rate of decomplexation. These decomplexation rates, in water at 25°, decreased in the order Ca^{2+}-(VI) > Ca^{2+}-(VII) > Ca^{2+}-(VIII). The observed complexation rates *also varied,* although the variance is not as pronounced as the change in the decomplexation rates.

Loyola *et al.* considered a mechanism in which the cryptand undergoes a conformational rearrangement prior to complexation. As has been previously noted (Lehn *et al.*, 1970), many of the cryptands and cryptates have endo-endo configurations (same as the in-in previously discussed). In the exo-exo form (same as the out-out configuration), where the nitrogen lone pairs are directed away from the cavity, appreciable complexation would not be expected. If the conformational equilibrium is established rapidly, then a mechanistic scheme similar to that noted earlier for the macrocyclic crown ethers and the antibiotics may apply:

$$\text{exo-exo} \underset{}{\overset{K_c}{\rightleftharpoons}} \text{endo-endo} \tag{7}$$

$$\text{endo-endo} + Ca^{2+} \underset{k_d}{\overset{k_{ca^{2+}}}{\rightleftharpoons}} Ca(\text{endo-endo})^{2+} \tag{8}$$

where $k_f \sim K_c k_{Ca^{2+}}$. If $k_{Ca^{2+}}$ were the specific rate of a diffusion-controlled reaction, then K_c would have to be $\sim 10^{-5}$ to produce the observed rate constant. If such a fast conformational change does exist prior to complexation (with only one form of the ligand complexing), the importance of evaluating the conformational equilibrium constant is obvious. Without the conformational equilibrium constant, one is limited to an observed complexation rate constant which can neither be compared to complexation rates with another ligand system nor be a value for the absolute complexation rate constant.

Utilizing their data and the preliminary data of Lehn *et al.* (1970), Loyola *et al.* suggested that a trend exists in the complexation rates of various cations with the cryptand (VI). This reactivity pattern is $Ca^{2+} \sim Sr^{2+} < Ba^{2+}$; $Na^+ < K^+ < Rb^+$. Loyola *et al.* then stated that this reactivity pattern does not support the idea of a rate-limiting conformational change associated with complexation. The rate constants for complexation of cryptand (VI) used in the foregoing reactivity pattern were taken from Lehn *et al.* (1973) (with the exception Ca^{2+}). Lehn in his 1973 paper points out that the complexation rate constants for (VI) with Na^+, K^+, Rb^+, Sr^{2+}, etc. (Table V) "are only approximate since they were calculated using the dissociation rates at the temperatures indicated and the stability constants at 25°C." Thus some caution is warranted in discussing trends based on these complexation rate constants. In addition, inspection of the rates of complexation of Na^+, K^+, and Rb^+ with (VI) (Lehn *et al.*, 1973) shows that the reactivity pattern is not $Na^+ < K^+ < Rb^+$, but rather $Na^+ < Rb^+ < K^+$. Finally, Loyola *et al.* stated that their activation entropies support the idea that formation of the transition state involves little loss of water bound to the metal ion. Additionally, $\Delta S°$ for complexation is positive in the three complexation cases, reflecting the fact that most, if not all, of the water is lost upon complex formation, while $\Delta S_f^{\neq}$ is negative and $\Delta S_d^{\neq}$ more negative.

We will conclude this section by considering some investigations that are not concerned with the kinetics of ion complexation. The first is a study by Lehn and Stubbs (1974) of the *intra*molecular cation exchange in [3]-cryptates (see IX) of the alkaline earth cations.

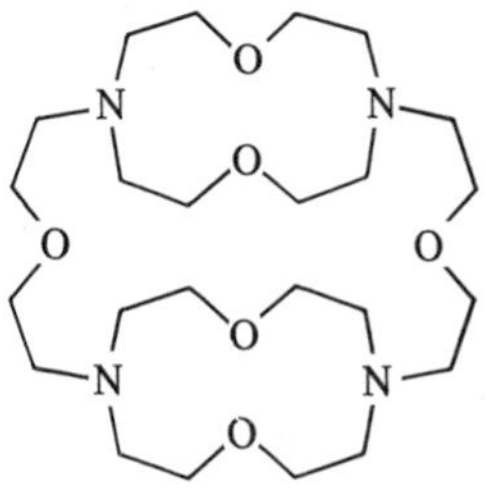

(IX)

Inclusion complexes are denoted by the mathematical sign of inclusion A⊂B (B includes A) (Lehn and Stubbs, 1974). These bivalent [3]-cryptates $[M^{n+}\subset(IX)]$, which display *intra*molecular cation exchange rates, were studied by ^{13}C nmr spectroscopy. The observations on the *intra*molecular cation jumping rates for $[M^{n+}\subset(IX)]$ (where M^{n+} was Ca^{2+}, Sr^{2+}, Ba^{2+}, and La^{3+}) were interpreted as follows:

(1) An intramolecular cation exchange process interconverts two species that have the cation located unsymmetrically in the molecular cavity.

(2) The intramolecular cryptates form a structure of the type (X) with the cation completing its coordination shell with one or two anions and/or water molecules. The spectral changes observed were attributed by Lehn and Stubbs to the *intra*molecular process (X) ⇌ (X′).

(3) The free energies of activation for (X) ⇌ (X′) decrease with increasing size and decreasing hydration energy of the cations; that is, $\Delta G^{\neq}$ is greater for

Ca^{2+} than for Sr^{2+}, etc. (Table VI).

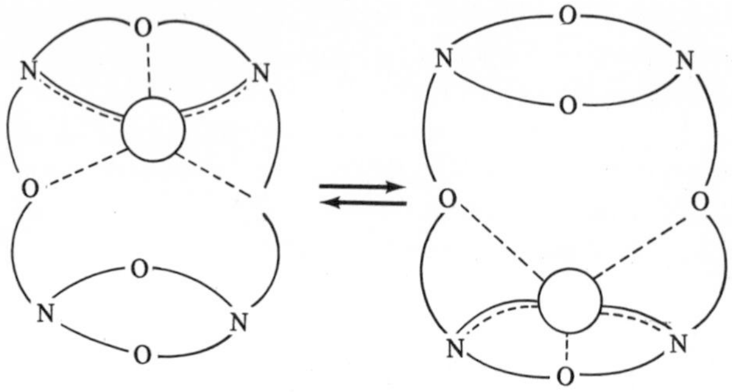

(X) (X′)

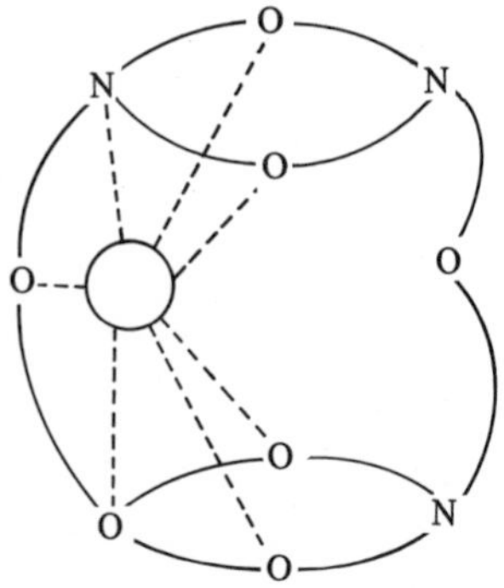

(XI)

(4) An *inter*molecular cation exchange also occurs, but with a much slower rate, and its free energy of activation is much higher (>19 kcal/mole) than those of the *intra*molecular process. The *inter*molecular cation exchange is very slow and the rates follow the sequence $Ca^{2+} > Sr^{3+} > Ba^{2+}$, which is the exact opposite of the trend for the *intra*molecular process.

(5) The *intra-* and *inter*molecular cation exchanges are both fast for weak complexes of (IX) with alkali cations.

At the conclusion of their article, Lehn and Stubbs suggest the feasibility of building synthetic molecules which could function as channels for the flow of metal cations.

A lithium-7 nmr study by Cahen *et al.* (1974) of lithium cryptates in different solvents led these workers to conclude that the donor properties of the solvent play an important role in the formation of a cryptate complex. A far-infrared and Raman spectroscopic study of lithium and sodium cryptates formed with ligands (VI)-(VIII) in nonaqueous solvents was carried out by Cahen and Popov (1975). From this latter study they concluded that alkali metal cations are completely enclosed in the cryptand cavity, the frequency of vibration of a cation within the cavity is proportional to the strength of the interaction with the cryptand (Li^+ vibrates more slowly than Na^+), and the cation-cryptand interaction is predominantly electrostatic in nature.

There is just one additional insight to which we wish to draw the reader's attention. Lehn *et al.* (1973) noted that Tl^+ complexes cryptand (VI) at an extraordinarily rapid rate, that is, $k_{23}^{obs} = 2 \times 10^8\ M^{-1}\ sec^{-1}$. Since the relationship between k_{23}^{obs} and a ligand conformational equilibrium constant K_c, as in reaction (7), is

$$k_{23}^{obs} = \frac{k_{23}K_c}{1+K_c} \tag{9}$$

the constant K_c cannot be less than 0.0204 if the true complexation-specific rate k_{23} is diffusion controlled, that is, $k_{23} = 1 \times 10^{10}\ M^{-1}\ sec^{-1}$. When similar k_{23}^{obs} values for Na^+, Rb^+, and alkaline earth metal cations reacting with cryptand (VI) are inserted into Eq. (9), it becomes apparent that none of the true complexation-specific rates k_{23} for these other ions approach the diffusion-controlled limit. They are, in fact, smaller by powers of ten.

The complexation rates decrease drastically in going from the alkali metal ions to the alkaline earth metal ions. Inner coordination sphere water loss may be rate determining in the complexation by cryptand (VI) of these cations. Since cryptands are comparatively inflexible, they probably force simultaneous loss of several first coordination sphere water molecules as complexation proceeds.

IV. MACROCYCLES CONTAINING NITROGEN

From a biological point of view, probably the most important of all the synthetic macrocyclic complexes are the octahedral macrocyclic tetramine complexes considered in this section. These are "model systems" devised to clarify natural complexation reactions in biological systems. There are many excellent review articles treating various aspects of this work (see e.g. Busch *et al.*, 1971; Poon, 1973). We will limit ourselves here to an outline of an important feature of this field, the macrocyclic effect, since this may lead to a better understanding of why these complexes are so stable and the manner in which kineticists may shed light on these enhanced stabilities.

Cabbiness and Margerum (1969) observed, in their investigation of copper-macrocyclic tetramine complexes, an enhanced stability over that of the similar noncyclic tetramine ligands, which they termed the "macrocyclic effect." Two important factors contribute to the macrocyclic effect: the *configuration* and *solvation* of a free macrocyclic ligand compared to its noncyclic analogue.

TABLE VII

Second-Order Rate Constants for the Reaction of Cu^{2+} with Various Ligands at 25.0°[a]

Ligand	Observed rate constants (M^{-1} sec^{-1})	
	0.5 *M* NaOH	pH 4.7
(XIII)	~10^7	*a, b*
(XIV)[a]	1.6×10^3	5.8×10^{-2} ($\mu = 1.0$ *M*)
		2.0×10^{-2} ($\mu = 0.1$ *M*)
(XIV)[b]	3.6×10^3	2.7×10^{-2} ($\mu = 1.0$ *M*)
(XV)	5.6×10^3	2.8×10^{-1} ($\mu = 1.0$ *M*)
(XVI)[a]	2.0×10^{-2}	*b*
(XVI)[b]	*b*	3.2 ($\mu = 0.5$ *M*)

[a]Data taken from Cabbiness and Margerum (1970).
[b]No reported complexation rates.

In a letter explaining the macrocyclic effect, Hinz, and Margerum (1974a) discussed the enhanced stability of the $[Ni\text{-}(XII)]^{2+}$ complex over that of the

TABLE VIII

Formation Rate Constants Resolved for Individual Species at 25.0°[a]

Reactants	Complexation rate constant
$Cu(OH)_3^-$ + (XIII)	~10^7 [b]
$Cu(OH)_3^-$ + (XIV)[a]	5×10^3
$Cu(OH)_3^-$ + (XIV)[a]	0.20
Cu^{2+} + H_2 $tetren^{2+}$	4.2×10^7
Cu^{2+} + $H(XIV)^{+}$[a]	7.6×10^3
Cu^{2+} + (XVI)[b]	4.3

[a]Data taken from Cabbiness and Margerum (1970).
[b]Reactivities of $Cu(OH)_4^{2-}$ and $Cu(OH)_3^-$ are nearly equal.

$[Ni\text{-}XIII)]^{2+}$ complex (a difference of ~10^7). From their equilibrium and

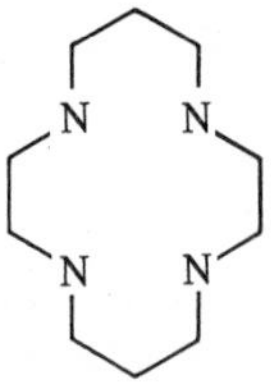

(XII)

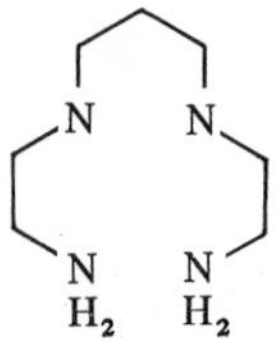

(XIII)

calorimetric studies (Table IX) Hinz and Margerum attributed the enhanced stability of the macrocyclic complex $[Ni\text{-}(XII)]^{2+}$ to a large negative difference in ΔH° (which overwhelms a less favorable ΔS°) for the formation reaction [scheme (10)]. The ΔH° for the $[Ni\text{-}(XII)]^{2+}$ complex is -31.0 ± 0.6 kcal/mole, whereas that for the noncyclic analogue, $[Ni\text{-}(XIII)]^{2+}$ (square planar), is -16.8 ± 0.5 kcal/mole. This is a favorable difference of -14 kcal/mole. The entropy difference is unfavorable. The $[Ni\text{-}(XII)]^{2+}$ has an entropy of complexation of -2 cal (mol °K)$^{-1}$, whereas that of the $[Ni\text{-}XIII)]^{2+}$ complex is 14 cal (mol °K)$^{-1}$. This difference of -16 cal (mol °K)$^{-1}$ corresponds to a $T\,\Delta S^\circ$ of about -4 kcal/mole at 25°.

TABLE IX

Stability Constants, Entropy, and Enthalpy of Formation of the Nickel Complexes of (XIII) and (XII) at 25.0° and Ionic Strength = 0.1[a]

Complex	Log K^b	ΔH° (kcal/mole)	ΔS° cal (°K mol)$^{-1}$
$[Ni\text{-}(XIII)]^{2+}$	15.8	-19.4 ± 0.1	7.4
$[Ni\text{-}(XIII)^{2+}]$ (square-planar)	15.4	-16.8 ± 0.5	14
$[Ni\text{-}(XII)]^{2+}$	22.2	-31.0 ± 0.6	-2

[a] Data taken from Hinz and Margerum (1974a).
[b] K denotes the stability constant.

In order to explain why $[Ni\text{-}(XII)]^{2+}$ has a more favorable ΔH°, Hinz and Margerum discussed ligand solvation in the complexation scheme

$$Ni(H_2O)_x^{2+} + L(H_2O)_y \rightarrow NiL(H_2O)_z^{2+} + (x+y-z)H_2O \qquad (10)$$

The ΔH° value for complexation will be a larger positive number as the enthalpy of solvation of the ligand increases. The ΔS° value for complexation will increase as the solvation of the ligand (y) increases, due to the increase in the number of water molecules desolvated. Hinz and Margerum reasoned that the macrocyclic ligand (XII) should be less solvated than its open-chain analogue (XIII), since the cyclic nature of (XII) physically prohibits it from being hydrated by the same number of hydrogen-bonded water molecules as (XIII). In addition, since ligands (XII) and (XIII) are wrapped around the ion, in the complexed state, both complexes have nearly the same hydration number. From these points Hinz and Margerum concluded that less enthalpy need be expended in desolvation of (XII) and thus (XII) has a more favorable complexation, that is, ΔH° is more negative than that for (XIII). Since ΔS° will increase in scheme (10) as ligand solvation increases, ΔS° should be less positive for lightly solvated (XII) compared to the complexation reaction of (XIII). This is not the case, since the difference for (XII) and (XIII) is less than one would predict for the release of two water molecules. Hinz and Margerum attributed this unexpectedly small difference to the greater loss of configurational entropy of the open-chain ligand (XIII).

Hinz and Margerum pointed out that although the effect is termed the macrocyclic effect, it is not limited to macrocyclic ligands, but applies to any ligand whose donor groups are forced to be close to one another or in some way shielded from solvation. Thus the terminology used by Busch *et al.* (1971) of "multiple juxtapositional fixedness" would be more appropriate.

In a follow-up to this letter, Hinz and Margerum (1974b) discussed the macrocyclic effect in greater detail with reference to a specific example: (Ni^{2+}-tetramine complexes). They observed that in the case of tetramine ligands the lower degree of solvation of the macrocycle is primarily responsible for the macrocyclic effect. Specifically, the ΔH° of complexation for the $[Ni\text{-}(XII)]^{2+}$ and $[Ni\text{-}(XIII)]^{2+}$ complexes, which differed by −14.2 kcal/mole, was equivalent to ligand (XII) being less solvated by at least two fewer water molecules than ligand (XIII). They used an average ΔH° in hydrogen bond formation with nitrogen bases of −7.3 kcal/mole to make this estimate. A smaller contribution for the macrocycle (XII) is the lower configurational entropy of the ligand, since it is already cyclic. Both Ni^{2+} and Cu^{2+} evidence enhanced stability with macrocyclic ligands, consistent with the idea that the macrocyclic effect should be independent of the metal ion as long as steric interactions attributable to the metal ion (such as the fit of the ion in the ligand cavity) do not come into play.

Ligand solvation is also important in other macrocyclic systems, as noted by Izatt *et al.* (1971) in the case of the cyclic polyethers. There is, for example, a pronounced difference in the association constants for the two isomers of dicyclohexyl-18-crown-6 complexing Na^+. However, the difference arises more from entropy than from enthalpy contributions. Clearly, if ligand solvation is important in the thermodynamics of metal complexation, then it must also be reflected in the *kinetics* of macrocyclic complexation or decomplexation, though little attention has thus far been directed to this effect.

As has been previously discussed, the enhanced stabilities of the macrocyclic polyethers with various cations is attributable to changes in the rates of dissociation of the complex. For the macrobicyclic polyethers, changes in the stability constants are reflected in both the complexation and decomplexation rates. Margerum and co-workers have investigated the kinetics of the macrocyclic tetramines and their open-chain analogues. As we will see later, the approach taken in kinetic studies of tetramines emphasizes changes in ligands rather than metal ions and has not demonstrated whether trends in complexation or decomplexation rates account for variations in stability constants.

Cabbiness and Margerum (1970) reported kinetic studies on the complexation of Cu^{2+} with ligands (XIII)-(XVI) in both acidic and basic solutions (Table X)..

TABLE X

Observed First-Order Dissociation Rate Constants in 6.1 *M* HCℓ at 25°[a]

Complex	Dissociation rate k_d (sec^{-1})
Cu-(XIV)a^{2+} (red)[b]	3.6×10^{-7}
Cu-(XIV)a^{2+} (unstable red)[c]	4.2×10^{-6}
Cu-(XIV)a^{2+} (blue)[d]	3.8×10^{-3}
Cu-(XV)$^{2+}$	1.2×10^{-3}
Cu-(XIII)$^{2+}$	4.1

[a]Data taken from Cabbiness and Margerum (1970).
[b]The Cu-(XIV)a^{2+} (red) complex has a square planar configuration found in pH 4.7 solution.
[c](XIV)a prepared in oxalate solution.
[d]Reaction product formed in basic solution.

From the kinetic studies carried out in aqueous acidic solution (pH = 4.7, 25°), Cabbiness *et al.* noted that aqueous Cu^{2+} reacted much slower with (XIV)a, (XIV)b, and (XV) than with the porphyrin ligand (XVI), which in turn complexed slower than the noncyclic analogue (XIII). In basic aqueous solution (0.5 *M* NaOH, 25°), where Cu^{2+} is present in two forms, $Cu(OH)_3^-$ and $Cu(OH)_4^{2-}$, they observed that (XIV)a, (XIV)b, and (XV) reacted much faster than the prophyrin ligand (XVI) with copper. This trend is opposite to that in acidic solution. However, in either acid or base, the noncyclic ligand (XIII) reacted faster with Cu^{2+} than any of the macrocyclic ligands studied.

In acidic or basic media one observes that the mechanism of complexation with Cu^{2+} differs between the noncyclic polyamine (XIII), the 14-membered rings (XIV) and (XV), and the more rigid porphyrin ligand (XVI). In a discussion of the mechanisms of the desolvation of the metal ion by the various ligands, Cabbiness *et al.* noted that the porphyrin structure, because of its rigidity, has the greatest tendency to force simultaneous multiple desolvation of the metal ion, and the open-chain polyamine (XIII) can cause a serial displacement of the coordinated solvent. Cabbiness *et al.* temporized with respect to the 14-membered rings [(XIV) and (XV)] and argued that both paths must be considered, since these ligands are not very flexible. Models of (XIV) and (XV) indicate that some degree of multiple desolvation is necessary in the coordination of the third and fourth ligand nitrogens. The rate-determining step may still occur earlier in the complexation process.

In addition to these differences, Cabbiness *et al.* treated the problem of protonation of the nitrogen donor atoms in acidic solution. The ligands (XIV)a, (XIV)b, and (XV) are strongly protonated in acidic solutions (pK_1 = 12.6, pK_2 = 10.4), so that during complexation, the two protons are forced much closer together in the macrocyclic structure than in the more flexible open-chain molecule. Thus, as observed, (XIV)a, (XIV)b, and (XV) should react slower with Cu^{2+} than with (XIII). As the acidity of the solution is increased, the porphyrin ligand (XVI) does not protonate until a pH of 4-5, so its rate of decomplexation is less drastically affected by acid. In basic solution one does not have this steric hindrance introduced by the protons.

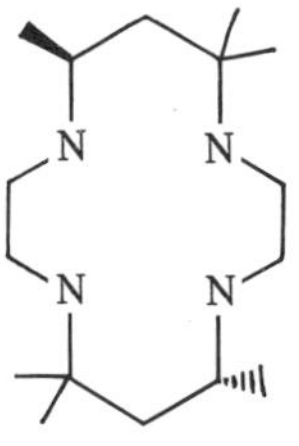

(XIV)

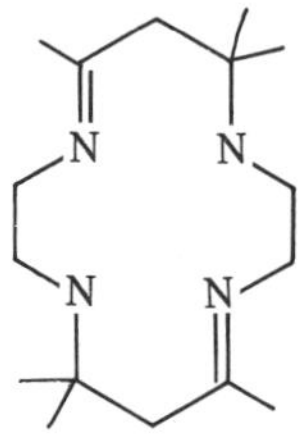

(XV)

a: 1 = ···ıı| CH_3
b: 1 = ◀ CH_3

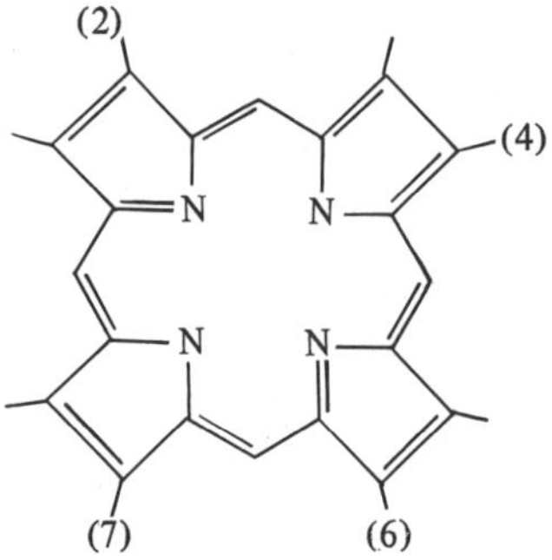

(XVI)

a: (2),(4) = $CHOHCH_3$
(6),(7) = CH_2CH_2COOH
b: (2),(4) = SO_3H
(6),(7) = $CH_2CH_2COOCH_3$

Lin *et al.* (1975) have also investigated the magnitude of the kinetic effects attributable to ligand cyclization. Kinetic measurements were made on the complexation of various cyclic and noncyclic polyamines with Cu^{2+} and carried out in basic aqueous media (25°) so as to eliminate steric effects between protonated nitrogens. Table XI contains the formation rate constants of the $Cu(OH)_3^-$ and $Cu(OH)_4^{2-}$ species with the various cyclic and noncyclic ligands.

TABLE XI

Resolved Formation Rate Constants for $Cu(OH)_3^-$ and $Cu(OH)_4^{2-}$ Species Reacting with Unprotonated Cyclic and Open-Chain Polyamines at 25°[a]

Ligand	$k^{obs}_{Cu(OH)_3}$	$k^{obs}_{Cu(OH)_4}$	$k^{obs}_{Cu(OH)_3}/k^{obs}_{Cu(OH)_4}$
(XIII)	$(1.0 \pm 0.7) \times 10^7$	$(4.3 \pm 0.2) \times 10^6$	2.4
Et_2-(XIII)[b]	$(3.0 \pm 0.3) \times 10^6$	$(2.9 \pm 0.6) \times 10^5$	10
(XII)	$(2.7 \pm 0.4) \times 10^6$	$(3.8 \pm 0.9) \times 10^4$	72
Me_2-(XII)[b]	5.6×10^5	0.9×10^4	60
(XIV)a	$\sim 10^4$	$<10^2$	$>10^2$
Et_4-dien[b]	4.0×10^5	$<10^3$	>400

[a]Data taken from Lin *et al.* (1975).
[b]See Fig. 4 for structures.

Utilizing the value for the rate constant in the reaction of Cu^{2+} with NH_3 (2×10^8 M^{-1} sec^{-1}, according to Diebler and Rosen, 1972) and accounting for steric and statistical factors, Lin *et al.* estimated that the rate of complexation of Cu^{2+} with (XIII) would be 8×10^7 M^{-1} sec^{-1}. "This is a factor of eight larger than the value obtained for $Cu(OH)_3^-$ and a factor of 20 larger than that for $Cu(OH)_4^{2-}$." In addition, the lability of axial water molecules is expected to increase in the order

$$Cu(H_2O)_6^{2+} << Cu(OH)_3(H_2O)_3^- < Cu(OH)_4(H_2O)_2^{2-}.$$

Lin *et al.* concluded that in the case of $Cu(OH)_3^-$ and $Cu(OH)_4^{2-}$ complexation by (XIII), the rate-determining step cannot be simple axial water dissociation. Thus either first or second coordinate bond formation or an associative inner sphere rearrangement must be rate determining. This is true since the formation rates are second order (first order in $Cu(OH)_4^{2-}$) and the coordination number of Cu^{2+} is four; thus third coordinate bond formation cannot be rate determining and still have a formation rate second order in $Cu(OH_4)^{2-}$ and ligand. In addition, if the rate-determining step of all ligands studied complexing Cu^{2+} involved second-bond formation, then "there would be no apparent reason for the ratio of $k^L_{Cu(OH)_3}/k^L_{Cu(OH)_4}$ to vary from 2.4 to 400 for the various ligands" (where $k^L_{Cu(OH)_3}$ is the complexation rate constant of $Cu(OH)_3^-$ with ligand L). The variation in this ratio suggests that in the complexation of $Cu(OH)_3^-$ and $Cu(OH)_4^{2-}$ with (XIII) first-bond formation was the rate-determining step, while for the cyclic ligands the rate-determining step shifts from first-bond formation to second-bond formation as the reactant changes from $Cu(OH)_3^-$ to Cu-$(OH)_4^{2-}$.

V. MACROCYCLES CONTAINING SULFUR

Jones *et al.* (1975) reported a kinetic and stability constant study of the solvated Cu^{2+} ion reacting with a third series of macrocyclic ligands, the cyclic polythioethers. The specific ligands investigated were 12-, 13-, 14-, 15-, 16- and 18-membered macrocyclic tetrathioethers [(XV)-(XXII)] and the open-chain ligand (XXIII). Table XII summarizes their kinetic results.

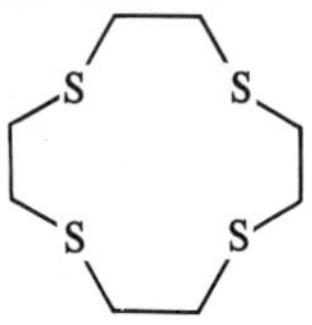

(XVII)

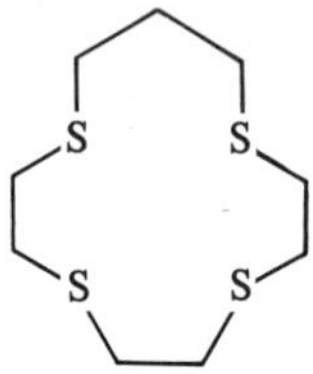

(XVIII)

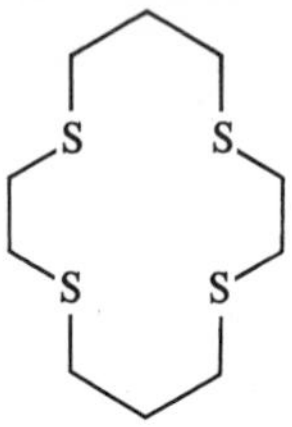

(XIX)

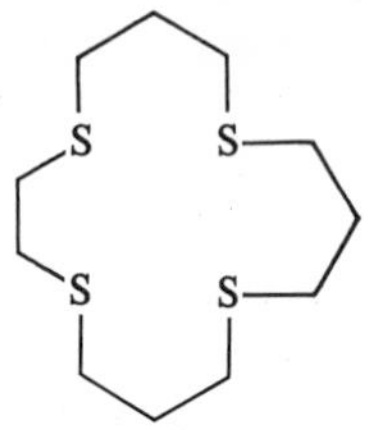

(XX)

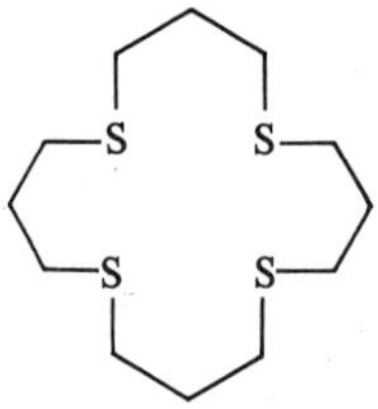

(XXI)

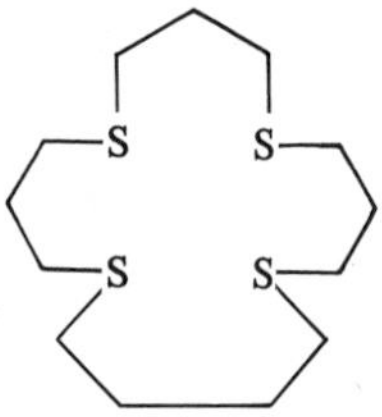

(XXII)

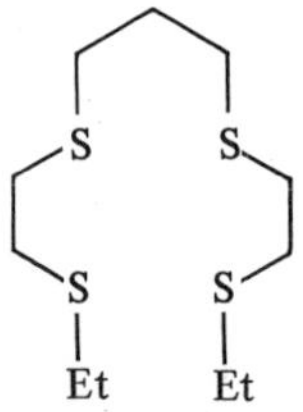

(XXIII)

As in the case of the macrobicyclic and macrocyclic polyethers, the rate constants for decomplexation are primarily responsible for variations in stability constants. These trends were not peculiar to the mixed solvent composition (CH_3OH–H_2O). Additionally, they noted that the solvent trends in the complexation rate are essentially independent of the ligand, being a characteristic instead of the Cu^{2+} solvation. Thus the trends in Table XII are not attributable to solvent-dependent conformational equilibria.

Jones *et al.* noted that the dependence of the dissociation rate constants and stability constants upon the ring size appears to be related to the bonding of the third and fourth donor atoms and associated ligand conformational changes. As ring size increases, the constraints of the ring diminish, with a consequent decrease in the macrocyclic effect. As for the relationship of the macrocyclic effect to kinetics, they stated that the similarity in specific rates of complexation exhibited by the corresponding open-chain and cyclic ligands implies that any ligand solvation effects contributing to the macrocyclic effect must be manifested in the dissociation rates.

TABLE XII

Kinetic and Equilibrium Data for Cu^{2+} Reaction with Several Macrocyclic Thioethers [a]
(in 80% Methanol at 25° and Ionic Strength = 0.1 M $HClO_4$ [a])

Ligand	Method	$k^{f}_{L\text{-}Cu(II)}$ (M^{-1} sec^{-1})	$k^{d}_{Cu(II)\text{-}L}$ (sec^{-1})	$K^{equil}_{M\text{-}L}$	$K^{kinetic}_{M\text{-}L}$
(XVII)	Stopped flow	1.2×10^3	4.4	3.3×10^3	2.7×10^2
(XVIII)	Stopped flow	1.4×10^4	51.0	[b]	2.7×10^2
(XIX)	Stopped flow, temperature jump	2.8×10^4	9.0	3×10^3	3.1×10^3
(XX)	Temperature jump	4.3×10^4	1.9×10^2	[b]	2.3×10^2
(XXI)	Temperature jump	2.9×10^4	3.2×10^3	11	9
(XXII)	[c]	[c]	[c]	<<10	[c]
(XXIII)	Temperature jump	4.1×10^5	3.0×10^4	13	14

[a]Data taken from Jones *et al.* (1975).
[b]No reported values.
[c]Complex too weak to allow a kinetic study.

TABLE XIII

Rate Comparisons for Liberation of *p*-Nitrophenol from 10^{-4} *M* Solutions of Amino Acid Ester Salts in the Presence of 5.0×10^{-3} *M* Cyclic Polyether at 25° C[a]

Medium[b]	Polyether present[c]	$RC^*H(NH_3^+)CO_2Ar$ R	$RC^*H(NH_3^+)CO_2Ar$ Config.	Rate constant[d] (10^3k, sec^{-1})	Rate factors
A	None	H	–	2.6	1
A	(S)-1a	H	–	>702	>270
A	(S)-2	H	–	5.5	2.1
A	(S)-1b	H	–	7.3	2.8
A	18-Crown-6	H	–	8.1	3.1
A	(S)-1a	Prolyl	L	16	0.8
A	(S)-2	Prolyl	L	21	
A	(S)-1a	H	L	>700	>130
A	(S)-2	H	L	5.4	
A	(S)-1a	CH_3	L	≳700	≳130
A	(S)-2	CH_3	L	5.4	
A	(S)-1a	$(CH_3)_2CH$	L	13	160
A	(S)-2	$(CH_3)_2CH$	L	0.08	
A	(S)-1a	$C_6H_5CH_2$	L	200	500
A	(S)-2	$C_6H_5CH_2$	L	0.41	

B	(S)-1a	$(CH_3)_2CHCH_2$	L	$\gtrsim 700$	$\gtrsim 1100$
B	(S)-2	$(CH_3)_2CHCH_2$	L	0.6	
B	(S)-1a	$C_6H_5CH_2$	D	25	69
B	(S)-2	$C_6H_6CH_2$	D	0.36	

[a]Data taken from Chao and Cram (1976).

[b]A, 20% EtOH in CH_2Cl_2 (v) buffered with 0.2 *M* AcOH and 0.1 *M* $(CH_3)_4N^+{}^-OAc$; B, 20% EtOH in CH_2Cl_2 (v) buffered with 0.3 *M* AcOH and 0.1 *M* $(CH_3)_4N^+{}^-OAc$; These buffers in water give pH 4.8.

[c]See Fig. 3 for the polyether structures.

[d]Pseudo-first-order rate constants, corrected for buffer solvolysis, made in triplicate runs of at least 14 points each, and followed for appearance *p*-$NO_2C_6H_4OH$ at 345 nm.

VI. HOST-GUEST CHEMISTRY

Host-guest chemistry is concerned with the structure and complexation requirements of highly structured molecular complexes. Cram (1976) states that the larger host compound, of the order typically of 2000 mass units, binds the smaller guest compound, which usually has a molecular weight in the hundreds of mass units or less. In order to form a complex, the host and guest must have binding sites and steric barriers located to complement one another's structure. Differentiation in molecular complexation thus depends on mutual structural recognition between potential hosts and guests. If ground-state molecules are complexed, their free energies are lower than or comparable to those of the uncomplexed state. Similarly, if transition states of molecules are complexed, their free energies are lower than or comparable to those of the uncomplexed state. Host compounds capable of lowering the free energies of rate-limiting transition states are catalysts. The macrocyclic compounds represent a beginning in the search for an understanding of the physicochemical basis of complexation, recognition, and catalysis by host compounds.

Recently, considerable synthetic organic research has been directed toward using these basic macrocyclic compounds to build stereospecific hosts (Lehn *et al.,* 1973; Cram and Cram, 1974; Cram, 1976) which are able to discriminate between enantiomeric guest molecules. Such systems are models of enzymes that are able to distinguish between enantiomeric substrates in reaction catalysis.

Cram and Cram (1974) have pointed out that "the first step in enzymatic catalysis is the formation of highly selective molecular *complexes* that orient the reactants and catalysts. Complexes form at very high rates and frequently are in equilibrium with their components. In later stages, covalent bonds are made and broken. Finally decomplexation occurs. Competitive complexation-decomplexation processes often account for enzyme inhibition and for enzymatic regulatory processes in cells."

A particularly interesting recent example of a crown ether-containing host molecule is the chiral host thiol (shown in Fig. 3), which selectively catalyzes the transacylation reaction between the chiral host and several amino acid ester salts [denoted as $RCH(NH_3^+)CO_2Ar$; see Chao and Cram (1976), and Table XIII].

Relaxation techniques such as the ultrasonic absorption method are ideally suited for the determination of rates and mechanisms of reversible complexation of substrates by these synthetic host molecules. Similarly, the stopped-flow method will likely prove useful in exploring the kinetics of the irreversible catalytic reactions of these same host molecules. Such future kinetic studies will probably be closely analogous methodologically to recent fast-reaction kinetic studies of enzyme catalysis (Hammes, 1974). It will be interesting to see if the mechanistic conclusions differ markedly from those obtained for enzymatic systems.

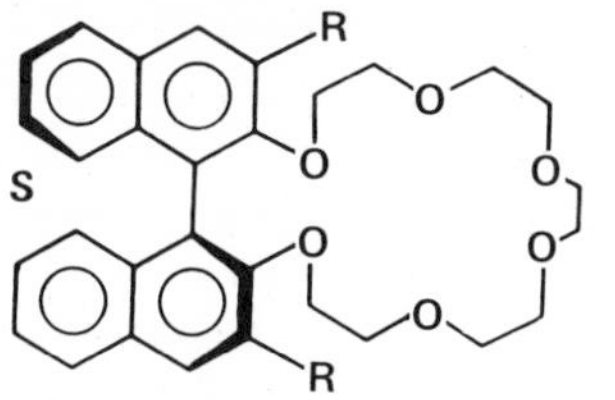

(S)-1a, R = CH_2SH

(S)-1b, R = CH_2OH

(S)-1c, R = H[2a]

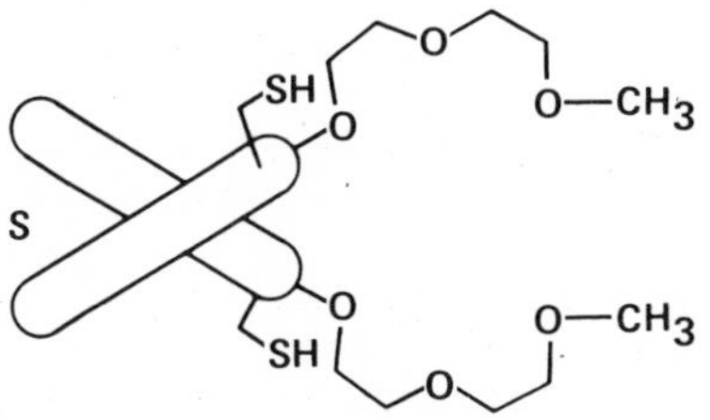

(S)-2

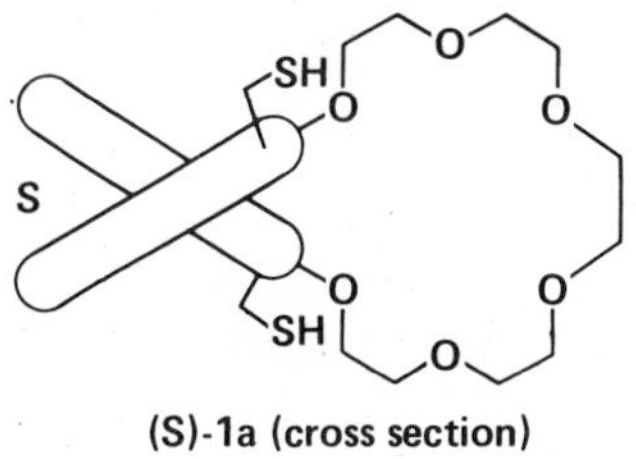

(S)-1a (cross section)

Fig. 3. Structures of several representative host molecules synthesized by Cram and co-workers. (Taken with permission from Chao and Cram, 1976.)

ACKNOWLEDGMENT

This work was sponsored by a contract from the Office of Naval Research.

REFERENCES

Burgermeister, W., Wieland, T., and Winkler, R. (1974). *Eur. J. Biochem.* **44**, 305-310, 311-316.

Busch, D. H. *et al.* (1971). *In* "Bioinorganic Chemistry" (R. F. Gould, ed.), Number 100, pp. 44-78. Am. Chem. Soc., Washington, D.C.

Cabbiness, D. K., and Margerum, D. W. (1969). *J. Am. Chem. Soc.* **91**, 6540-6541.

Cabbiness, D. K., and Margerum, D. W. (1970). *J. Am. Chem. Soc.* **92**, 2151-2153.

Cahen, Y. M., and Popov, A. I. (1975). *J. Solut. Chem.* **4**, 599-608.

Cahen, Y. M., Dye J. L., and Popov, A. I. (1974). *Inorg. Nucl. Chem. Lett.* **10**, 899-902.

Cahen, Y. M., Dye, J. L., and Popov, A. I. (1975). *J. Phys. Chem.* **79**, 1292-1295.

Ceraso, J. M., and Dye, J. L. (1973). *J. Am. Chem. Soc.* **95**, 4432-4434.

Chao, Y., and Cram, D. J. (1976). *J. Am. Chem. Soc.* **98**, 1015-1917.

Chock, P. B. (1972). *Proc. Nat. Acad. Sci. U.S.* **69**, 1939-1942.

Chock, P. B., and Titus, E. P. (1974). Current research topics in bioinorganic chemistry, *Progr. Inorgan. Chem.* **18**, 287-382.

Cram, D. J. (1976). *Pure Appl. Chem.* **43**, 327-349.

Cram, D. J., and Cram, J. M. (1974). *Science* **183**, 803-809.

Diamond, J. M. (1975). *J. Exp. Zool.* **194**, 227-240.

Diebler, H., and Rosen, P. (1972). *Ber. Bunsenges. Phyw. Chem.* **76**, 1031-1034.

Diebler, H., Eigen, M., Ilgenfritz, G., Maass, G., and Winkler, R. (1969). *Pure Appl. Chem.* **20**, 93-115.

Dietrich, B., Lehn, J. M., and Sauvage, J. P. (1969a). *Tetrahedron Lett.* **34**, 2885-2888.

Dietrich, B., Lehn, J. M., and Sauvage, J. P. (1969b). *Tetrahedron Lett.* **34**, 2889-2892.

Eigen, M., and Maass, G. (1966). *Z. Phys. Chem. N. F.* **49**, 163-177.

Eigen, M., and Wilkins, R. G. (1965). *Adv. Chem. Ser.* **49**, 55-80.

Eigen, M., and Winkler, R. (1970). *In* "The Neurosciences: Second Study Program" (F. O. Schmitt, ed.-in-chief), pp. 685-696. Rockefeller Univ. Press, New York.

Frensdorff, H. K. (1971). *J. Am. Chem. Soc.* **93**, 600-606.

Fuoss, R. M. (1958). *J. Am. Chem. Soc.* **80**, 5059-5061.

Grell, E., and Funck, T. (1973). *J. Supramol. Struct.* **1**, 307-335.

Grell, E., Funck, T., and Eggers, F. (1972). *In* "Molecular Mechanisms of Antibiotic Action on Protein Biosynthesis and Membranes" (E. Muñoz, F. Garcia-Ferrández, and D. Vazquez, eds.), pp. 646-686. Elsevier, Amsterdam.

Grell, E., Funck, T., and Eggers, F. (1975). *In* "Membranes" (G. Eisenman, ed.), Vol. III, pp. 1-171. Dekker, New York.

Hammes, G. G. (1974). *Pure Appl. Chem.* **40**, 523-547.

Hammes, G. G., and Steinfeld, J. I. (1962). *J. Am. Chem. Soc.* **84**, 4639-4643.

Hemmes, P. (1972). *J. Am. Chem. Soc.* **94**, 75-76.

Hinz, F. P., and Margerum, D. W. (1974a). *J. Am. Chem. Soc.* **96**, 4993-4994.

Hinz, F. P., and Margerum, D. W. (1974b). *Inorg. Chem.* **13**, 2941-2949.

Izatt, R. M., Nelson, D. P., Rytting, J. H., Haymore, B. L., and Christensen, J. J. (1971). *J. Am. Chem. Soc.* **93**, 1619-1623.

Jones, T. E., Zimmer, L. L., Diaddario, L. L., Rorabacher, D. B., Ochrymowycz, L. A. (1975). *J. Am. Chem. Soc.* **97**, 7163-7165.

Krasne, S. J., and Eisenman, G. (1973). *In* "Membranes, A Series of Advances" (G. Eisenman, ed.), Vol. II, pp. 277-328. Dekker, New York.

Laprade, R., Ciani, S., Eisenman, G., and Szabo, G. (1975). *In* "Membranes" (G. Eisenman, ed.), Vol. III, pp. 127-214. Dekker, New York.

Lehn, J.-M. (1973). *Structure and Bonding* **16**, 1-69.

Lehn, J.-M., and Stubbs, J. M. (1974). *J. Am. Chem. Soc.* **96**, 4011-4012.

Lehn, J.-M., Sauvage, J. P., and Dietrich, B. (1970). *J. Am. Chem. Soc.* **92**, 2916-2918.

Lehn, J.-M. Simon, J., and Wagner, J. (1973). *Angew. Chem. Int. Ed. Engl.* **12**, 578-579, 579-580.

Liesegang, G. W., Farrow, M. M., Purdie, N., and Eyring, E. M. (1976a). *J. Am. Chem. Soc.* 98, 6905-6908.

Liesegang, G. W., Farrow, M. M., Vazquez, A., Purdie, N., and Eyring, E. M. (1977). *J. Am. Chem. Soc.* 99, 3240-3243.

Lin C. T., Rorabacher, D. B., Cayley, G. R., and Margerum, D. W. (1975). *Inorg. Chem.* **14**, 919-925.

Loyola, V. M., Wilkins, R. G., and Pizer, R. (1975). *J. Am. Chem. Soc.* 97, 7382-7383.

Moreno, J. H., and Diamond, J. M. (1974). *J. Membr. Biol.* **15**, 277-318.

Neupert-Laves, K., and Dobler, M. (1975). *Helv. Chim. Acta* **58**, 432-442.

Park, C. H., and Simmons, H. E. (1968). *J. Am. Chem. Soc.* **90**, 2428-2429.

Parks, J. E., Wagner, B. E., and Holm, R. H. (1970). *J. Am. Chem. Soc.* **92**, 3500-3502.

Phillies, G. D. J., Asher, I. M., and Stanley, H. E. (1975). *Science* **188**, 1027-1029.

Poon, C. K. (1973). *Coord. Chem. Rev.* **10**, 1-35.

Purdie, N. (1975). Private communication.

Rodriguez, L. J., Liesegang, G. W., White, R. D., Farrow, M. M., Purdie, N., and Eyring, E. M. (1977). *J. Phys. Chem.*, in press.

Shchori, E., Jagur-Grodzinski, J., Luz, Z., and Shporer, M. (1971). *J. Am. Chem. Soc.* **93**, 7133-7138.

Shchori, E., Jagur-Grodzinski, J., and Shporer, M. (1973). *J. Am. Chem. Soc.* **95**, 3842-3846.

Shemyakin, M. M. *et al.* (1969). *J. Membr. Biol.* **1**, 402-430.

Shporer, M., and Luz, Z. (1975). *J. Am. Chem. Soc.* 97, 665-666.

Shporer, M., Zemel, H., and Luz, Z. (1974). *FEBS Lett.* **40**, 357-360.

Simmons, H. E., and Park, C. H. (1968). *J. Am. Chem. Soc.* **90**, 2429-2432.

Simon, W., and Morf, W. (1973). *In* "Membranes, A Series of Advances" (G. Eisenman, ed), Vol. II, pp. 329-375. Dekker, New York.

Szabo, G., Eisenman, G., and Ciani, S. (1969). *J. Membr. Biol.* **3**, 346-382.

Talekar, S. V. (1975). *Stud. Biophys.* **49**, 1-11.

Tsatsas, A. T., Stearns, R. W., and Risen, W. M., Jr. (1972). *J. Am. Chem. Soc.* **94**, 5247-5253.

Weatherburn, D. C., Billo, E. J., Jones, J. P., and Margerum, D. W. (1970). *Inorg. Chem.* 9, 1557-1559.

Winkler, R. (1972). *Structure and Bonding* **10**, 1-24.

6 DEVELOPING THE COMMERCIAL POTENTIAL OF MACROCYCLIC MOLECULES

Roger A. Schwind

Monsanto Research Corporation
Miamisberg, Ohio

Thomas J. Gilligan and E. L. Cussler

Department of Chemical Engineering
Carnegie-Mellon University
Pittsburgh, Pennsylvania

I. Potential of Macrocyclic Molecules 289
II. Liquid-Liquid Extractions 291
A. Isotope Separation 292
B. Other Promising Separations 294
III. Membrane Separations 294
A. How Liquid Membranes Work 295
B. How Liquid Membranes Are Made 297
C. Examples 298
D. Other Types of Membranes 301
IV. Other Practical Possibilities 302
A. Catalysis by Macrocyclic Compounds 302
B. Ion-Selective Electrodes 304
References 306

I. POTENTIAL OF MACROCYCLIC MOLECULES

Because macrocyclic molecules can selectively solubilize ions, they have considerable potential for ionic separations, nonaqueous catalysis, and selective analysis. The selectivity characteristic of these molecules is illustrated in Fig. 1 (Christensen *et al.*, 1971). These data represent the relative strength of the interaction between dicyclohexyl-18-crown-6 and the metal ions.

This selectivity of macrocyclic molecules is routinely higher than that commonly encountered with ordinary organic solutes. Moreover, these molecules show a systematic variation based on ion size rather than on an unknown plethora of specific chemical interactions. This behavior is particularly interesting because of the anticipated long-term shortages of metals like chromium and copper.

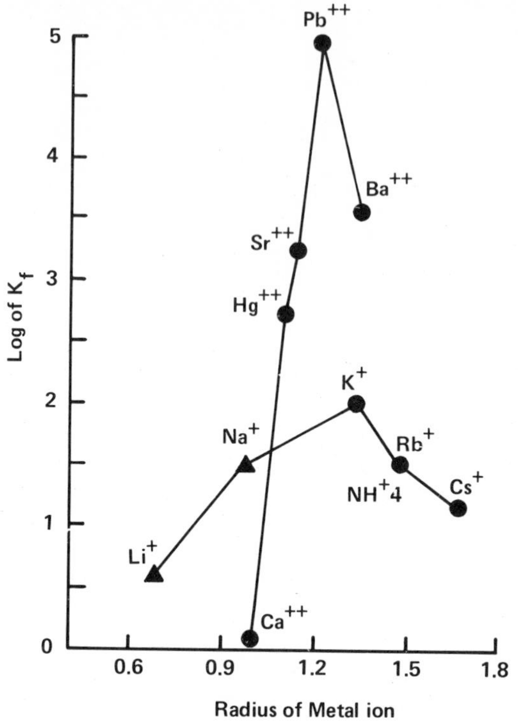

Fig. 1. Selectivity of Dicyclohexo-18-crown-6 versus ion size. The selectivity is exemplified by an equilibrium constant K for formation of the cation-polyether complex. (Data taken from Christensen *et al.*, 1971.)

In addition, macrocyclic molecules can often greatly increase the solubility of ions in organic solvents, as illustrated in Fig. 2 (Pederson, 1967) and in Pedersen (Chapter 1) and Liotta (Chapter 3). The amount of salt solubilized is roughly proportional to the amount of macrocyclic compound present. In this solubilization, however, the ionic complexes may not always have the same stoichiometric form. Initially, there may be one ion for two solubilizing molecules, but at higher ionic concentrations there may be two ions in every complex.

These solubilizations are superficially similar to those effected by detergents. In both cases, solutes which ordinarily are insoluble are incorporated into a complex aggregate and protected from the surrounding solution. The solubilized ions discussed here, however, are considerably less solubilized from the surrounding environment than the solutes contained within detergent micelles. This lack of shielding can cause ionic catalysis in organic solvents.

Molecules like the macrocyclic polyethers currently show much more potential than accomplishment; they are expensive to synthesize, and therefore have

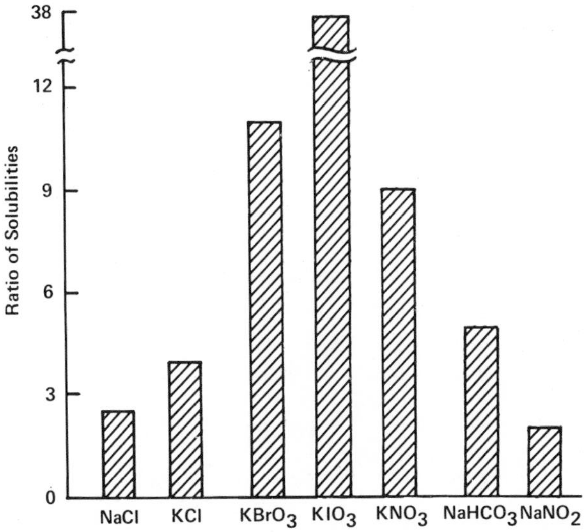

Fig. 2. Increase in the solubility of alkali salts in methanol. The solubility of these salts is greatly increased by the presence of dibenzo-18-crown-6. (Data taken from Pedersen, 1967.)

limited commercial application. Most research has tended to focus on new types of complexes of higher and higher selectivity, rather than on methods of cheap synthesis suitable for industrial development. We must now direct our efforts to decreasing the water solubility of these molecules by attaching side chains or by polymerizing them.

As a result, this chapter is more a guide for development than a summary of accomplishment. It contains three sections. In the first, we describe liquid-liquid extractions using these molecules. These extractions are the most completely developed for large-scale use. In the next two sections, we discuss membrane separations, catalysis, and specific ion electrodes. Membranes and catalysis are promising areas as yet publicly described only at bench scale. Specific ion electrodes allow selective analysis rather than large-scale application. We hope this summary will spur future developments of this fascinating area.

II. Liquid-Liquid Extractions

Liquid-liquid extractions consist of the contacting of two immiscible liquids so that the desired specie(s) is preferentially extracted into one of the liquids. We use traditional countercurrent contacting to obtain several stages of separation. The liquid-liquid system usually consists of an aqueous phase in contact with an organic phase containing the macrocyclic ligand.

Because of the great expense of these macrocyclic ligands, it is necessary to separate materials of sufficient value to justify the expense. In this section, we emphasize isotope separation, the most promising area of application. Others are reviewed more briefly.

A. Isotope Separation

Although the majority of the work to date concerns separation of calcium isotopes, liquid-liquid extraction studies at the Mound Laboraotry include several processes for the separation of the metal isotopes (Jepson, 1974). This process is applicable to the isotopes of essentially all the metals that form complexes with macrocyclic polyethers. Table I gives a partial list of these metals.

TABLE I

Possible Metal Isotopes to Be Separated

Barium	Iron	Potassium
Cadmium	Lanthanum	Rubidium
Calcium	Lead	Silver
Cerium	Lithium	Strontium
Cesium	Magnesium	Titanium
Chromium	Mercury	Zinc

The process is based upon the exchange between a metal ion in the aqueous phase and the metal ion-macrocyclic polyether complex in the organic phase. For example, for the separation of calcium-48 from calcium-40, this reaction is

$$^{40}\mathrm{Ca}^{2+}_{(aq)} + {}^{48}\mathrm{CaL}^{2+}_{(org)} \rightleftharpoons {}^{48}\mathrm{Ca}^{2+}_{(aq)} + {}^{40}\mathrm{CaL}^{2+}_{(org)}$$

where L represents a macrocyclic polyether. This reaction tends to go preferentially to the right as written; therefore, calcium-40 is preferentially enriched in the organic phase. For the polyether dicyclohexyl-18-crown-6, the equilibrium separation factor of this reaction has been measured to be 1.0080 ± 0.0016 (Jepson and DeWitt, 1976).

We can use one of several commercially available countercurrent extractors to obtain multiple-stage separations. For example, six columns 60 ft long and 1½-6 in. in diameter can enrich approximately 1000 g of calcium per year from natural abundance to the enrichments shown in Table II.

Closed reflux of these columns can be accomplished by removing the calcium from the calcium polyether complex at one end of the column and re-forming the complex at the other end of the column in the system consisting of

TABLE II

Calculated Calcium Isotope Enrichments from a Six-Column Production Cascade

Isotope	Natural abundance (%)	Product conc. (%)
Calcium-48	0.185	10.0
Calcium-46	0.0033	0.1
Calcium-44	2.06	40.0
Calcium-43	0.145	1.8
Calcium-42	0.64	2.0
Calcium-40	96.97	46.0

a calcium chloride aqueous solution and a chloroform solution of dicyclohexyl-18-crown-6 calcium complex. The organic phase is more dense. Therefore, the bottom reflux reaction is

$$CaL^{2+}_{(org)} + H_2O \longrightarrow L_{(org)} + Ca^{2+}_{(aq)}$$

This reaction is possible because the calcium ion prefers the dilute aqueous environment to that of the polyether complex in the organic solution. The equilibrium ratio of the aqueous phase calcium concentration to the calcium-polyether complex in the organic phase (distribution ratio) was found to be in the range of 20-130; the exact value depends upon the aqueous phase concentration and the organic phase polyether concentration. The calcium-polyether complex is formed in the separation column by the use of relatively high calcium concentrations (approximately 5 M) in the aqueous phase compared to the organic phase (approximately 0.2 M).

The reflux reaction at the top of the column is

$$Ca^{2+}_{(aq)} + L_{(org)} \longrightarrow CaL^{2+}_{(org)} + Ca^{2+}_{(aq)}$$

Some reconcentration of the aqueous calcium chloride stream from the bottom of the column is required before contacting with the polyether (L).

One might improve this isotope separation system by the appropriate choice or design of a different polyether. In it, one would want to increase the concentration range and enhance selectivity of the polyether. The first objective requires changing the side chains or external groups on the polyether; the second depends on modification of the hole size and of the nature of the coordinating bonds in the macrocyclic molecule.

Even in its present form, the polyether system has several advantages over other calcium isotope separation systems. It is considerably easier to operate than the mercury amalgam system (Zucker and Drury, 1964; Klinskii and Knyazev, 1974). The polyether system is simpler, and avoids the toxicity of mercury. It is also superior to the ion exchange system: the reflux is simpler, and the single-stage separation factor is considerably larger (Aaltonen, 1971; Klinskii *et al.*, 1974).

B. Other Promising Separations

Besides isotope separation, liquid-liquid extraction has possible applications for expensive chemical separations such as separation of lanthanides (Heckley and King, 1973), optically active isomers, and amino acids (Cram and Cram, 1974), and the partitioning of radioactive waste streams. Both selectivity and distribution coefficients are important considerations when evaluating potential separation processes.

The distribution coefficient for several of the most readily available polyether-metal systems is such that concentrated aqueous solutions are required to obtain adequate complex concentrations in the organic phase when the two phases are contacted. The metal-polyether systems are, therefore, not applicable to the removal of undesirable cations from dilute waste streams. By proper design of a polyether, however, more favorable distribution coefficients may be obtained.

We may also apply several of the liquid-liquid extraction systems to chromatographic separations. For example, Cram and Cram (1974) successfully used liquid-liquid chromatography to separate the enantiomers of α-phenylethylamine salts. They also suggested the resolution of racemic amino acids with an optically active cyclic ether by either liquid-liquid extraction or chromatography.

We can use polyethers in either the stationary phase or the mobile phase. they can be dissolved in a liquid which is adsorbed on a solid phase for liquid-liquid chromatography or they can be bonded directly to a solid phase for liquid-solid or gas-solid chromatography. Macrocyclic compounds, with exchangeable atoms such as hydrogen, may also be prepared; these can form the basis for more selective ion exchange.

III. MEMBRANE SEPARATIONS

Membranes containing macrocyclic polyethers can effect rapid selective separations of ionic solutes. These membranes can concentrate the solute of interest, converting a dilute mixture into a concentrated solution of the one desired component. However, they require juggling both an empirical membrane technology and a relatively sophisticated membrane mechanism. Because of this, they have not been as completely developed as the more conventional extractions that we have just discussed.

There are two basic types of membranes that are macrocyclic molecules: liquid membranes and solid membranes. Liquid membranes currently offer high selectivity, but solid membranes promise greater stability. Most research to date has concentrated on the development of liquid membranes and they are now suitable for large-scale application. Much less has been done on solid membranes, which require more careful chemical formulation. Both types are discussed in this section.

A. How Liquid Membranes Work

A typical liquid membrane consists of an organic solution of a macrocyclic molecule which forms a thin layer between two aqueous solutions containing different ionic concentrations. The macrocyclic molecules alter the permeability of the membrane to the ions and thus facilitate selective diffusion across the thin layer. This facilitated diffusion produces, in turn, selective separations. The basic mechanism by which these separations operate is idealized in Fig. 3 (Reusch and Cussler, 1973; Cussler, 1976). The solute to be separated first reacts with the macrocyclic molecule to form a complex which is soluble in the membrane

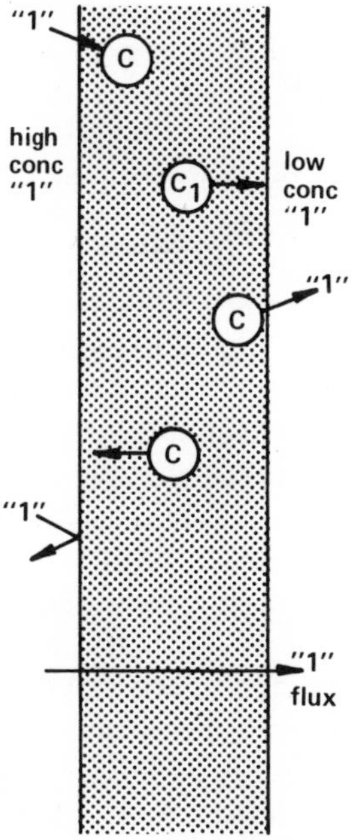

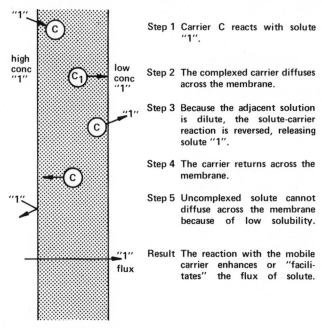

Step 1 Carrier C reacts with solute "1".

Step 2 The complexed carrier diffuses across the membrane.

Step 3 Because the adjacent solution is dilute, the solute-carrier reaction is reversed, releasing solute "1".

Step 4 The carrier returns across the membrane.

Step 5 Uncomplexed solute cannot diffuse across the membrane because of low solubility.

Result The reaction with the mobile carrier enhances or "facilitates" the flux of solute.

Fig. 3 The basic mechanism for liquid membrane separation. The selectivity of the separation depends on the reaction of solute "1" with the macrocyclic carrier C.

liquid, but not in the adjacent solutions. This complex then diffuses across the membrane. At the other side, the complex decays into the original macrocyclic molecule and into the original solute. Selective separations result when the reaction between the macrocyclic molecule and the carrier is in turn selective.

Mechanisms like this often represent a chemically well-defined analogue of the mobile carrier mechanism frequently postulated to explain facilitated diffusion in biological systems (Stein, 1967; Ward, 1970). The mechanism is, however, an idealization in two important ways. First, it implicitly assumes that the macrocyclic molecule always reacts very rapidly with the solute. This is approximated only when the second Damköhler number is large, that is, when

$$\frac{l^2}{t_{1/2}D} >> 1$$

where l is the membrane thickness, $t_{1/2}$ the half-life of the chemical reaction, and D the diffusion coefficient of the complex within the membrane. Second, it implies that the macrocyclic molecule reacts with the solute only near the membrane surface. In fact, this reaction takes place throughout the membrane, approaching the equilibrium

$$\text{macrocyclic molecule} + \text{solute} \rightleftharpoons \text{complex}$$

Removing these two idealizations requires heady doses of mathematics and of careful chemical insight (Schultz *et al.*, 1974). However, neither idealization compromises the primary message that diffusion and reaction combine to produce selective separations.

When the macrocyclic molecule reacts with two separate solutes, the membrane can both separate and concentrate the solute of interest. There are two different ways in which this can happen. First, the macrocyclic molecule reacts competitively with two solutes. In this case, a flux of one solute against its gradient in one direction can be generated by the flux of a second solute in the opposite direction. The first solute is most commonly a metal ion; the second is a proton. This mechanism, an analogue of "countertransport" observed in biological systems, is effectively a form of ion exchange (Cussler *et al.*, 1971). It will most often occur when the macrocyclic molecule is charged and the complex is not.

Alternatively, the macrocyclic molecule can react cooperatively with two different solutes. In this case, the flux of one solute against its gradient can be generated by the flux of a second solute in the same direction. Often, the two solutes react with the macrocyclic molecule to form a modified type of ion pair. This mechanism, similar to "co-transport" in biological systems, is more common for uncharged macrocyclic molecules (Schiffer *et al.*, 1974). Clearly, separations based on this mechanism again depend on combinations of diffusion and selective chemical reaction. Their design is described in the next subsection.

B. How Liquid Membranes Are Made

Developing a liquid membrane separation requires two basic steps: choosing the right macrocyclic molecule, and choosing the right geometry for the liquid membrane. The first requires chemical intuition; the second is necessary if the membrane is to useful.

Choosing the right macrocyclic molecule for a liquid membrane is exactly like choosing one for a liquid-liquid extraction. Indeed, it is much easier to screen systems for potential membrane applications by performing liquid-liquid extractions first. This is especially true when one wishes to simultaneously separate and concentrate a specific solute (Cussler and Evans, 1974).

Choosing the right geometry for the liquid membrane depends on how the liquid membrane can be stabilized. Obviously, making the liquid membrane very thin will both increase the rate at which separations can occur and decrease the stability of the membrane. If the membrane ruptures, no separation is possible.

The best way of stabilizing liquid membranes is to re-form them as the small bubbles shown in Fig. 4 (Cahn and Li, 1974; Matulevicius and Li, 1974; Li and

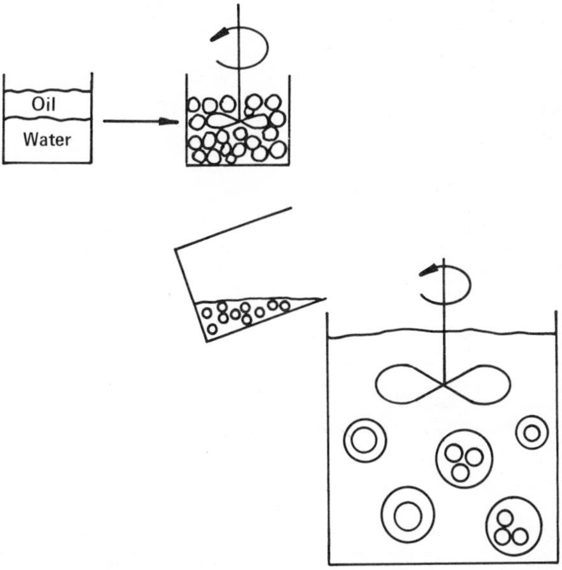

Fig. 4. Schematic procedure for making liquid surfactant membranes. The result is commonly a water-in-oil-in-water emulsion.

Shrier, 1972; Hochhauser and Cussler, 1976). These bubbles, called *liquid surfactant membranes,* are more accurately regarded as a water-in-oil-in-water emulsion. The two water phases represent the two solutions adjacent to the membrane, while the oil phase, which contains the macrocyclic molecules, represents the membrane itself.

Liquid surfactant membranes are easy to make, but hard to make well. As Fig. 4 suggests, one first uses rapid stirring to make a water-in-oil emulsion. This emulsion is stabilized by an oil-solvable surfactant. When it is added, with moderate stirring, to a second aqueous phase, one obtains the desired liquid surfactant membranes. To be stable, these membranes must have a high viscosity, a neutral buoyancy, and a large concentration of surfactant. The combination of these factors is an art and has been truly successful in only a few cases.

C. Examples

Two examples of the foregoing generalizations will illustrate how these liquid

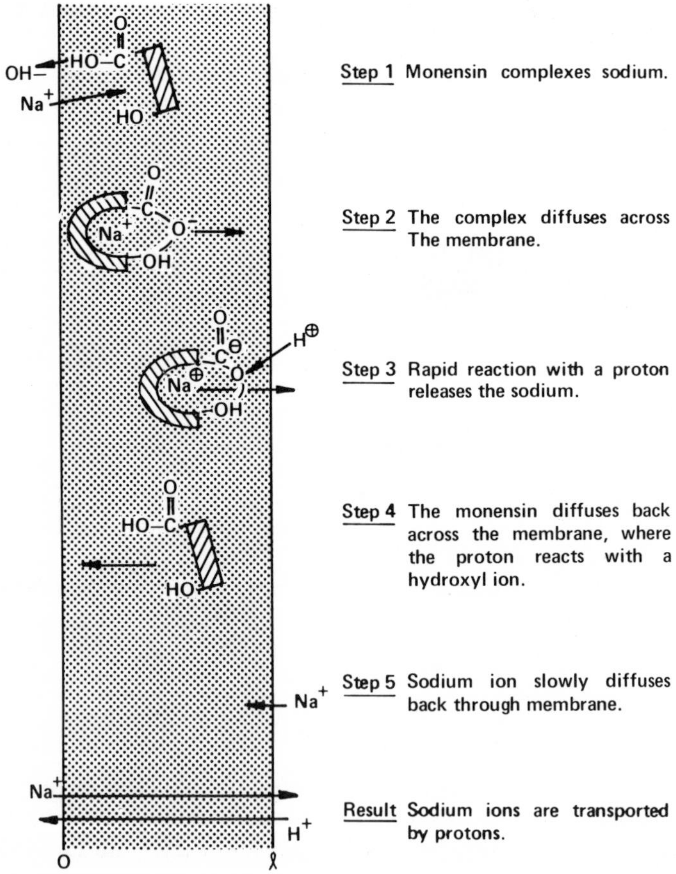

Fig. 5. Sodium transport mechanism using the macrocyclic antibiotic monensin. Sodium can be moved against its gradient by a flux of protons.

membranes work. Although neither of these examples has been developed to give commercial separations, both clearly illustrate the basic effects involved.

The first example, shown in Fig. 5, uses a flux of protons to selectively transport sodium ion against its gradient (Choy *et al.,* 1974). The membrane consists of an octanol solution of the macrocyclic antibiotic monensin (Chamberlin and Agtarap, 1970). In a typical experiment, shown in Fig. 6, the membrane originally separates two well-stirred solutions containing equal concentrations of sodium

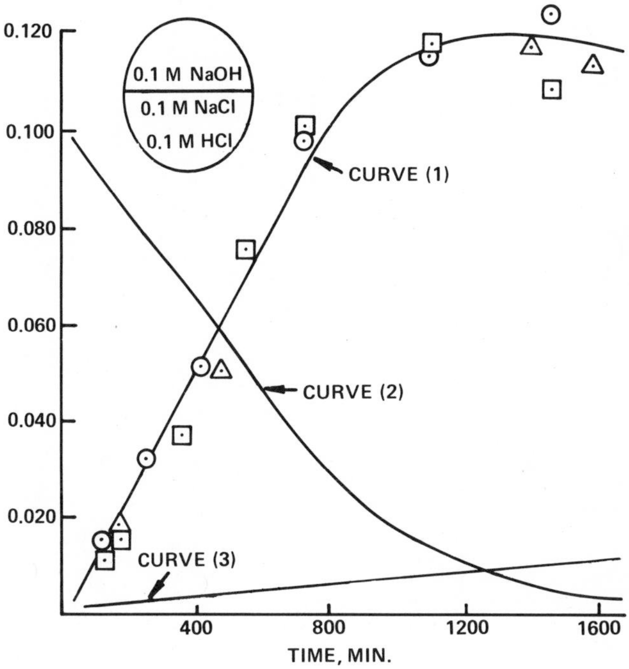

Fig. 6. Experimental results with a membrane containing monensin. As the concentration difference of protons decreases (curve 2), the flux of sodium ion against its gradient also decreases (curve 1). The sodium ion transport without monensin is shown in curve 3. (Data taken from Choy *et al.,* 1974.)

ions but unequal concentrations of protons. As the experiment proceeds, the large sodium ion concentration difference shown by curve 1 develops. This concentration difference is not altered by changes in the electrical potential across the membrane because the diffusing species (MCOOH and MCOONa) have no charge. Moreover, when the proton concentration difference shown by curve 2 is exhausted, the sodium ion concentration difference reaches its maximum.

The second example uses a gradient of chloride ions to selectively concentrate potassium (Caracciolo *et al* ., 1975; Wong *et al.,* 1975). This membrane consists of a solution in dichloroethane of the macrocyclic polyether, dibenzo-18-crown-6.

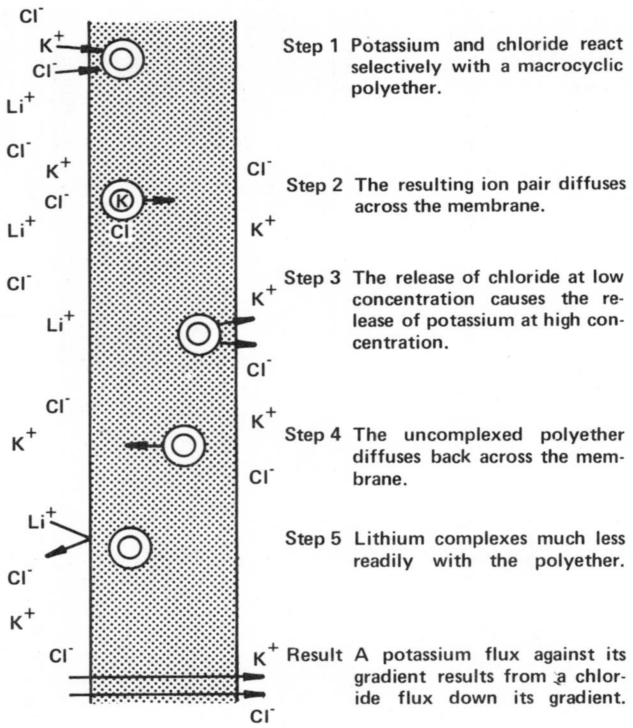

Cl⁻
K⁺
Cl⁻
Li⁺
Cl⁻
K⁺
Cl⁻
Li⁺
Cl
Cl⁻
Li⁺
Cl⁻
K⁺
Li⁺
Cl⁻
K⁺
Cl⁻
Cl⁻
K⁺
K⁺
Cl⁻
K⁺
Cl⁻
K⁺
Cl⁻
Step 1 Potassium and chloride react selectively with a macrocyclic polyether.
Step 2 The resulting ion pair diffuses across the membrane.
Step 3 The release of chloride at low concentration causes the release of potassium at high concentration.
Step 4 The uncomplexed polyether diffuses back across the membrane.
Step 5 Lithium complexes much less readily with the polyether.
Result A potassium flux against its gradient results from a chloride flux down its gradient.

Fig. 7. Potassium transport mechanism using the crown ether dibenzo-18-crown-6. Because ions are transported as pairs, the flux across this membrane depends on the product of potassium and chloride concentrations.

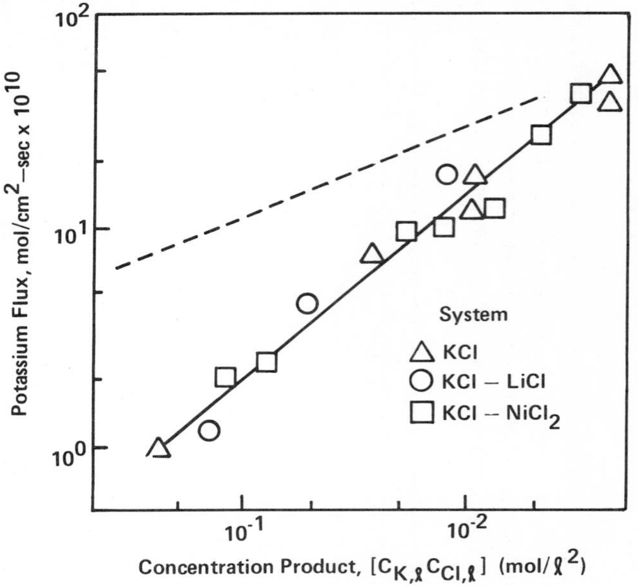

10²
10¹
10⁰
Potassium Flux, mol/cm²—sec x 10¹⁰
10⁻¹
10⁻²
Concentration Product, [C_K,ℓ C_Cl,ℓ] (mol/ℓ²)
System
KCl
KCl – LiCl
KCl – NiCl₂

Fig. 8. The flux versus concentration product across a membrane containing dibenzo-18-crown-6. The potassium ion concentration shown is that on one side of the membrane; that on the other side is zero. The broken and solid lines are theoretical predictions for transport of ions and of ion pairs, respectively. (Data taken from Caracciolo *et al.*, 1975).

The mechanism by which this membrane functions is shown in Fig. 7 and some typical experiments are summarized in Fig. 8. In all these experiments the membrane just described separates two well-stirred solutions. The top solution initially contains pure water; the bottom solution contains both potassium and chloride ions, not necessarily at equal concentrations. After the start of the experiment, complexed potassium chloride ion pairs diffuse across the membrane. As a result, the flux depends not on the potassium concentration difference, but on the difference of the product of potassium and chloride concentrations.

Three characteristics of these examples deserve special mention. First, the effects involved do not originate from electrostatics. More specifically, these ionic separations do not depend on Donnan equilibria and would exist even when all ions involved had the same charge and the same permeability. This is because both experiments involve ion transport as uncharged species. Second, both experiments use a thick liquid membrane supported by a piece of glass fiber filter paper. Thick membranes mean slow separations. Nevertheless, these two experiments can be converted into liquid membranes in a straightforward fashion (Li *et al.*, 1974; Hochhauser and Cussler, 1976).

The final, more serious, problem is the loss of the macrocyclic molecules from the membrane solutions. For example, mercury and nickel can be separated using toluene membranes containing the macrocyclic polyether dicyclohexo-18-crown-6, but this polyether is significantly soluble in aqueous salt solutions. A membrane separation which rapidly separates these cations with this polyether will also tend to rapidly lose the polyether. The obvious cure is to add large hydrocarbon side chains to the polyether molecules in an effort to decrease their water solubility. Unfortunately, decreasing their solubility also decreases their diffusion, so that the separation will become slower. The optimum compromise between polyether solubility and diffusion will clearly depend on the cost of the carrier.

D. Other Types of Membranes

The obvious alternatives to the liquid membranes discussed in the foregoing are solid membranes. However, if macrocyclic polyethers are trapped in thin polymer sheets, they have almost no effect on ion fluxes across these sheets. More specifically, the large selective flux increases which occur when polyethers are added to liquid membranes do not occur in solid ones.

The reason for this change is that the mechanism shown in Fig. 3 cannot effectively operate in solid membranes. The mobile carrier cannot diffuse rapidly enough to significantly enhance solute diffusion. In a few cases, enhanced transport in these solid membranes does occur by means of selective pores (Eisenman *et al.*, 1973) but these cases are not as yet commercially adaptable.

The most promising alternatives to liquid membranes are the so-called solid-liquid membranes. These membranes can be cut and handled without damaging their properties; one of us even kept one in his pocket for a week and it still worked. They consist of porous polymer films impregnated with viscous solutions of polyethers or of submicron polymer sheets plasticized with solutions of macrocyclic molecules. In the latter case, the macrocyclic molecules can sometimes function as mobile carriers even when they are sterically trapped within the polymer matrix.

IV. OTHER PRACTICAL POSSIBILITIES

A. Catalysis by Macrocyclic Compounds

Catalysis reduces the cost of commercial chemical processes by increasing the efficiency of chemical reactions. In the presence of catalysts, reactions can occur under milder conditions and at faster reaction rates. Catalysts are, in general, either homogeneous or heterogeneous. Most heterogeneous catalysts are metals, present either at an interface or as a finely divided dispersion. Because they have the advantage of being readily separable or allowing flow-through processing, they are widely used for industrial applications. Homogeneous catalysts, normally more efficient than heterogeneous ones, are used in the liquid phase reactions. They can, however, be more difficult to separate and recover than heterogeneous catalysts.

The major chemical changes effected by catalysis are listed in Table III. Cyclic polyethers will be of little industrial importance for the first three types of catalysis because satisfactory catalysts are already in use. Macrocyclic catalysts may prove useful for nitrogen fixation. The main objective would be to supplement the enzyme, nitrogenase in the soil. At present, several groups (van

TABLE III

Major Types of Catalysis

Type	Examples of catalysts
Acid-base	H^+, OH^-
Hydrogenation-dehydrogenation	Platinum metal
Polymerization	Free radicals
Nitrogen fixation	Enzyme, transition metal complexes
Oxidation-reduction (electron transfer)	Tetraazas, polyelectrolytes
Synthesis-decomposition	Enzymes, crown ethers

Tamelen, 1970) have succeeded in this by using macrochelates of ruthenium, titanium, and iron, but the yield is low because the reaction is not reversible. The ammonia remains bound to the metal ion.

Electron transfer is a second promising area for possible commercial applications of macrocycles as catalysts. Electron transfer reactions can be accelerated with a macrochelated transition metal ion (Rillema *et al.*, 1972; Lovecchio *et al.*, 1974). In most cases, the macrochelates employed are cyclic tetraaza compounds or similar heterocyclic macrocompounds where the coordination occurs between the transition metal ion and the nitrogens. Table IV shows accelerations

TABLE IV

Rates of Electron Transfer of Macrocyclic Co(III) Complexes with $Ru(NH_3)_6^{2+}$

Oxidant	k (M^{-1} sec^{-1})	Reference[a]
$Co(NH_3)_5NH_3^{3+}$	0.01	*1*
Co(*trans*-(14)-diene)$(NH_3)_2^{+3}$	3.0	*2*
Co(*trans*-(14)-diene)$(OH_2)_2^{+3}$	8×10^2	*2*
$Co(NH_3)_5(OH_2)^{+3}$	3.0	*1*

[a]References *1* Endicott and Taube (1964), *2* Rillema *et al.* (1972).

of 300 for the oxidation of Co(III) compounds when tetraaza ligands replace ammonias. The acceleration of the electron transfer results from the alteration of the electronic energy levels of the metal through the coordination. These chemical changes parallel those in biological systems where bound metal ions can activate the protein. Ferrodoxin, hemoglobin, and carbonic anhydrase are some examples of this.

The greatest potential for macrocyclic catalysts lies in the synthesis of specialized chemicals that otherwise could not be made, or could be made only at prohibitive cost. Because catalysts are recoverable, the cost of the macrocycles does not dominate their use in commercial production. The ability of these materials to solubilize charged species in apolar solvents makes them especially useful in this regard. Many organic reactions make use of alkali and alkaline earth salts, but their efficiency is impaired by low solubility. Cyclic polyethers have been used to solubilize permanganate salts (Sam and Simmons, 1972). By complexing the cation, macrocycles can also enhance the reactivity of the counterion. One example is the use of a cyclic polyether to catalyze the reaction of carbanions (Makosza and Ludwikow, 1974; Liotta and Harris, 1974). Other possible areas of application will be found in alkyllithium compounds, Grignard reagents, and alkali metals (Pederson and Frensdorff, 1972). In the presence of macrobicycles (cryptates), alkali metal solutions give solvated electrons (Lehn, 1973). Macrocycles could also serve as catalysts by altering the solvation of a compound or by causing a conformational change once the ligand is in the microcavity.

B. Ion-Selective Electrodes

Ion-selective electrodes can be divided into roughly three classes: solid ion exchangers, liquid ion exchangers, and neutral sequestering agents. The solid ion exchangers, which include glass electrodes and solid-state crystal electrodes, are the most familiar type. The glass electrodes are made from mixtures of silicon oxides with other metal oxides. The selectivity of the electrode results from the type of defect (caused by the oxides other than silicon) in the glass structure and is varied by varying the glass composition (Eisenman, 1967). Macrocyclic additives will not commonly affect these electrodes.

Liquid ion exchange electrodes are based on a liquid membrane containing a mobile carrier. Examples of mobile carriers include phosphate esters, fatty acids, and amines dissolved in water-insoluble membrane (Ross, 1969). The first commercial liquid membrane electrode was calcium selective; it used didecyl phosphate (Ross, 1967). The selectivity of these electrodes is largely based on the specific reaction between the carrier and the solute being studied. The mobile carriers in this case are charged. Macrocyclic molecules may be effectively used as mobile carriers in these electrodes.

Neutral sequestering electrodes also use liquid membranes containing a mobile carrier, but this carrier is uncharged. Their selectivity depends totally on the specificity of the reaction between this carrier and the solute being studied. Macrocyclic additives which act as mobile carriers are most useful for this type of electrode.

1. Mechanical Considerations

Whether they are the ion exchange or neutral carrier type, liquid membrane ion-selective electrodes have two major problems: mechanical stability and reasonable response times. We can overcome the problem of stability by using an inert support for the liquid (like a porous glass frit or a porous plastic membrane). Figure 9 shows a simplified diagram of such a liquid membrane electrode. The support is thin to allow for rapid establishment of the potential across the membrane. To minimize contamination, the electrode is usually designed for some outward leakage from the membrane. The electrical contact for the membrane electrode is noramlly established by having an internal aqueous electrolyte and an internal reference electrode (0.1 *M* KCl and a Ag/AgCl

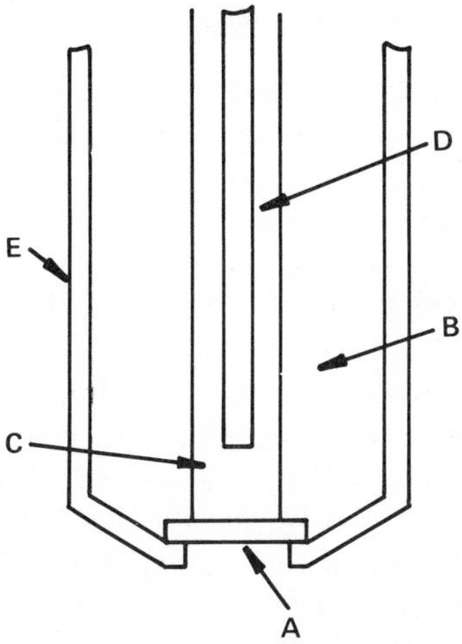

Fig. 9. Schematic construction of a liquid membrane electrode. A is the inert support membrane. B contains the mobile carrier dissolved in a suitable solvent. C is the internal aqueous electrolyte, usually KCl; D the internal reference electrode (Ag/AgCl); E the electrode body..

electrode is one possible arrangement). Orion and Corning make electrode bodies suitable for such use. Microelectrodes have also been constructed (Walker, 1971).

2. Electrodes Using Macrocyclic Antibiotics

Two classes of mobile carriers for electrodes, natual antibiotics and synthetic macrocycles, have been used in liquid membrane specific ion electrodes. Most of the work with natural antibiotics has centered on developing cation-selective electrodes for the alkali and alkaline earth ions. The earliest electrode of this type, based on valinomycin, shows a selectivity for potassium over sodium 10-100 times greater than the commercial glass electrodes (Mueller and Rudin, 1968; Eyal and Rechnitz, 1971). This selectivity is independent of the membrane solvent used. Other neutral antibiotics, actin, enniatins, gramicidin, and tetralactones, also show potassium selectivity. For example, Stefanac and Simon (1967) found a selectivity of 750 for potassium over sodium using nonactin homologs with a glass frit support membrane. Rubidium and cesium can interfere because their selectivity is quite similar to that of potassium. An ammonium ion electrode, using macrotetrolides, has also been studied (Cosgrove *et al.*, 1970). Charged antibiotics, like monensin and nigericin, behave in a more elaborate fashion; nigericin is selective for sodium and monensin is selective for potassium (Lutz *et al.*, 1971).

3. Electrodes Using Synthetic Macrocyclic Compounds

Liquid membrane electrodes can also be based on synthetic macrocyclic molecules. The first class of compounds used in this way was the crown ethers, originally synthesized by Pedersen (1967). Using these molecules as a template, other workers have made new classes by replacing the oxygen molecules with nitrogen and sulfur, and by making bicyclic and tricyclic structures, commonly called cryptates. Different classes of these compounds are selective for different types of ions (Dietrich *et al.*, 1974).

Since synthetic macrocycles selectively complex ions, they can be substituted for the natural antibiotics described earlier. As in the case of antibiotics, the selectivity of the resulting electrode depends largely on two factors. The first and most important is the relative equilibrium constant for formation of the ion-macrocycle complex. The second main factor is the membrane solvent: in some cases, changing the membrane solvent can reverse the selectivity.

Electrodes containing the crown ethers constitute the only class which has been studied extensively. Rechnitz and Eyal (1972) studied nitrobenzene solutions of dicyclohexyl-18-crown-6 in both Orion series 92 and Corning Model 476041 electrodes. They found that the selectivity relative to potassium for the electrodes agreed with the equilibrium constants for complexation: the selectivity of potassium over sodium is much less than that of valinomycin electrodes. Mascini and Pallozi (1974), using polyvinyl chloride (PVC) support membranes, found essentially the same result when they compared valinomycin and dibenzo-18-crown-6. They had to add small amounts of plasticizer to obtain stable behavior. Finally, Simon's group (Morf *et al.*, 1973) developed a calcium-selective electrode that can be used in the presence of potassium, sodium, and magnesium. The carrier is not cyclic or a polyether, but the mode of complexation is the same as the cyclic polyethers. While other macrocyclic ligands have not been used, their potential is greater than the natural antibiotics because they can be synthesized with properties specific to the ion to be measured.

ACKNOWLEDGMENTS

R. A. S. works at the Mound Laboratory, which is operated by Monsanto Research Corporation for the U.S. Energy Research and Development Administration under Contract No. E-33-i-GEN-53. The research of T.J.G. and E.L.C. is supported by National Science Foundation Grant No. GK-32313 and by National Institutes of Arthritis, Metabolic and Digestive Disease Grant No. 1-R01-AM-16143. E.L.C. is supported by National Institutes of Health Research Career Development Award 1KO-AM-70461.

References

Aaltonen, J. (1971). *Soumen Kemistilehti* **44B**, 1-4.

Cahn, R. P., and Li, N. N. (1974). Paper presented at Am. Chem. Soc. Meeting, Los Angeles, March.

Caracciolo, F., Cussler, E. L., and Evans, D. F. (1975). *AIChE J.* **21**, 160-167.
Chamberlin, J. W., and Agtarap, A. (1970). *Org. Mass Spectrom.* **3**, 271-285.
Choy, E., Evans, D. F., and Cussler, E. L. (1974). *J. Am. Chem. Soc.* **96**, 7085-7090.
Christensen, J. J., Hill, J. O., and Izatt, R. M. (1971). *Science,* **174**, 459-467.
Cosgrove, R. E., Mask, C. A., and Krull, I. A. (1970). *Anal. Lett.* **3**, 457-464.
Cram, D. J., and Cram, J. M. (1974). *Science* **183**, 803-810.
Cussler, E. L. (1976). "Multicomponent Diffusion." Elsevier, Amsterdam.
Cussler, E. L., and Evans, D. F. (1974). *Separat. Purificat. Methods* **3**, 399-421.
Cussler, E. L., Evans, D. F., and Matesich, M. (1971). *Science* **172**, 377-379.
Dietrich, B., Sauvage, J. P., and Lehn, J.-M. (1973). *J. Chem. Soc. Chem. Commun.* 15-16.
Eisenman, G. (1967). *In* "Glass Electrodes for Hydrogen and Other Cations" (G. Eisenman, ed.), pp 268-283. Dekker, New York.
Eisenman, G., Szabo, G., McLaughlin, S. G. A., and Ciani, S. M. (1973). *J. Bioenerg.* **4**, 93-148.
Endicott, J. F., and Taube, H. (1964). *J. Amer. Chem. Soc.* **86**, 1686-1691.
Eyal, E., and Rechnitz, G. A. (1971). *Anal. Chem.* **43**, 1090-1093.
Heckley, P. R., and King, R. B. (1973). *Proc. Rare Earth Res. Conf., 10th Conf.* 730402-PI. Carefree, Arizona, April 30-May 3.
Hochhauser, A. M., and Cussler, E. L. (1975). *CEP Symp. Ser.,* **71**, 136-142.
Jepson, B. E., (1974). U.S. Patent No. 3,914,373.
Jepson, B. E., and DeWitt, R. (1976). *J. Inorg. Nucl. Chem.,* **38**, 1175-1177.
Klinskii, G. D., and Knyazev, D. A. (1974). *Russ. J. Phys. Chem.,* 48, 890-891.
Klinskii, G. D. Knyazev, D. A., and Vlasova, G. I. (1974). *Russ. J. Phys. Chem.* **48**, 380-382.
Lehn, J. M. (1973). *Struct. Bond.* **16**, 1-69.
Li, N. N., and Shrier, A. L. (1972). *In* "Recent Developments in Separation Science" (N. N. Li, ed.), Vol. I, pp. 163-174. CRC Press, Cleveland, Ohio.
Li, N. N., Cahn, R. P., and Shrier, A. L. (1974). U.S. Patent No. 3,799,907.
Liotta, C. L., and Harris, H. P. (1974). *J. Am. Chem. Soc.* **96**, 2250-2252.
Lovecchio, F. V., Gore, E. S., and Busch. D. H. (1974). *J. Am. Chem. Soc.* **96**, 3109-3118.
Lutz, W. K., Fruh, P. U., and Simon, W. (1971). *Helv. Chim. Acta* **54**, 2767-2770.
Makosza, M., and Ludwikow, M. (1974). *Angew. Chem.* **86**, 744-745.
Mascini, M., and Pallozi, F. (1974). *Anal. Chim. Acta* **73**, 375-384.
Matulevicius, E. S., and Li, N. N. (1974). Paper presented at AIChE Meeting, Pittsburgh, Pennsylvania, June.
Morf, W. E., Ammann, D., Pretch, E., and Simon, W. (1973). *Pure Appl. Chem.* **36**, 421-439.
Mueller, P., and Rudin, D. O. (1967). *Biochem. Biophys. Res. Commun.* **26**, 398-404.
Pedersen, C. J. (1967). *J. Am. Chem. Soc.* **89** 7017-7036.
Pedersen, C. J., and Frensdorff, H. K. (1972). *Angew. Chem. Int. Ed. Engl.* **11**, 16-25.
Rechnitz, G. A., and Eyal, E. (1972). *Anal. Chem.* **44**, 370-372.
Reusch, D. F., and Cussler, E. L. (1973). *AIChE J.* **19**, 736-741.
Rillema, D. P., Endicott, J. F., and Patel, R. C. (1972). *J. Am. Chem. Soc.,* **94**, 394-401.
Ross, J. W. (1967). *Science* **156**, 1378-1379.
Ross, J. W. (1969). *In* "Ion-selective Electrodes" (R. A. Durst, ed.), pp. 57-88. Nat. Bur. Std. Publ. 314, Washinton, D.C.
Sam, D. J., and Simmons, H. F. (1972). *J. Am. Chem. Soc.* **94**, 4024-4025.
Schiffer, D. K., Hochhauser, A. M., Evans, D. F., and Cussler, E. L. (1974). *Nature (London)* **250**, 484-486.
Schultz, J. S., Goddard, J. D., and Suchdeo, S. R. (1974). *AIChE J.* **20**, 417-445.
Stefanac, Z., and Simon, W. (1967). *Microchem. J.* **12**, 125-132.

Stein, W. D. (1967). "Movement of Molecules Across Cell Membranes." Academic Press, New York.

van Tamelen, E. E. (1970). *Ann. Chem. Res.* **3**, 361-367.

Walker, J. L. Jr. (1971). *Anal. Chem.* **43**, 89A-93A.

Ward, W. J. (1970). *AIChE J.* **16**, 405-410.

Wong, K. H., Efagi, K., and Smid, J. (1975). *J. Am. Chem. Soc.* 97, 3642.

Zucker, D., and Drury, J. D. (1964). *J. Chem. Phys.* **41**, 1678-1681.

AUTHOR INDEX

Numbers in parentheses are page numbers that indicate that an author's work is referred to although his name is not cited in the text. Numbers in italics show the page on which the complete reference is listed.

A

Aaltonen, J., 294, *306*
Ackman, R. G., 10, *49*
Adams, R., 135, *201*
Agtarap, A., 299, *307*
Ah-Kow, G., (120), *205*
Aldanova, N. A., (6), *51*
Allan, A. R., (56), *108*
Allen, F. H., 135, *201*
Almy, J., 189, 194, *201*
Ammann, D., (306), *307*
Ave, D. H., (113), *202*
Arnold, C., 135, *204*
Asher, I. M., (253), *287*
Ashby, J., 55, *107*
Atkins, T. J., 56, *109*

B

Bailey, M. F., (236), *243*
Baillargeou, M. H., 47, *49*
Barefield, E. K., (236), *243*
Barton, D. H. R., 141, *201*
Bartsch, R. A., 151-155, 157, 159, *202*
Bauer, R. A., 236, *241*
Belelskaya, I. P., (197), (200), *203*
Belford, R. L., (235), *242*
Biellmann, J. F., 188, *202*
Biggi, G., (147), *202*
Bioch, D. R., 191, *203*
Black, D. S. C., 53, *107*
Blanzat, J., (56), *107*
Boden, R. M., (128), 177, *202, 203*
Bohme, D. K., 113, *202, 205*
Boileau, L., (201), *203*
Bombieri, G., 230, 231, *241*
Borchardt, J. K., 163, *202*
Borgen, G., (55), *107*
Borrows, E. T., (10), *51*
Bosnich, B., 236, *241*
Bowers, A., 141, *202*
Bowers, M. T., 113, *202*
Bradshaw, J. S., 36, *49*, 54, (55), 56, 57, *107, 108*
Bram, G., (198), *202*
Brandstrom, A., 113, *202*
Brauman, J. I., 113, *202*
Bright, D., 16, 26, *49*, 227, 237, *242*
Brockmann, H., 6, *49*
Bromels, M. H., 37, 43, *50*
Broussard, J. B., (55), *108*
Brown, W. H., (10), *49*
Burgemeister, W., 257, *286*
Bush, M. A., 16, 26, 27, *49*, 226-228, 237, 241, *242*
Busch, D. H., (15), (26), *51*, 273, 276, *286*, (303), *307*
Bushman, D. G., (153), *204*

C

Cabbiness, D. K., 273, 274, 277, *286*
Cahen, Y. M., 253, 268, 272, *286*
Cahn, R. P., 297, (301), *306, 307*
Cambillau, C., 198, *202*

Caracciolo, F., 299, 300, *307*
Cardillo, G., 176, *202*
Cassol, A., (230), (231), *241*
Cayley, G. R., (278), *287*
Ceraso, J. M., (56), *107*, 266, *286*
Chamberlin, J. W., 299, *307*
Chan, Y., (57), *107*
Chao, Y., 283-285, *286*
Cheney, J., 57, *107*
Chenoweth, M. B., (48), *50*
Chern, C., (167), (168), (170), *204*
Chiesa, P. P., (57), *107*
Chock, P. B., 254, 258, 267, *286*
Choy, E., 299, *307*
Christensen, J. J., 18, (28), (32), 36, 39, *49, 50,* 53, (55), (57), *107, 108,* 111, 113, *202,* 228, (229), (237), *242*, (276), *286*, 290, *307*
Ciani, S., 18, (28), *49, 50,* (246), (253), *286, 287,* (301), *307*
Cima, F. Del, 147, *202*
Cinquini, M., 123, 141, 175, 180, 183, 192, *202*
Cockerill, A. F., (153), *204*
Coetzee, J. F., 118, *202*
Collet, A., (201), *203*
Cook, F. L., (15), (33), *49,* (55), *107*
Cooper, M. J., (55), *107*
Cope, A. C., 125, 135, *203*
Corbaz, R., 6, *49*
Corey, E. J., 165, 171, *202*
Cosgrove, R. E., 305, *307*
Cotton, F. A., 45, *51*
Couvillion, J. L., (157), *205*
Coxon, A. C., 57, *107*
Cram, D. J., (10), 11, (15), 33, *49, 50,* 53, 55, 56, 58, 59, *107, 108, 109,* 111, 189, (194), *201, 202, 204,* 229, *242,* 283-285, *286,* 294, *307*
Cram, J. M., 11, *49*, 111, *202*
Curci, R., 149, *202*
Curtis, A. B., 163, *205*
Curtis, N. F., 235, 236, *242, 243*
Curtis, W. D., 59, *107*
Cussler, E. L., 295-297, (299), (300), 301, *307*

D

Daasvatn, K., (55), *107*
Dale, J., 15, *49,* 55, *107,* 237, *242*
Dalley, N. K., 229, 237, *242*
Danesh-Khoshboo, F., (55), *108*
D'Aniello, Jr., M. J., (236), *243*
Dann, J. R., 57, *107*
Davis, M., 240, *242*
Davis, P. H., 235, *242*
DeBacker, M. G., (30), *49*, (201), *202*
Dehm, D., (128), 129, *202, 203*
Dehmlow, E. V., 113, *202*
de Jong, F., 58, *107*
DePaoli, G., (230), (231), *241*
DeSimone, R. E., 235, 237, *242*
DeWitt, R., 292, *307*
Diaddario, L. L., (235), *242*, (280), (281), *286*
Diamond, J. M., 251, *286,* 252, *287*
Diebler, H., 246, 250, 251, 253, 254, 267, 279, *286*
Dietrich, B., 42, 43, *49,* 56, 57, *107*, 144, 150, 177, *202*, (254), 260, (261), (266), (270), *286*, 306, *307*
Dietz, R., 163, *202*
Difuria, F., 149, *202*
Dobler, M., 6, 27, *49*, 216, (224), (225), (227), (241), *242, 243*, 246, *287*
Dockx, J., 113, *202*
Dotsevi, G., 58, *107*
Down, J. L., 10, 29, *49*
Drury, J. D., 294, *308*
Dunbar, B. I., (134), (141), *205*
Dunitz, J. D., (6), (27), *49*, 216, (224), 225, (227), 236, 237, 241, *242, 243*
Durst, H. D., 111, 127, (128), (134), (141), 175, *202, 203, 205*
Dye, J. L., 30, *49*, 56, *107,* 201, *202, 203,* 266, (268), (272), *286*

E

Eatough, D. J., (36), (39), *49*, (53), *107*, (111), (113), *202*
Edwards, J. O., 117, *202*
Efagi, K., (299), *308*
Eggers, F., (246), (247), (252), (256), *286*
Eigen, M., 246, 250, 251, (253), (254), (267), *286*
Eisenman, G., 18, (28), 29, *49, 50*, (246), 253, *286, 287,* 301, 304, *307*
Ellingsen, T., (145), *205*
Endicott, J. F., 303, *307*
Erdman, J. P., 184, *204*

Ettlinger, L., (6), *49*
Evans, D. A., 190, *202*
Evans, D. F., (296), (297), (299), (300), *307*
Evans, R. E., 10
Eyal, E., 305, 306, *307*
Eyring, E. M., (256)-(259), (266), *287*

F

Farrow, M. M., (256)-(259), (266), *287*
Feigenbaum, W. M., 33, 34, *49*
Feigina, M. Yu., (6), *51*
Fiandanese, V., 162, *202*
Fishman, M., (141), *205*
Fishman, J., 141, *203*
Flitter, D., 32, *51*
Ford, W. T., 190, 194, *202, 203*
Forno, A. E. J., (163), *202*
Fraenkel, G., 183, 184, *202*
Freidman, L., 135, *202*
Frensdorff, H. K., 21, 23, 25, 26, 28-30, 36, 37, *49, 50,* 53, 55, *107, 109,* 111, 143, 146, *204,* 231, *242,* 304, *307*
Fruh, P. U., (305), *307*
Fujimoto, M., (141), *204*
Funck, T., (246), 247, (252), (256), *286*
Fuoss, R. M., 248, *286*
Furtado, D., (56), *108*

G

Garwood, D. C., (189), (194), *201*
Gates, Jr., J. W., (57), *107*
Gaumann, E., (6), *49*
Gavel, D. P., (235), *242*
Girodeau, J. M., 59, *107*
Glick, M. D., 235, 237, *242*
Goddard, J. D., (296), *307*
Gokel, G. W., 15, 33, *49,* 55, 58, *107, 108,* 111, *202*
Goldberg, I., 229, 237, 241, *242*
Golob, A. M., 190, *202*
Gonzalez, T., (21), *50,* (125), (131), (134), *203*
Gore, E. S., (303), *307*
Graf, E., 57, *108*
Gray, R. T., 55, *109*
Greene, R. N., 15, 16, *50,* 55, *108*
Grell, E., 246, 247, 252, 256, *286*
Grisdale, E. E., (116), (117), (180), *203*
Grovenstein, E., 186, *203*
Grunwald, E., 113, 117, *203*
Grushka, E., 127, *203*
Grütze, J., (55), (56), *109*
Gueta, F., 59, *109*

H

Hall, Jr., H. K., 121, *205*
Hammes, G. G., 248, 284, *286*
Handel, H. 172, 173, *203, 204*
Hanson, M. P., 194, *204*
Hapala, J., (156), *205*
Harman, M. E., 229, *242*
Harris, H. P., (15), 21, (33), *49, 50,* (55), *107,* 118, (125), (131), 134, *203,* 303, *307*
Hart, F. A., (229), *242*
Hartshorn, 53, *107*
Hassel, O., 240, *242*
Hayase, Y., (141), *204*
Haymore, B. L., (28), (32), *50,* (55), (57), *107, 108,* (276), *286*
Heckley, P. R., 294, *307*
Helgeson, R. C., 58, *108,* (237), *243*
Hemmes, P., 248, *286*
Henbest, H. B., 141, *203*
Herceg, M., 209, 231, 237, *242, 243*
Herriott, A. W., 180, *203*
Hill, J. O., (18), *49,* (53), *107,* (111), *202,* (289), (290), *307*
Hinz, F. P., 274-276, *286*
Hirai, S., (141), *204*
Hochhauser, A. M., (296), 297, 301, *307*
Hogberg, S. A. G., 56, *108*
Hogen-Esch, T. E., (34), *50*
Holm, R. H., (260), *287*
Holmes, H. L., 125, 135, *203*
Hopkins, H. P., (116), (117), (180), *203*
House, H. O., 125, 135, *203*
Hughes, D. L., 226, 227, *242*
Hui, J. Y., 36, *49,* 54, 57, *107*
Hull, R., (55), *107*
Hunter, D. H., 130, *203*
Hurd, C. D., 45, *50*
Hursthouse, M. B., (229), *242*

I

Iitaka, Y., 231, *242*
Ilgenfritz, G., (246), (250), (251), (253),

(254), (267), *286*
Immirzi, A., (230), (231), *241*
Ivanov, V. T., (6), *50*
Izatt, R. M., (18), 28, 32, (36), (39), *49, 50,* 53, 55, (57), *107, 108*, (111), (113), *202,* (229), (237), *242*, 276, *286*, (289), (290), *307*

J

Jackson, W. R., 141, *203*
Jagur-Grodzinski, J., 33, *51*, 179, *204,* (253), 254, (257), (258), (266), *287*
Jepson, B. E., 292, *307*
Jirkovsky, L., (141), *205*
Johnson, C. K., 224, 225, *242*
Johnson, R. A., 163, 164, 167, *203*
Johnson, W. S., 141, *203*
Johnston, N. J., 141, *204*
Jones, G. H., (59), *107*
Jones, T. E., 280, 281, *286*
Joyce, M. A., (181), (183), *204*

K

Kaempf, B., 201, *203*
Kapoor, P. N., 53, *108*
Kappenstein, C., 53, *108*
Kaura, C. K., (56), *108*
Kayser, R. H., 155, *202*
Keane, J. F. W., 141, *203*
Kebarle, P., 113, 118, *203, 205*
Keller-Schierlein, W., (6), *49, 51*
Kieczykowski, G. R., (127), (175), *202*
Kikkawa, K., (141), *204*
Kikta, E. J., (127), *203*
Kilbourn, B. T., (6), (27), *49*
Kimura, E., (231), *242*
King, A. P., 56, *108*
King, R. B., 294, *307*
Klinskii, G. D., 294, *307*
Knöchel, A., 115, 128, 142, *203*
Knyazev, D. A., 294, *307*
Koga, K., (58), (59), *108*
Koida, K., 171, *203*
Konizer, G., (28), *51*, (194), *205*
Kopolow, S., 34, *50*
Kormarynsky, M. A., 201, *203*
Kosower, E. M., 117, *203*
Kostikov, R. R., 183, *203*
Kradolfer, F., (6), *49*
Krasne, S. J., 253, *286*
Krespan, C. G., 56, *108*
Kristiansen, P. O., 15, *49*, 237, *242*
Krull, I. A., (305), *307*
Kurts, A. L., 197, 200, *203*
Kyba, E. P., 10, *50*, 55, 58, 59, *108*

L

Laidler, D. A., (59), *107*
Landini, D., 123, 177, *203*
Laprade, R., 246, *286*
Larcombe, B. E., (163), *202*
Lardy, H. A., 18, *50*
Larson, S. B., (237), *242*
Lee, W., (130), *203*
Lee-Ruff, E., (113), *205*
Leffler, J. E., 113, 117, *203*
Lehn, J. M., (42), 43, *49, 50*, 53, (56), 57, 59, *107, 108,* 111, 144, 150, 177, (201), *202, 203,* 216, *242,* 246, 253, 254, (260), 261, 266, 267, 270, 271, 273, 284, *286, 287,* 304, (306), *307*
Leonard, J. E., (195), *205*
Leong, B. K. J., 48, *50*
Lewis, J., (10), (29), *49*
Li, N. N., 297, 301, *306, 307*
Lieder, C. A., (113), *202*
Lier, E. F., (141), *201*
Liesegang, G. W., (256), 257-259, 266, *287*
Lin, C. T., 278, *287*
Lindoy, L. F., 53, *108*
Liotta, C. L. (15), 21, (33), *49, 50*, (55), *107*, 116-118, 125, 131, 134, 180, *203*, 303, *307*
Lockhart, J. C., 56, *108*
Lok, M. T., (56), *107*, 201, *203*
Louis, R., 56, *109*
Lovecchio, F. V., 303, *307*
Loyola, V. M., 268, 269, *287*
Ludwikow, M., 192, *203*, 303, *307*
Luteri, G. F., 190, *203*
Luttringhaus, A., 10, *50*, 57, *108*
Lutz, W. K., 305, *307*
Luz, Z., (253), (254), 258, (266), *287*

M

Maas, G., (246), 250, (251), (253), (254), (267), *286*
Machida, Y., (165), (171), *202*

Mack, M. M., 128, *203*
Machay, G. I., (113), *202*
Makosza, M., 113, 192, *203*, 303, *307*
Mallinson, P. R., 26, *50*, 227, 228, 237, *242*
Mandan, K., 58, *108*
Marchese, G., (162), *202*
Margerum, D. W., (236), *241*, 273-277, *286*, (278), *287*
Marshall, J. A., 141, *203*
Martin, J. C., 191, *203*
Marvel, C. S., 135, *203*
Marx, J., 41, *51*
Mascini, M., 306, *307*
Mask, C. A., (305), *307*
Maskornick, M. J., 149, *203*
Mason, R., (236), *241*
Matesich, M., (296), *307*
Matheson, K. L., (237), *242*
Mathieu, F., 215, *242*
Matsuda, T., 171, *203*
Matulevicius, E. S., 297, *307*
Maverick, E., (237), *243*
Mehrotra, R. C., 53, *108*
Mercer, M., 10, 26, *50*, 226, 227, 229, 241, *242*
Merritt, M. V., 163, *203*
Metz, B., (209), 215, 216, 231, 234, (236), 241, *242, 243*
Meyer, W., 141, *204*
Michel, R. H., 33, 34, *49*
Milligan, D. V., 192, *204*
Mintz, E. A., (154), *202*
Mocella, M. T., (236), *243*
Molchanov, A. P., 183, *203*
Montanari, F., (123), (141), (175), (177), (180), (183), 192, *202, 203*
Montavan, F., 57, *108*
Moore, B., (10), (29), *49*
Moras, D., 209, 215, 231, (234), 236, 241, *243*
Moreau, J. M., (58), *108*
Moreno, J. H., 252, *287*
Morf, W., 251, *287*, 306, *307*
Moss, G. P., (229), *242*
Moss, R. A., 180, 181, 183, *203, 204*
Mowry, D. J., 135, *204*
Moyer, J. M., (58), *108*
Mueller, P., 5, *50*, 305, *307*
Murakami, M., (141), *204*
McClure, G. L., (55), *108*
McColm, E. M., 135, *203*
McDermott, M., (21), *50*, (125), (131), (134), *203*
McGhie, J. F., (141), *201*
McIver, R. T., (113), *202, 203*
McLaughlin, S. G. A., 28, *50*, (301), *307*

N

Nagata, W., 141, *203, 204*
Nager, U., 6, *51*
Narisada, M., (141), *204*
Naso, F., 121, (162), *202, 204*
Nätscher, R., (55), (56), *109*
Neipp, L., (6), *49*
Newkome, G. R., 55, *108*
Nelson, D. P., (28), *50*, (276), *286*
Nelson, G. V., 201, *204*
Neumann, P., 53, *109*, 111, *205*
Neupert-Laves, K., 246, *287*
Newcomb, M., 55, 58, *108*
Nicely, V. A., (30), *49*, (201), *202*
Nicolaou, K. C., (165), (171), *202*
Nidy, E. G., 163, 164, 167, *203*

O

Ochrymowycz, L. A., (280), (281), *286*
Oehler, J., (115), (128), *203*
Olmstead, W. N., (113), *202*
Orena, M., (176), *202*
Ovchinnikov, Yu. A., 6, *50*
Owen, J. D., 229, *243*
Owens, R. W., 135, *205*

P

Padwa, A., 129, *202*
Pallozi, F., 306, *307*
Pankova, M., (158), *205*
Park, C. H., 41, *51*, 260, 261, *286, 287*
Parker, A. J., 117, *204*
Parker, K., 237, *243*
Parlman, R. M., (154), *202*
Parks, J. E., 260, *287*
Parsons, D. G., 10, 29, *50*
Parsons, D. J., 55, *109*
Patel, R. C., (303), *307*
Pauling, P. J., (236), *241*
Payzant, J. D., (113), *202*
Peacock, S. C., (58), *108*
Pearson, R. G., 21, 24, *50*, 117, *204*

Pechhold, E., 183, 184, *202*
Pederson, C. J., 2, 4-6, 10-26, 28, 30, 32-37, 42, 43, 46, 47, *50, 51,* 53-55, 57, *109,* 111, 120, 143, 146, *204*, 208, 236, *243*, 290, 291, 304, 306, *307*
Pelissard, D., 56, *109*
Peover, M. E., (163), *202*
Phillies, G. D. J., 253, *287*
Phillips, S. E. V., 45, *50*
Phizackerley, R. P., 216, (225), (241), *242, 243*
Picker, D., 180, *203*
Pierre, J. L., 172, 173, *203, 204*
Pietra, F., (147), *202*
Pilkiewicz, F. G., 180, (181), (183), *203, 204*
Pirisi, F. M., (123), (177), *203*
Pizer, R., (268), (269), *287*
Plattner, P. A., 6, *51*
Poon, C. K., 273, *287*
Poon, L., (141), *205*
Poonia, N. S., 23, 24, *51*
Popov, A. I., 253, (268), 272, *286*
Pouet, M., (120), *205*
Powell, H. M., 32, *51*
Prelog, V., (6), *49*
Pressman, B. C., 18, 26, *51*
Pretch, E., (306), *307*
Prewitt, C. T., 215, *243*
Purdie, N., (256)-(259), (266), (267), *287*

Q, R

Quitt, P., (6), *51*
Raithby, P. R., (229), *242*
Ramage, E. M., (55) *107*
Ratajczyk, J. F., (195), *205*
Raynal, S., (201), *203*
Rechnitz, G. A., 305, 306, *307*
Reinhoudt, D. N., 55, *109*
Reusch, D. F., 295, *307*
Reutov, O. A., (197), (200), *203*
Richey, F. A., 125, *203*
Richman, J. E., 56, *109*
Rillema, D. P., 303, *307*
Risen, W. M., Jr., (253), *287*
Ritchie, C. D., 117, *204*
Robertson, G. B., (236), *241*
Robinson, J. M., 55, *108*
Robinson, W. R., (236), *241*
Robson, A. C., (56), *108*
Rodig, O. R., 141, *204*
Rodriguez, L. J., 256, *287*
Roitman, J. N., 189, *204*
Romano, L. J., 170, *204*
Ronzini, L., 121, *204*
Rorabacher, D. B., (235), *242*, (278), (280), (281), *286, 287*
Rosalky, J., (216), *243*
Rosen, P., 279, *286*
Ross, J. W., 304, *307*
Rudin, D. O., 5, *50*, 305, *307*
Rudolph, G., (115), (128), 142, *203*
Rytting, J. H., (28), *50*, (276), *286*

S

Sakembaeva, S. M., 197, 200, *203*
Sam, D. J., 10, 21, *51,* 34, 120, 122, 134, 144, 175, *204*, 303, *307*
Sandri, S., (176), *202*
San Filippo, J., 167, 168, 170, *204*
Sarthou, P., (198), *202*
Saunders, W. H., 153, 163, *202, 204*
Sauvage, J. P., (42), (43), *49*, (56), (57), (59), *107*, (216), *242*, (254), (260), (261), (266), (270), *286*, (306), *307*
Sawyer, D. T., 163, *203*
Schaefer, A. D., (195), *205*
Schecter, J., 135, *202*
Scherer, K. V., (183), *204*
Schiessler, R. W., 32, *51*
Schiffer, D. K., 296, *307*
Schmidt-Kastner, G., 6, *49*
Schmitt, J. F., 188, *202*
Schnautz, N., 141, *204*
Schué, F., (201), *203*
Schultz, J. S., 296, *307*
Sciacovelli, O., (162), *202*
Scott, C. R., 117, *205*
Seiler, P., 216, 224, 225, 227, 236, 237, (241), *242, 243*
Sengstad, J., 117, *204*
Sepp, D. T., 183, *204*
Shannon, R. D., 215, *243*
Sharpe, W. R., 118, *202*
Shchori, E., 33, *51*, 179, *204,* 253, 254, 257, 258, 266, *287*
Shelly, T. A., 157, *202*
Shemyakin, M. M., 6, *51*, 248, *287*
Shibasaki, M., (165), (171), *202*
Shina, M., (231), *242*
Shiner, C. S., (165), (171), *202*
Shkrob, A. M., (6), *50*

Shporer, M., (253), (254), (257), 258, (266), *287*
Shrier, A. L., 297, (301), *307*
Sichert-Modrow, I., 10, *50*
Sicher, J., 154, *204*
Siegel, M. G., (10), *50*, (55), (58), 59, *107, 108*
Simmons, H. E., 21, 41, *51*, 120, 122, 134, 144, 175, *204*, 260, 261, *286, 287*, 303, *307*
Simon, J., (57), *107, 108*, (246), (253), (266), (267), (271), (273), (284), *287*
Simon, W., 251, *287*, 305, (306), *307*
Simonnin, M., (120), *205*
Sims, R. J., 45, *50*
Sims, S. K., (130), *203*
Smid, J., (28), (34), *50, 51*, (194), *205*, (299), *308*
Smiley, R. A., 135, *204*
Smith, D. E., (229), *242*
Smith, J. S., (237), *242*
Smith, K., (21), *50*, (125), (131), (134), *203*
Smith, S. G., 192, 194, *204*
Sogah, G. D. Y., (10), *50*, (55), (58), *107, 108*
Sousa, L. R., (10), *50*, (55), (58), (59), *108*
Staley, S. W., 184, *204*
Stanley, H. E., (253), *287*
Starks, C. M., 113, 135, *204, 205*
Stearns, R. W., (253), *287*
Stefanac, Z., 305, *307*
Stein, W. D., 296, *308*
Steinfeld, J. I., 248, *286*
Stetter, H., 41, *51*
Stewart, D. G., 10, *51*
Stoddart, J. F., 55, 57, 59, *107, 109*
Stone, D. B., (157), *205*
Stott, P. E., *12*
Streitweiser, Jr., A., 117, 119, *205*
Strike, D. P., 141, *203*
Stubbs, J. M., 271, *286*
Stubbs, M. E., (57), *107*
Studer, R. O., (6), *51*
Su, A. C. L., 26, 31, 32, *51*
Suchdeo, S. R., (296), *307*
Sugasawa, T., (141), *204*
Sugimoto, N., 135, *205*
Svoboda, M., 156, (158), *205*
Swain, C. G., 117, *205*
Swann, D. A., (235), (236), *242*
Szabo, G., (28), *50*, (246), 253, *286, 287*, (301), *307*

T

Tabner, B. J., 192, *205*
Taft, R. W., 113, *205*
Takekoshi, T., 34, *51*
Talekar, S. V., 253, *287*
Taube H., 303, *307*
Taylor, L. T., 15, 26, *51*
Tehan, F. J., (56), *107*, (201), *203*
Terasawa, T., (141), *204*
Terrier, F., 120, *205*
Thal, A. F., 135, *201*
Thiem, J., (142), *203*
Thomassen, L. M., 145, *205*
Thompson, M. E., (56), *108*
Timko, J. M., 55, 58, *108, 109*
Tobe, M. L., (236), *241*
Torigoe, M., 141, *203*
Tosteson, D. C., 18, *51*
Traynham, J. G., 157, *205*
Trueblood, K. N., (237), *243*
Truter, M. R., 10, 16, 23, 24, 26-28, (29), 42, 43, 45, *49, 50, 51*, 54, *109*, 207, 226-229, 236, 237, 241, *242, 243*
Tsatsas, A. T., 253, *287*
Ts'o T. O. T., (48), *50*
Tundo, P., (123), (141), (175), (180), (183), (192), *202*

U, V

Ugelstad, J., (145), *205*
Valentine, J. S., 163, (167), (168), (170), *204, 205*
Van Tamelen, E. E., 303, *308*
Vazquez, A., (258), (259), (266), *287*
Vergez, S. C., (15), (26), *51*
Vinogradova, E. I., (6), *51*
Virtanen, P. O. I., 117, *204*
Vlasova, G. I., (294), *307*
Vogler, K., (6), *51*
Vögtle, F., 53, 55, 56, *109*, 111, *205*
Von Zelewsky, A., 201, *204*
Voorhees, K. J., (56), *107*

W

Waddam, D. Y., (10), *51*

Wagner, B. E., (260), *287*
Wagner, J., (57), *108*, 236, *243*, (246), (253), (266), (267), (271), (273), (284), *287*
Wakabayashi, T., (141), *204*
Walker, J. L., Jr., 305, *308*
Walker, T., 192, *205*
Wang, A. H. J., (236), *243*
Ward, W. J., 296, *308*
Waters, T. N., (235), (236), *242*
Webb, H. M., (113), *202*
Webb, J., 34, *51*
Weber, E., 55, (56), *109*
Weber, W. P., (183), *204*
Wehner, W., (55), (56), *109*
Weiher, J. F., 26, 31, 32, *51*
Weiss, R., 209, 215, (216), 231, (234), 236, 237, 241, *242, 243*
Weissman, S. I., 201, *203*
Wenkert, E., 141, *203*
Wheatley, C. M., 55, *109*
Whimp, P. O., 236, *243*
White, L. K., (235), *242*
White R. D., (256)-(259), (266), *287*
Wiegers, K. E., 154, *202*
Wieland, T., (257), *286*
Wieser, K., 141, *205*
Wilkins, R. G., 250, (268), (269), *286, 287*
Wilkinson, G., (10), (29), 45, *49, 51*
Williamson, R. E., 186, *203*
Wingfield, J. N., (29), *50,* 229, *243*
Winkler, R., 246, (250), 251, (253), (254), (257), (267), *286*
Wolfe, J., 141, *204*
Wong, K. H., 20, 28, *51*, 194, *205*, 299, *308*
Wright, G. F., (10), *49*
Wudl, F., 59, *109*

Y

Yamdagni, R., 118, *205*
Ykman, P., 121, *205*
Yoshioka, M., (141), *204*
Young, L. B., 113, *205*

Z

Zahner, H., (6), *49*
Zaugg, H. E., 195, *205*
Zavada, J., (156), 158, *205*
Zemel, H., (258), *287*
Zimmer, L. L., (280), (281), *286*
Zubrick, J. W., (127), 134, 141, (175), *202, 205*
Zucker, D., 294, *308*

SUBJECT INDEX

A

Alkali metal amides, activation by dibenzo-18-crown-6, 185
Alkali metal anion, formation of, 201
Alkenes, activation by 18-crown-6, 181-182
Alkylation
of phenylacetone catalyzed by dicyclohexyl-18-crown-6, 192
rates of, 196
Amines, *see* Macrocyclic polyether amines
Anions, formation of alkali metal anions, 201
Antaminide, rate constants for reaction of cations with, 255
Aza crown compounds, *see* Macrocyclic polyether amines

B

Benzo-12-crown-4, 13
Benzo-15-crown-5, 13
activation of hydroxide ions with, 146
average C-C bond lengths of, 240
cavity size, 217
complex formation with alkali cations, 20
crystalline complexes of, 23
structure, 112
Benzo-18-crown-6, 13
activation of hydroxide ions with, 146
Benzo-cryclohexyl-18-crown-6, 13
Bicyclic diamines, 41
chloride encapsulation by, 41
Biphenyl-17-crown-5, 10
Bis (*t*-butylbenzo)-18-crown-6, activation of hydroxide ions with, 146
Boiling points, macrocyclic and macropolycyclic ring systems, 60-105
Bond lengths
in benzo-15-crown-5, 240
in 18-crown-6, 240
in cryptates, 212-214
in diaza-18-crown-6, 240
in dibenzo-18-crown-6, 240
in dibenzo-30-crown-10, 240
in dithia-18-crown-6, 240
in macrocyclic polyether complexes, 218-222
in macrocyclic polyether molecules, 238-240
in substituted macrocyclic polyethers, 232-233
Bromide ion, reaction of organic substrates with, 122-124
Bromination reaction, activation by dibenzo-18-crown-6, 179

C

Cage, macrocycle, 208
Carbanions, generation by use of dibenzo-18-crown-6, 192
Carboxylate ion
activation by 18-crown-6, 125-131
activation by dibenzo-18-crown-6, 129
activation by dicyclohexo-18-crown-6, 127-131
activation by macrocyclic polyether diamines, 129
reaction of organic substrates with, 125-131
Catalysis, macrocyclic compounds used as, 302-304
Cation diameters, values of, 24, 114, 215
Cations
sizes of 24, 114, 215
solubilization of, 114-116
thermodynamic cycle of, 115
Cavity, macrocycles, 208
Cavity size

of benzo-15-crown-5, 217
of 18-crown-6, 114, 217
of 21-crown-7, 217
of dibenzo-18-crown-6, 217
of dicyclohexo-18-crown-6, 217
of macrocyclic polyethers, 25, 114, 217
of tetramethyl dibenzo-18-crown-6, 217
Cesium ion, crystal structure of 18-crown-6 with, 225
Chiral crown compounds, *see* Chiral macrocyclic polyethers
Chiral macrocylic polyethers, 10, 58-60
synthesis of, 58-60
Club sandwich complex, 26-27
Complex formation, macrocyclic polyethers-alkali cations, 20
Coordination template effect, *see* Template effect
Copper ion
crystal structure of tetraamine complex with, 236
kinetics of complexation with macrocyclic thioethers, 281
kinetics of complexation with tetraamines, 274, 277, 279
Courtauld model, of dibenzo-18-crown-6, 6
12-Crown-4
cavity size, 114, 217
structure, 112
toxicity, 48
15-Crown-5
cavity size, 114, 217
crystal structure of sodium ion with, 228
structure, 112
18-Crown-6
activation of alkenes with, 181-182
activation of carboxylate ions with, 125-131
activation of cyanide ion with, 132-133, 136-138
activation of hydroxide ions with, 144
activation of metal hydride ions with, 175
activation of nitrite ion with, 142
activation of oxidation reactions with, 178
activation of superoxide anion with, 164, 167, 168, 169
average C-C bond lengths of, 240
cavity size, 114, 217
crystal structure of cesium ion with, 225
crystal structure of sodium ion with, 224-225
crystal structure of UO_2 ion with, 230
potassium acetate solubilized by, 127
potassium fluoride solubilized by, 120
potassium salts solubilized by. 115
production of radical metal ions M^- with, 201
rate of anionic oxy-cope rearrangement effect of, 190-191
rate constants for reaction of metal cations with, 259
reaction of ethyl acetoacetate with ethyl iodide effect of, 200
reaction of ethyl acetoacetate with ethyl tosylate effect of, 199
reaction of fluoride with organic substrates promoted by, 119
rearrangement of 2,2,3-triphenylpropy alkali metal compounds with, 187
stability constants of cation complexes with, 29
structure of, 112
structure of potassium complex of, 30
synthesis of, 15,55
21-Crown-7
cavity size, 217
stability constants of cation complexes, 29
24-Crown-8
stability constants of cation complexes, 29
structure, 112
30-Crown-10, crystal structure of potassium ion with, 228
Crown compounds, *see* Macrocyclic polyethers
nomenclature, 13
Cryptates, 41-43, 57
activation of carboxylate ions with, 129
activation of cyanide ions with, 141
activation of metal hydride ions with, 174
activation of nitrate ions with, 143
activation of oxidation reactions with, 177
comprehensive list of, 60-105
crystal structure of, 209-216
endo-endo configuration of, 270
equilibrium constant for metal interaction with, 216
exo-exo configuration of, 270
in-in stereoisomer of, 260
kinetic parameters for complexation of, 262-265
out-in stereoisomer of, 260
out-out stereoisomer of, 260
production of radical metal ions M^- with, 201

reaction of ethyl acetoacetate with ethyl iodide effect of, 200
reaction of ethyl acetoacetate with ethyl tosylate effect of, 199
Stevens-Sommelet type rearrangement effect of, 188
structures, 112
synthesis of, 42-43
Crystal structure
of 15-crown-5-potassium ion complex, 228
of 18-crown-6-cesium ion complex, 225
of 18-crown-6-$PdCl_2$ complex, 234
of 18-crown-6-potassium ion complex, 225
of 18-crown-6-sodium ion complex, 224-225
of 18-crown-6-UO_2 ion complex, 230
of 30-crown-10-potassium ion complex, 228
cryptate-cation complexes, 209-216
of dibenzo-24-crown-8-two sodium ions complex, 227
of dithia-18-crown-6-$PdCl_2$ complex, 234
of substituted macrocyclic polyethers, 237-240
of tetraamine macrocyclic ligand-Cu^{2+} ion complex, 236
of tetrathia macrocyclic ligand-$NbCl_5$ complex, 235
of uncomplexed cryptates, 236-237
of uncomplexed macrocyclic polyethers, 236-237
Cyanide ion
activation by 18-crown-6, 132-133, 136-138
activation by cryptates, 141
activation by dicyclohexo-18-crown-6, 141
reaction of organic substrates with, 131-141
Cyclohexyl-15-crown-5
solvent extraction of alkali earth metal ions using, 22
stability constants of cation complexes, 29
Cyclohexyl-18-crown-6
stability constants of cation complexes, 29
4H-Cyclopental[def] phenanthrene, effect of dibenzo-18-crown-6 on stability of salts of, 192

D

Dehydrocyclopentadienyl anion, generation using dicyclohexyl-18-crown-6, 191
cis-4,4′-Diaminodibenzo-18-crown-6, rate constants for reaction of metal cations with, 255
Diaza-18-crown-6, average C-C bond lengths of, 240
Dibenzo-12-crown-4, 13
activation of hydroxide ions with, 146
Dibenzo-14-crown-4, 13
complex formation with alkali cations, 20
Dibenzo-15-crown-5, 13
activation of hydroxide ions with, 146
crystalline complexes of, 23
Dibenzo-18-crown-6, 13
activation of alkali metal amides with, 185
activation of bromination reaction with, 179
activation of carboxylate ions with, 129
activation of hydroxide ions with, 146, 149
activation of metal hydride ions with, 171, 172
activation of methyl iodide with, 194
activation of methyl tosylate with, 194
activation of nitrate ions with 142-143
activation of oxidation reactions with, 176-177
asym-dibenzo-18-crown-6, 13
average C-C bond lengths of, 240
cavity size, 217
complex formation with alkali cations, 20
Courtauld model, 6
crystalline complexes of, 23
discovery, 2-6
ethylation rate constant effect of, 197
generation of carbanions with, 192
generation of halocarbenes with, 192
increased solubility in methanol, 20
infrared spectrum, 17
ion selective electrodes using, 306
isomer B, 10
potassium ion transport in liquid membranes using, 299-301
production of radical metal ions M^- with, 201
rate constants for reaction of cations with, 255
solubility of alkali salts enhanced by, 291
solvent extraction of alkali earth metal ions, using, 22
stability constants of cation complexes, 29

stability of salts of 4H-cyclopental[def]-phenanthrene effect of, 192
structure of, 112
synthesis of, 15, 55
transport of potassium ion in liquid membranes using, 299, 301
Dibenzo-20-crown-4, 13
Dibenzo-21-crown-7, 13
complex formation with alkali cations, 20
stability constants of cation complexes, 29
Dibenzo-24-crown-8, 13
activation of hydroxide ions with, 146
crystal structure of sodium ion with, 227
crystalline complexes of, 23
stability constants of cation complexes, 29
Dibenzo-28-crown-4, 13
Dibenzo-30-crown-10, 13
average C-C bond lengths of, 240
complex formation with alkali cations, 20
crystalline complexes of, 23
rate constants for reaction of cations with, 255
stability constants of cation complexes, 29
Dibenzo-60-crown-20, stability constants of cation complexes, 29
2,3,11,12-Dibenzo-1,4,7,10,13,16-hexaoxacycloocta-2,11-diene, *see* Dibenzo-18-crown-6
Dibiphenyl-28-crown-8, 10
Dicyclohexo-, *see* Dicyclohexyl-
Dicyclohexyl-14-crown-4
solvent extraction of alkali earth metal ions using, 22
stability constants of cation complexes, 29
Dicyclohexyl-18-crown-6, 13
activation of carboxylate ions with, 127-131
activation of cyanide ions with, 141
activation of halide ions with, 118-124
activation of hydroxide ion with, 146, 148, 152, 153, 154, 155, 156
activation of oxidation reactions with, 177
activation of superoxide anion with, 164
alkylation of phenylacetone catalyzed by, 192
alkylation rates effect of, 196
cavity size, 217
crystalline complexes of, 23
crystalline complex of $CoCl_2$, 31
equilibrium constants for metal ions interaction with, 290
generation of dehydrocyclopentadienyl anion with, 191
infrared spectrum, 17
ion selective electrodes using, 306
isomer A, 10
isotope separation using, 292
rate constants for reaction of metal cations with, 255
sandwich complex of, 31-32
solubilization of metal salts, 20-21
solvent extraction of alkali earth metal ions using, 22
stability constants of cation complexes, 29
stereochemical studies using, 189
structure, 112
synthesis of, 55
toxicity, 47-48
value of isotope separation factor, 292
Dicyclohexyl-21-crown-7, solvent extraction of alkali metal ions using, 22
Dicyclohexyl-24-crown-8, solvent extraction of alkali metal ions using, 22
cis-4,4′-Dinitrodibenzo-18-crown-6, rate constants for reaction of metal cations with, 255
Dithia-18-crown-6
average C-C bond lengths of, 240
crystal structure of $PdCl_2$ with, 234

E

Electron transfer
macrocycles as catalysts for, 303
rates of, 303
Emperical formulas, crown compounds, 11-12
Enthalpy, Ni^{2+}-macrocyclic tetraamine complexes, values of, 275
Entropy, Ni^{2+}-macrocyclic tetraamine complexes, values of, 275
Ether esters, *see* Macrocyclic polyether esters
Ethylation rate constants, effect of dibenzo-18-crown-6, 197
Equilibrium constant
for cryptate-metal ion interaction, 216, 262-265
for dicyclohexo-18-crown-6-metal ion

interaction, 5, 290
for macrocyclic thioethers-Cu^{2+} interaction, 281
Ethyl acetoacetate
reaction with ethyl iodide, effect of 18-crown-6, 200
reaction with ethyl iodide, effect of cryptate, 200
reaction with ethyl tosylate, effect of 18-crown-6, 199
reaction of ethyl tosylate, effect of cryptate, 199

F

Fluoride ion, reaction of organic substrates with, 119-122
Free energies of activation, cryptate complexation, values of, 264-265

H

Halide ion
activation by dicyclohexo-18-crown-6, 118-124
reaction of organic substrates with, 118-124
Halocarbenes, generation by use of dibenzo-18-crown-6, 192
Host-guest chemistry, 284-285
Host-guest reactions, rate constants for, 282, 283
Host-guest compounds, 229
structures of, 229
Hydride ion, reaction of organic substrates with, 171-175
Hydroxide ions
activation by benzo-15-crown-5, 146
activation by benzo-18-crown-6, 146
activation by bis(*t*-butyl-benzo)-18-crown-6, 146
activation by 18-crown-6, 144
activation by dibenzo-12-crown-4, 146
activation by dibenzo-15-crown-5, 146
activation by dibenzo-18-crown-6, 146, 149
activation by dibenzo-24-crown-8, 146
activation by dicyclohexo-18-crown-6, 146, 148, 152, 153, 154, 155, 156
activation by tetramethyl-12-crown-4, 155

I

Iodide ion, reaction of organic substrates with, 122-124
Ion exchange electrodes, liquid, 304
Ion selective electrodes, 304-306
using dibenzo-18-crown-6, 306
using dicyclohexyl-18-crown-6, 306
using macrocyclic antibiotics, 305
using macrocyclic molecules, 306
Infrared spectra
of dibenzo-18-crown-6, 17
of dicyclohexyl-18-crown-6, 17
of macrocyclic polyethers, 16-17
Inorganic salts, solubilization
in organic solvents, 20-21
Isotope separation
macrocyclic polyethers used in, 292-294
using dicyclohexyl-18-crown-6, 292
values for calcium enrichment using dicyclohexyl-18-crown-6, 293
Isotope separation factor, for dicyclohexyl-18-crown-6, 292

K

Katapinosis, 41
Kinetics
of macrobicyclic ligand complexation, 260-273
of macrocyclic polyether complexation, 253-259
of sulfur contained macrocycles, 280-283
Kinetic parameters
of antaminide-metal cation complexation, 255
of 18-crown-6-metal cation complexation, 259
of cryptate-metal cation complexation, 262-265
of cis-4,4′-diaminodibenzo-18-crown-6-metal cation complexation, 255
of dibenzo-18-crown-6-metal cation complexation, 255
of dibenzo-30-crown-10- metal cation complexation, 255
of dicyclohexyl-18-crown-6-metal cation complexation, 255
of *cis*-4,4′-dinitrodibenzo-18-crown-6-metal cation complexation, 255
of host-guest reactions, 282-283

of macrocycle-metal cation complexation, 255
of macrocyclic tetraamine-Cu^{2+} ion complexation, 274, 277, 279
of macrocyclic thioethers-Cu^{2+} ion complexation, 281
of monactin-metal cation complexation, 255
of murexide-metal cation complexation, 250
of uramildiacetate-metal cation complexation, 250
of valinomycin-metal cation complexation, 246,249

L

Lanterns, 40
Liquid ion exchange electrodes, 304
Liquid-liquid extractions,
macrocyclic polyethers used in, 291-294
Liquid membranes, 295-301
basic mechanism for separation using, 295
potassium ion transport using dibenzo-18-crown-6 in, 299-301
procedure for making, 297
sodium ion transport using monensin in, 298, 299
stabilization of, 297
transport of potassium ion using dibenzo-18-crown-6 in, 299-301
Liquid surfactant membranes, 297-298

M

Macrobicyclic ligands, *see* Cryptates
Macrocyclic antibiotics, naturally occurring, 5
Macrocyclic compounds, discovery, 2-9
Macrocyclic effect, of tetraamine complexes, 273-276
Macrocyclic polyethers
cavity diameters, 114
comprehensive list of, 60-105
crystalline complexes of, 23
formula and physical properties, 11-12
hole diameters, 25
infrared spectra of, 16-17
nomenclature, 13
properties, 16-18
structural formulas of, 6-9
substituted compounds preparation of, 33
synthesis of, 13-16, 54-56
synthesis yields, 15
toxicity, 47
uses of, 46-47
Macrocyclic polyether amines, 37-39
comprehensive list of, 60-105
synthesis of, 56-57
Macrocyclic polyether diamines, *see* Cryptates
Macrocyclic polyether esters, 56
Macrocyclic polyether sulfides, 34-36
comprehensive list of, 60-105
synthesis of, 57
Macrocyclic tetraaza compounds, electron transfer using, 303
Macrocyclic thioethers
equilibrium data for Cu^{2+} reaction with, 281
kinetic data for Cu^{2+} reaction with, 281
Macropolycyclic compounds, synthesis of, 57-58
Melting points
crown compounds, 11-12
macrocyclic and macropolycyclic ring systems, 60-105
Membrane separations, 294-302
Metal ions, *see* Cations
Metal hydride ions
activation by 18-crown-6, 175
activation by cryptates, 174
activation by dibenzo-18-crown-6, 171-172
Metal radii, values of, 215
Methyl iodide, activation by dibenzo-18-crown-6, 194
Methyl tosylate, activation by dibenzo-18-crown-6, 194
Molecular weights, crown compounds, 11-12
Monactin, rate constants for reaction of metal cations with, 255
Monensin
in liquid membranes, 298-299
sodium ion transport in liquid membranes using, 298-299
Murexide, kinetic parameters for binding metal ions with, 250

N

Naked anions, 21,116
Neodymium, crystal structure of tetrathia macrocycle complex with, 235

Neutral sequestering electrodes, 304
Nickel, stability constants for macrocyclic tetraamine complexes of, 275
Nitrate ion
activation by cryptates, 143
activation by dibenzo-18-crown-6,142-143
activation by macrocyclic polyether diamines, 143
reaction of organic substrates with, 142-143
Nitrite ion,
activation by 18-crown-6, 142
reaction with organic substrates with, 141-142
Nomenclature
crown compounds, 13
macrocyclic polyethers, 13

O

Open-chain polyethers, 43-46
complexing of potassium ion, 45
Out-in isomerism, 41
Oxidation reaction
activation by 18-crown-6, 178
activation by cryptates, 177
activation by dibenzo-18-crown-6, 176, 177
activation by dicyclohexo-18-crown-6, 177
activation by macrocyclic compounds, 175-178
Oxy-cope rearrangement, 18-crown-6 effect on, 190-191
Oxygen nucleophiles, reaction of organic substrates with, 143-163

P

Palladium ion, crystal structure of dithia-18-crown-6 with, 234
Polyether diamines, *see* Cryptates
Polymeric macrocyclic polyethers, 34
Potassium acetate, solubilized by 18-crown-6, 127
Potassium fluoride, solubilized by 18-crown-6, 120
Potassium ion
complexing with open-chain polyethers, 45
crystal structure of 15-crown-5 with, 228
crystal structure of 18-crown-6 with, 225
crystal structure of 30-crown-10 with 228
transport in liquid membranes, 299-301
Potassium salts, solubilized by 18-crown-6, 115

R

Radical metal ions
production using 18-crown-6, 201
production using cryptates, 201
production using dibenzo-18-crown-6, 201

S

Salt complexes, macrocyclic polyethers, 18-30
Sandwich structure
of 15-crown-5 with potassium ion, 227-228
of dicyclohexyl-18-crown-6 with Co^{2+}, 31-32
Sodium ion
crystal structure of 18-crown-6 complex with, 224-225
crystal structure of dibenzo-24-crown-8 with, 227
transport in liquid membranes using monensin, 298-299
Solid membranes, 301-302
Solid ion electrodes, 304
Solubility, alkali salt in methanol, enhancement of, 291
Solubilization, of inorganic salts in organic solvents, 20-21
Solvent extraction of alkali earth metal ions
using cyclohexyl-15-crown-5, 22
using dibenzo-18-crown-6, 22
using dicyclohexyl-14-crown-4, 22
using dicyclohexyl-18-crown-6, 22
using dicyclohexyl-21-crown-7, 22
using dicyclohexyl-24-crown-8, 22
macrocyclic polyethers used in, 22-23
Stability constant
cryptate-metal cation complexes, values of, 216, 264-265
macrocyclic polyether-alkali earth metal ion complexes, values of, 29
macrocyclic tetraamine complexes-Ni^{2+} complexes, values of, 275
Stereospecific hosts, 284-285
Stevens-Sommelet rearrangement, effect of cryptates, 188
Structure

of cation complexes, 25-27
of club sandwich complex, 26-27
of 24-crown-8, 112
of cryptate-metal complex, 212-214
of dibenzo-18-crown-6, 112
of host-guest compounds, 229
of macrocyclic and macropolycyclic ring systems, 60-105
of macrocyclic polyether-metal complexes, 218-219
of sandwich complex, 26-27
of substituted polyether-metal complexes, 232-233
of two metal-one ligand complex, 226-227
Sulfur containing macrocycles, kinetics of, 280-283
Sulfur nucleophiles, 180
Superoxide anion
activation by 18-crown-6, 164, 167, 168, 169
activation by dicyclohexo-18-crown-6, 164
organic substrates reaction with, 163, 171
Synthesis
of dibenzo-18-crown-6, 15, 55
of dicyclohexo-18-crown-6, 55
of macrocyclic polyether amines, 56-57
of macrocyclic polyether diamines, 42-43
of macrocyclic polyether sulfides, 57
methods and yield for macrocyclic polyethers, 13-16, 33, 54-56
Synthesis of specialized chemicals, use of macrocyclic catalysis in, 303-304
Synthesis yields, macrocyclic and macropolycyclic ring systems, 60-105

T

Template effect, 15-16, 46, 55
Tetrabenzo-24-crown-8, 13
Tetramethyl-12-crown-4
activation of hydroxide ions with, 155
stability constants of cation complexes, 29
Tetramethyl-dibenzo-18-crown-6, 10
cavity size, 217
Thermodynamic parameters, cryptate complexation, 264-265
Thia crown compounds, *see* Macrocyclic polyether sulfides
Torsion angle, 208-209
cryptates, 212-214
macrobicyclic molecules, 212-214
macrocyclic polyether complexes, 218-222
noncomplexed macrocyclic molecules, 238-239
substituted macrocyclic polyethers, 232-233
Toxicity, of macrocyclic polyethers, 47-48
Tribenzo-18-crown-6, 13

U

Ultraviolet spectrum
of dibenzo-18-crown-6, 5
of dibenzo-18-crown-6-salt complexes, 19
of macrocyclic polyether-salt complexes, 16, 18-20
Uramildiacetate, kinetic parameters for binding metal ions with, 250
Uranium ion, crystal structure of 18-crown-6 with, 230

V, W

Valinomycin
kinetic parameter for binding metal ions with, 246, 249
rate constants for reaction of metal cations with, 255
structure of K^+ with, 246
Wrap-around structure, 228

A
B
C 8
D 9
E 0
F 1
G 2
H 3
I 4
J 5